Biology and Ecology of Aphids

Biology and Ecology of Aphids

Editor

Andreas Vilcinskas
Institute of Phytopathology and Applied Zoology
Justus-Liebig University of Giessen
Giessen
Germany

CRC Press
Taylor & Francis Group
Boca Raton London New York

CRC Press is an imprint of the
Taylor & Francis Group, an **Informa** business

A SCIENCE PUBLISHERS BOOK

Cover illustration reproduced by kind courtesy of Henrike Schmidtberg (author of Chapter 2) and Marisa Skaljac (author of Chapter 5)

CRC Press
Taylor & Francis Group
6000 Broken Sound Parkway NW, Suite 300
Boca Raton, FL 33487-2742

First issued in paperback 2021

© 2016 by Taylor & Francis Group, LLC
CRC Press is an imprint of Taylor & Francis Group, an Informa business

No claim to original U.S. Government works

Version Date: 20151209

ISBN 13: 978-0-367-78318-1 (pbk)
ISBN 13: 978-1-4822-3676-7 (hbk)

Library of Congress Cataloging-in-Publication Data

Names: Vilcinskas, Andreas.
Title: Biology and ecology of aphids / [edited by] Andreas Vilcinskas.
Description: Boca Raton : Taylor & Francis, 2016. | Includes bibliographical references and index.
Identifiers: LCCN 2015041280 | ISBN 9781482236767 (hardcover : alk. paper)
Subjects: LCSH: Aphids.
Classification: LCC QL527.A64 B545 2016 | DDC 595.7/52--dc23
LC record available at http://lccn.loc.gov/2015041280

Visit the Taylor & Francis Web site at
http://www.taylorandfrancis.com

and the CRC Press Web site at
http://www.crcpress.com

Preface

Aphids, commonly known as plant lice, are important agricultural pests in temperate regions, causing both direct damage to crops as phloem feeders, and indirect yield losses by the transmission of plant pathogens. However, aphids are also fascinating model organisms in evolutionary biology and ecology, because they have evolved complex life cycles encompassing sexual and asexual reproduction as well as the development of multiple distinct phenotypes. The formation of one of several distinct phenotypes defined by the same genotype (polyphenism) is triggered by environmental stimuli and reflects adaptations to ecological challenges. Another prominent feature of aphids is their association with obligate and facultative microbial symbionts, which allow them to feed exclusively on phloem sap and which provide ecological benefits such as resistance to pathogens, parasites or environmental stress. The rapidly expanding scientific literature focusing on the biology and ecology of aphids calls for a textbook that summarizes the recent progress of research in this field, particularly for students in biology, ecology and the agricultural sciences. Accordingly, this volume offers a collection of chapters, written by experts, focusing on recent advances in distinct aspects of aphid biology and ecology.

Aphids belong to the superfamily Aphidoidea (~4400 species) which is part of the insect order Hemiptera representing a large taxon of hemimetabolous insects. The first chapter, entitled "Phylogeny of the aphids", summarizes current opinion concerning the placement of aphids in the tree of life. The reconstruction of the phylogenetic relationship among aphids and their hemipteran relatives has been supported recently by the rapidly increasing availability of transcriptome datasets and complete genome sequences, particularly the genome of the pea aphid *Acyrthosiphon pisum*.

The second chapter, entitled "The Ontogenesis of the Pea Aphid *Acyrthosiphon pisum*", focuses on the developmental biology of aphids and provides an illustrated guidance summary of oogenesis, embryogenesis, viviparous development and the complex life cycle of aphids. We are convinced that a text book for students should include such a chapter summarizing the basic processes of aphid development. The illustrations were drafted by Dr. Henrike Schmidtberg for teaching purposes in our agricultural biotechnology Master's programs at the Justus-Liebig University of Giessen in Germany.

The third chapter, entitled "Functional and Evolutionary Genomics in Aphids", highlights the rapidly developing field of aphid genome biology. The first completely sequenced genome of a hemipteran species was that of the pea aphid *Acyrthosiphon pisum*. Denis Tagu and his colleagues describe the background of this pioneering project and have extracted the most important information for analysis in this chapter. They also consider potential future developments in aphid genome biology.

Chapter 4, entitled "Epigenetic Control of Polyphenism in Aphids", summarizes recent insights into the development of distinct morphological phenotypes of aphids in response to environmental stimuli. Epigenetics is a rapidly expanding research discipline addressing hereditary mechanisms that are not mediated by changes in the DNA sequence but which instead involve environmental mechanisms that reprogram transcriptional responses. Although the translation of environmental stimuli into particular aphid phenotypes is not fully understood, the current body of evidence suggests that epigenetic mechanisms participate in the regulation of aphid polyphenism.

The role of "Bacterial Symbionts of Aphids" is discussed in Chapter 5. The coevolution of obligate and facultative aphid symbionts along with their hosts can be thought of as a holobiont which fills an ecological niche as a highly-adapted phloem feeder. Phloem sap is rich in sugars, but lacks a balanced mixture of amino acids. The dependency of aphids on their microbial symbionts, which compensate for their unbalanced diet, is reflected by striking adaptations of their immune system, as highlighted in Chapter 6, entitled "Aphid immunity". In Chapter 7, Laramy Enders and Nicholas Miller provide a complementary analysis of "Aphid Molecular Stress Biology". Aphids must cope with environmental stressors such as heat, starvation and plant-derived secondary metabolites, and the corresponding adaptations are regulated by molecular mechanisms which now form an exciting research area.

The ecology of aphids is also determined by their host plants, which can be used for overwintering or reproduction. Chapter 8, entitled "The Effect of Plant Within-Species Variation on Aphid Ecology", by Sharon Zytynska and Wolfgang Weisser, introduces the reader to community genetics and multitrophic interactions. Communication among aphids, or between aphids and host plants or aphidophagous insects, is mediated by infochemicals known as semiochemicals or pheromones. The corresponding research field, the "Chemical Ecology of Aphids", is discussed by Antoine Boullis and François Verheggen in Chapter 9. A key feature of aphids is their ability to produce honeydew. This excretion product was originally considered as a waste product, but current insights suggest that honeydew not only removes surplus sugars from the diet but also plays a role in the chemical ecology of aphids. Klaus Hoffmann summarizes the state of the art in this area in Chapter 10, entitled "Aphid Honeydew: Rubbish or Signaler". Another key trait of aphids is their ability to produce saliva that facilitates their adaptation as phloem sap feeders. Recent biochemical and physiological studies have revealed a number of constituents and their functions, which are explained and discussed in Chapter 11, entitled "Function of Aphid Saliva in Aphid-Plant Interaction".

Aphids rapidly develop resistance against insecticides. Consequently, there is a pressing demand for sustainable control measures that benefit the environment and consumers. The development of such "Biotechnological Approaches to Aphid Management" is subject of Chapter 12 contributed by Benjamin Deist and Bryony Bonning. The final chapter, entitled "Aphid Techniques", focuses on the description of methods and technologies which are widely used in aphid research, such as the electrical penetration graph technique, styloctomy and artificial feeding.

Although a single book is not sufficient to provide exhaustive coverage of the burgeoning research on aphids, this volume covers an unprecedented spectrum of basic knowledge and recent research developments. I am convinced that this book will be a valuable source of information for researchers working with aphids and for students in biology, ecology and agriculture.

Autumn 2015   **Andreas Vilcinskas**

Contents

1

Phylogeny of the Aphids

Lars Podsiadlowski

Introduction

Aphids (Aphidoidea) are a part of the Hemiptera, a species-rich branch of the hemimetabolous insects (Fig. 1). Reflected in their alternative name "Rhynchota" all members of Hemiptera have mouthparts transformed to a "sucking beak". Mandibles and maxillae are used as *stechborsten*, the labium forms a sheath around them, maxillary and labial palps are reduced. This innovation initially evolved to optimize the exploitation of plant saps for feeding. The sistergroup of Hemiptera is most probably the Thysanoptera (thrips), which also suck plant saps, but their mouthparts are derived in a different way. Among the hemipteran subtaxon Heteroptera several branches secondarily switched to predatory behavior. Retaining sucking mouthparts, these bugs feed on invertebrate hemolymph or vertebrate blood.

The currently described about 100,000 species of Hemiptera were traditionally split into Heteroptera ("typical bugs") and "Homoptera", but the latter group appears to be a paraphylum in most molecular analyses (e.g., von Dohlen and Moran 1995, Sorensen et al. 1995, Ouvrard et al. 2000, Song et al. 2012, Cui et al. 2013). "Homoptera" is now usually split into the clades Sternorrhyncha (plant lice, including the aphids), Cicadomorpha (cicadas, leaf hoppers, tree hoppers), Fulgoromorpha (plant hoppers) and Coleorrhyncha (moss bugs or beetle bugs). The old name "Auchenorrhyncha" for the combination of Cicadomorpha and Fulgoromorpha is no longer supported as it seems to be a paraphyletic group united by shared plesiomorphic characters. Nevertheless, molecular analyses differ in the exact placement of these taxa to each other. Early studies, based on 18s rRNA data, placed Cicadomorpha as the sistergroup to a combined clade of Fulgoromorpha and the sistergroups Coleorrhyncha and Heteroptera (Sorensen et al. 1995), another one with the same marker set found

Institute of Evolutionary Biology and Zooecology, Rheinische Friedrich-Wilhelms-Universität Bonn, An der Immenburg 1, 53121 Bonn, Germany.
E-mail: lpodsiadlowski@evolution.uni-bonn.de

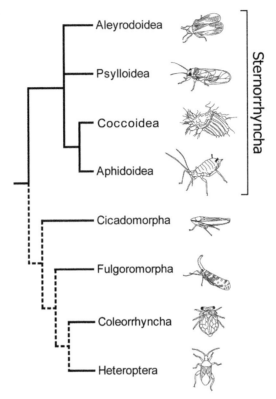

Fig. 1. Hemipteran relationships. In most phylogenetic analyses the Sternorrhyncha are supported as monophylum, and aphids (Aphidoidea) are the sistergroup to scale insects (Coccoidea). Relationships of the other groups vary between different studies, thus the dotted lines depict just one of several alternative hypotheses. Line drawings of insects are not to scale.

only support for the clade combining all those four taxa (Ouvrard et al. 2000). More recent studies with mitochondrial genome sequences confused the overall picture by combining Cicadomorpha, Fulgoromorpha and Sternorrhyncha, with the latter two as sistergroups (Song et al. 2012), or supporting a sistergroup relation between Cicadomorpha and Heteroptera (Cui et al. 2013). In contrast, a recent phylogenomic study, making use of 1,500 orthologous genes and covering all insect orders, placed Coleorrhyncha as sistergroup to a clade combining Fulgoromorpha and Cicadomorpha, while Heteroptera is sistergroup to these three taxa (Misof et al. 2014).

Almost all studies favour monophyly of the clade Sternorrhyncha, which includes the aphids (Aphidoidea), along with scale insects (Coccoidea), white flies (Aleyrodoidea), and jumping plant lice (Psylloidea). All Sternorrhyncha feed on plant saps by tapping the phloem. Many species have economical impact by damaging plants directly or acting as vectors of plant pathogens. In phylogenetic analyses of molecular datasets aphids (Aphidoidea) and scale insects (Coccoidea) usually appear as sistergroups (e.g., von Dohlen and Moran 1995, Xie et al. 2008, Misof et al. 2014). Most of the 7,700 species of scale insects feed on plant saps, a few exploit fungal food sources. In most species the adult females are wingless and stay permanently attached to their food plant. The bodies of many of these sessile females are covered by wax secretions as a defense against predators. The adult males have wings, but are short-lived and take up no food. The relationship of the sistergroups Aphidoidea and Coccoidea to the remaining two clades of Sternorrhyncha varies between different analyses.

General Features of Aphids

Typical morphological characteristics include a head that is immovably joined to the thorax, which itself is broadly conjoined with the abdomen, thus the body has a compact oval shape. As in other Sternorrhyncha the base of the "sucking beak" is located between the coxae of the first walking legs. Species of the largest family Aphididae have a noticeable pair of siphons on the back. While morphology is rather uniform throughout the aphids, their life cycles vary strongly, often involving alternating generations, e.g., of winged (alate) and wingless (apterous), sexual and parthenogenetic forms, with oviparous or viviparous females.

Several bacterial endosymbionts are described from aphids. In the largest family Aphididae there is a long known mutual relationship (obligatory, "primary" endosymbionts) with *Buchnera* bacteria (Buchner 1965), which supplement their hosts' poor diet of phloem sap with essential aminoacids (Ramsey et al. 2010, Wilson et al. 2010). In addition some species also harbour facultative (secondary) symbionts (Moran et al. 2005a, 2005b). Aphids from the two other families Adelgidae and Phylloxeridae have no association with *Buchnera* strains, but other kinds of endosymbionts (Havill and Foottit 2007, Toenshoff et al. 2012a, 2012b, Michalik et al. 2013, Toenshoff et al. 2014). Here the host-symbiont interactions are poorly investigated. The grape phylloxera seems to have no endosymbionts at all (see Chapter 5 "Bacterial Symbionts of Aphids (Hemiptera: Aphididae)" in this volume).

The earliest fossils from Coccoidea and Aphidoidea date back to the jurassic (Grimaldi and Engel 2005), thus preceding the evolution of angiosperms. The aphidoid family Adelgidae may represent the plesiomorphic host range by exploiting only gymnosperm trees. The quick radiation and high divergence of the aphidoid family Aphididae is often mentioned to reflect the parallel radiation of the angiosperm lineages (von Dohlen and Moran 2000).

Aphids and Plants

Aphid evolution is strongly shaped by their dependency on their host plants. About 99% of the species are specialists, associated with one or just a few closely related plant species. About 10% of the species regularly switch between two host species (primary and secondary host) during the seasons. The primary host is usually a woody plant, which is inhabited in autumn, winter and spring, while the secondary host may be a herbaceous plant, exploited during summer season. Life cycles with host alternation evolved several times independently and are found in species of Adelgidae,

Phylloxeridae, Hormaphidinae, Eriosomatinae, Anoeciinae and Aphididae (Blackman and Eastop 2006).

The majority of all aphid species are found in temperate regions of the northern hemisphere. For unknown reasons tropical trees are not often inhabited by aphids. The complex life cycles of aphids with alternation between flightless parthenogenetic generations and winged sexual morphs seem to be adaptations to the seasonal changes of plant food supply in temperate regions. Probably aphid life cycle evolution has to overcome some hurdles to adapt to the tropical absence of seasons.

Besides the classical mode of allopatric speciation, sympatric ecological speciation was proposed for specialized phytophagous insects, e.g., by reduced interactions between "ecotypes" exploiting different host plants (reviewed in Berlocher and Feder 2002). Aphid research provided several examples demonstrating that speciation processes coincide with host plant switches, e.g., in genus *Aphis* (e.g., Hawthorne and Via 2001, Peccoud et al. 2010). Nevertheless, in many cases the allopatric mode of speciation contributes substantially to biodiversity, e.g., shown in the genus *Cinara* (Jousselin et al. 2013). Here as in many other biological fields, general conclusions of trends and features do not well fit to the complicated ecological interactions and historical processes occurring in real populations.

Phylogeny of Aphidoidea

The basal splits of Aphidoidea are between the Adelgidae, Phylloxeridae, and Aphididae, while the interrelations of these three groups are not finally resolved. Most studies favoured a sistergroup relation between Adelgidae and Phylloxeridae (e.g., Heie and Wegierek 2009), but alternatively Adelgidae may be the sistergroup of all other Aphidoidea (Phylloxeridae and Aphididae), which have reduced an ovipositor in contrast to Adelgidae (Grimaldi and Engel 2005). The Aphididae contain by far the majority (> 97%) of all aphid species. This group is clearly different from the other two families by the presence of *Buchnera* endosymbionts and by life cycles involving viviparous generations. Typical morphological features include a pair of siphuncular pores on the fifth tergite, used for chemical communication (e.g., alarm pheromones) (see Chapter 9: "Chemical Ecology of Aphids (Hemiptera: Aphididae)" in this volume).

Heie and Wegierek proposed a classification of living and fossil aphids using a small set of morphological characters and life style features (Heie and Wegierek 2009). But morphological characters used for delimiting subtaxa of aphids seem to have a great plasticity in evolution. Molecular phylogenetic studies yielded support for several subfamilies and often resolved internal relationships of subfamilies, tribes and genera. However, a well supported phylogeny of Aphididae is far from reality. One reason for this unsatisfying situation may be the fast radiation of aphids running in parallel with the angiosperm evolution in the cretaceous (von Dohlen and Moran 2000).

The biggest recently used molecular dataset derived from aphids included four genes (Ortiz-Rivas and Martinez-Torres 2010), two mitochondrial (COII, ATP6) and two nuclear encoded genes (elongation factor 1 alpha, long-wavelength opsin). While aphid mitochondrial genes are of limited value in resolving phylogeny on deeper (= subfamily) level (see Novakova et al. 2013), there is growing effort to use genomic sequences from *Buchnera* endosymbionts, which are vertically transmitted to the offspring, and therefore co-evolve with their hosts. The most comprehensive approach used five genes derived from *Buchnera* strains of 70 aphid species, covering 15 of the 23 subfamilies of Aphididae (Novakova et al. 2013). Nevertheless these two currently broadest approaches both failed to get good resolution at the base of Aphididae and at some points even contradict each other. Besides the lack of well-supported clades above subfamily level in the analyses of *Buchnera* genes, another problematic issue is the rooting of the tree. Adelgidae and Phylloxeridae do not bear *Buchnera* endosymbionts, thus the chosen outgroup representatives have to be the closest relatives from other bacteria. The currently best choice is *Ishikawaella*, a symbiont of true bugs (Husnik et al. 2011), but it is quite far away from *Buchnera* by means of sequence evolution. A long branch to the outgroup questions the informative value of the basal split of the ingroup (see also Chapter 5 "Bacterial Symbionts of Aphids (Hemiptera: Aphididae)" in this volume).

A tree combining evidences from the two biggest molecular studies shows not much resolution above subfamily level (Fig. 2). Seven subfamilies were not at all represented in the molecular analyses. A clade comprising the subfamilies Phyllaphidinae, Calaphidinae and Saltusaphidinae found good support by one of the molecular datasets (Novakova et al. 2013) and appeared here as sistergroup to Aphidinae. Heie and Wegierek (2009) placed those three subfamilies inside of their "Drepanaphididae" clade. In the other mentioned study (Ortiz-Rivas and Martinez-Torres 2010) none of the relationships between subfamilies found good bootstrap support. Hopefully, future phylogenomic approaches based on whole genomes and transcriptomes may help to get a better resolution of the basal parts of the aphid tree of life.

Detailed Accounts of the Families, Subfamilies and Tribes of Aphidoidea

Adelgidae, the Conifer Woolly Aphids

There are two genera in this family, *Adelges* and *Pineus*, with about 55 gall-inducing species from the Palaearctic. The basal life cycle in Adelgidae involves a sexual stage parasitizing a primary host (spruce in both genera) and asexually generations living on a secondary host, pine (*Pinus*) in *Pineus*, spruce (*Picea*), fir (*Abies*) or other conifers in *Adelges* (Havill and Foottit 2007). This complete cycle takes two years. Several species from both genera independently switched to a complete asexual life style involving only one host tree species. In some of these cases the primary host trees became the solitary feeding source, in others the secondary host (Havill et al. 2007). As in Phylloxeridae all female morphs are oviparous (plesiomorphic condition in aphids). Adelgids are the only aphids retaining an ovipositor.

Phylloxeridae

Phylloxeridae currently comprise eight genera with about 80 gall-inducing species, living on trees and one species on vine stocks. All female morphs are oviparous. *Acanthochermes*, two Palaearctic species on oak trees (*Quercus*, Fagaceae); *Aphanostigma*, two Palaearctic species from *Pyrus* (Rosaceae); *Foaiella danesii*, on

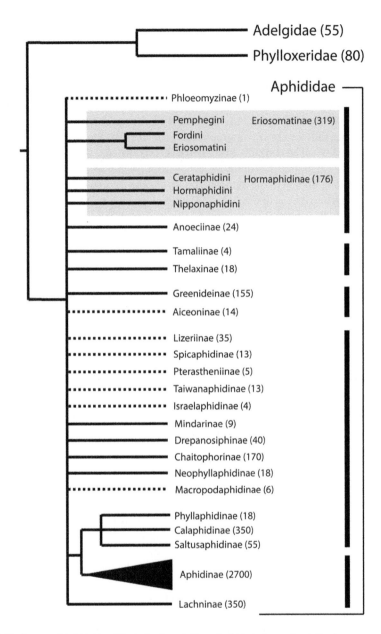

Fig. 2. Phylogeny of Aphidoidea. The lack of resolution of aphid phylogenetic studies becomes apparent when extracting information for good support from the recent most comprehensive analyses (Ortiz-Rivas and Martinez-Torres 2010, Novakova et al. 2013). Interrelationships of the 23 subfamilies of Aphididae are almost without resolution in these two studies. Eriosomatinae and Hormaphidinae (shaded boxes) are not supported as monophyla and very probably have to be split into separate subfamilies. Numbers accompanying subfamily names correspond to the amount of described extant species. Clades with dotted lines depict clades which were not represented by multi-gene datasets in the phylogenetic analyses mentioned above. The black dashes on the right reflect higher ranking taxa suggested by Heie and Wegiereck (2009) based on a few shared morphological and life style characters.

the roots of *Quercus* (Fagaceae); *Moritziella* with at least three Palaearctic species from Fagaceae; *Olegia ulmifoliae*, from Japanese *Ulmus* (Ulmaceae); *Phylloxera*, 60 Palaearctic species living on Fagaceae and Juglandaceae; *Phylloxerina*, with nine palaearctic species from Salicaceae. Finally, there is one North American genus with only one species, the grape phylloxera (*Daktulosphaira vitifoliae*). It was incidentally introduced to Europe around 1850. This species feeds on roots and leaves of grapevine, thereby promoting secondary fungal infections of the roots. The grape phylloxera had a major impact on the vine industry. About 70–90% of all European vine stocks were destroyed by the end of the 19th century, forcing vine producers to switch to resistant types or plants grafted to resistant root stocks. There is still no definite cure against the grape phylloxera.

Aphididae

The remaining aphids are here placed in one family Aphididae *sensu lato*, subdivided into a set of 23 subfamilies, thus following Remaudiere and Remaudiere (1997), with the revisions made by Nieto Nafria et al. (1998). In contrast Heie and Wegierek (2009) suggested family status for most of these subfamilies, except for a set of 14 subfamilies combined to one family "Drepanosiphidae" (Mindarinae, Drepanosiphinae, Neophyllaphidinae, Spicaphidinae, Lizeriinae, Pterastheniinae, Israelaphidinae, Taiwanaphidinae, Phyllaphidinae, Calaphidinae, Saltusaphidinae, Macropodaphidinae, Chaitophorinae, Parachaitophorinae). In the following treatment *Parachaitophorus* is included in the subfamily Drepanosiphinae (Sugimoto 2011) instead of having its own subfamily.

1. Phloeomyzinae with only one species, *Phloeomyzus passerinii*, which lives on poplar trees, *Populus* spp. (Salicaceae), without host alternation. Occurs in Eurasia, with a few records from North Africa, USA and South America.

2. Eriosomatinae (syn. Pemphiginae) with about 320 species in 56 genera from the Holarctic. This is a taxon of gall-inducing lice. Mouthparts and gut are simplified in comparison to other aphids. Among this group a complex social system with a soldier caste was firstly described (Aoki 1977), similar traits are also found in Hormaphidinae. Monophyly of this subfamily is not supported in any of the recent broad range analyses, as the three tribes Eriosomatini, Pemphigini and Fordini appear separated from each other in most of the published trees (Ortiz-Rivas and Martinez-Torres 2010, Novakova et al. 2013). A detailed phylogenetic test for monophyly of Eriosomatinae revealed that other subfamilies (Hormaphidinae, Anoeciinae, Mindariinae and Thelaxinae) interfere with the tribes of Eriosomatinae, thus this subfamily is very probably not a monophylum (Li et al. 2014).

   Members of the tribe Eriosomatini (14 genera) are found in the northern hemisphere, forming leaf-galls on Ulmaceae (*Ulmus* or *Zelkova*) as primary host. Secondary hosts vary between species. A phylogenetic study based on morphological characters exists (Sano and Akimoto 2011). The tribe Pemphigini combines 20 genera occurring throughout the northern hemisphere. Monophyly of this tribe is not well supported (Zhang and Qiao 2008). Many species, e.g., the about 70 species of *Pemphigus*, form leaf galls on *Populus* (Salicaceae) as primary host, and use various herbaceous plants as secondary hosts. The genus *Prociphilus* (45 species) is unusual in that it switched to other primary hosts (e.g., Rosaceae, Caprifoliaceae, Oleaceae) but got stuck with conifer roots as secondary hosts. The tribe Fordini comprises 20 genera of gall-inducing aphids from the northern hemisphere, with most species alternating between primary hosts of Anacardiaceae (*Pistacia, Rhus*) and secondary hosts of grasses or mosses. There is a detailed molecular phylogenetic analysis of this tribe (Zhang and Qiao 2007).

Two species with unsure affinities are sometimes placed here in a fourth tribe named Phyllaphidini, e.g., by Blackman and Eastop (2006) (they otherwise appear also as members of Calaphidinae, Panaphidini). One is *Appendiseta robiniae* from North America, feeding on *Robinia*, the other one is, *Apulicallis trojanae* from Italy, which feeds on *Quercus*.

3. Hormaphidinae 46 genera with 176 species of gall-inducing aphids; these galls can harbor more than thousand clones. A few species are even eusocial, with a sterile soldier caste (Aoki and Kurosu 2010). Most species are found in East and Southeast Asia. The genera *Hamamelistes* and *Hormaphis* both show a disjunctive distribution in North America and Japan following the distribution of their host plants (von Dohlen et al. 2002). Recent analyses of molecular and morphological datasets covering almost 80% of the genera shed light on the phylogenetic relationships of this subfamily (Huang et al. 2012, Chen et al. 2014). Of the three distinguished tribes, Nipponaphidini with about 30 genera turns out to be well supported by this analysis. From the formerly recognized 13 genera of Cerataphidini, three were not combined with the others: *Doraphis* was closest to the third tribe, Hormaphidini (with genera *Hormaphis* and *Hamamelistes*). *Protohormaphis* and *Tsugaphis* each formed clades by its own, the latter as sister to a clade combining Cerataphidini and Nipponaphidini. Primary hosts of Cerataphidini are *Styrax* species (Styracaceae), while the remaining Hormaphidinae live on Hamamelidae as primary host, *Hormaphis* and *Hamamelistes* on *Hamamelis*, Nipponaphidini on *Distylium*. Other host plants are associated with the genera *Protohormaphis* (*Picea*, Pinaceae), *Tsugaphis* (*Tsuga*, Pinaceae), and *Doraphis* (*Populus*, Salicaceae). The secondary hosts are more diverse throughout the subfamily.

4. Anoeciinae, with only one Holarctic genus *Anoecia*, with about 24 species. Some species show annual alternation between *Cornus* (Cornaceae) and grass roots, others live their entire life cycle subterranean on grass roots (Poaceae or Cyperaceae).

5. Tamaliinae, with only genus *Tamalia* comprising at least five species from the Nearctic. These gall-inducing species live their entire life cycle on *Arctostaphylos* species (manzanitas and bearberries, Ericaceae). Some undescribed species of this subfamily may live on other Ericaceae. There is a study on phylogeny, host-plant use and the evolution of inquilinism in *Tamalia* (Miller and Crespi 2003), as one species (*T. inquilinus*) does not induce galls by itself, but uses galls induced by another species.

6. Thelaxinae comprises four genera with 18 species: *Glyphina*, Holarctic on *Alnus* and *Betula* in Europe and North America; *Kurisakia* from Eastern Asia, on Juglandaceae and Fagaceae; *Thelaxes* from Europe and North America on *Quercus* (Fagaceae); *Neothelaxes* from India on *Parthenocissus* (Vitaceae).

7. Greenideinae comprises 14 genera with 155 species. Three tribes are distinguished: (a) Cervaphidini, five genera with 17 species living on various trees and shrubs of Asia and Australia; (b) Shoutedeniini contains three genera with seven species from Africa, living on Euphorbiaceae and Phyllanthaceae; (c) Greenideini comprise six genera with more than 130 species, living mainly on Fagaceae and some other trees in Asia. A phylogenetic analysis of this subfamily revealed paraphyly of Cervaphidini with respect to Shoutedeniini and Greenideini, which appear as sistergroups (Liu et al. 2015).

8. Aiceoninae, with only one genus *Aiceona* comprising 14 species from East and Southeast Asia living on Lauraceae.

9. Lizeriinae comprises three genera and 35 species: *Ceriferella*, two species from Australia living on Epacridaceae. *Lizerius*, 11 South American species living on Lauraceae and Combretaceae. *Paoliella*, a mainly African genus (one species is known from India) with 22 species living on Burseraceae and Combretaceae.

10. Spicaphidinae, which comprises two genera (*Neosensoriaphis, Neuquenaphis*) with 13 species living on *Nothofagus* (Nothofagaceae) in South America.

11. Pterastheniinae are five species in two African genera (*Neoantalus, Pterasthenia*) living on trees and shrubs of Fabaceae.

12. Taiwanaphidinae, with two genera and 13 species: *Sensoriaphis* with four species on *Nothofagus* (Nothofagaceae) and *Melaleuca* (Myrtaceae) in Papua New Guinea, Australia and New Zealand, *Taiwanaphis* with 9 species, mainly from Myrtaceae in Asia.

13. Israelaphidinae. A single genus *Israelaphis*, with four species living in the European Mediterranean region, feeding on grasses (Poaceae).

14. Mindarinae, genus *Mindarus* with at least nine species from Holarktis. Parasites of coniferous trees (mainly *Picea* and *Abies*).

15. Drepanosiphinae, six genera, 40 species from Holarctic, mainly on Aceraceae. *Drepanaphis* comprises about 20 species from Nearctic, living on Aceraceae. *Drepanosiphoniella aceris*, a Palaearctic species from *Acer*. *Drepanosiphum* with eight Holarctic species on *Acer*. *Shenahweum minutum* from North America feeds on *Acer saccharum*. *Yamatocallis* comprises nine Oriental species from *Acer. Parachaitophorus spiraeae*, from Eastasian *Spiraea* species is an enigmatic species, sometimes placed in its own subfamily (e.g., Heie and Wegierek 2009).

16. Chaitophorinae. Twelve genera with about 170 Holarctic species. Two tribes: (a) Chaitophorini comprises five genera and about 140 species living on Salicaceae and Aceraceae; (b) Siphini contains five genera with 24 species which spend their complete life cycle on grass roots (Poaceae and Cyperaceae). A phylogenetic study of Siphini covering all genera combined morphological and molecular data (Wieczorek and Kajtoch 2011). Two separate genera with no affinity to the two

tribes are *Yamatochaitophorus albus* (on *Acer* in Japan) and *Pseudopterocomma* (two species on *Populus* in North America).

17. Neophyllaphidinae; the single genus *Neophyllaphis* contains 18 species from tropical mountains of the southern hemisphere and north to China and Japan. Host plants are Podocarpaceae and Araucariaceae.

18. Macropodaphidinae; this subfamily contains a single genus *Macropodaphis* with about 6 Asian species. Life cycles are unknown, specimens were mainly found on *Potentilla* (Rosaceae), *Artemisia* (Asteraceae) and *Carex* (Cyperaceae).

19. Phyllaphidinae with four genera and 18 mainly Holarctic species (one species occurs also in India), living on Fagaceae (*Fagus*, *Quercus*), except for the two Asian *Machilaphis* species living on Lauraceae.

20. Calaphidinae with 60 genera and about 350 species in Holarctic, divided into two tribes: (a) Calaphidini comprises 17 genera with about 70 species. These live mainly on Betulaceae (*Betula*, *Alnus*), with a few exceptions on Magnoliaceae and Fagaceae. (b) Panaphidini combines 43 genera and about 275 species. A great variety of trees is used as host plants, e.g., Fagaceae, Betulaceae, Ulmaceae, Tiliaceae, Juglandaceae. Some genera do not live on trees, but on bamboo (*Bambusa*, Poaceae) or Fabaceae.

21. Saltusaphidinae with 12 genera and about 55 species, mainly living on sedges (*Carex*, Cyperaceae) in the Holarctic region. One of the few higher taxa which spend their entire life cycle on herbaceous plants. A phylogenetic analysis according to morphological characters is available (Zhang and Qiao 1998).

22. Aphidinae. This group is found in the Holarctic, South and Southeast Asian regions and contains the highest amount of biodiversity of all aphid lineages: about 2,700 species in 177 genera, in other words about 60% of the currently described aphid species. There is a great variety of simple and host-alternating life cycles. Phylogenetic results suggest that a complex life cycle is a basal trait of Aphidinae and simple life cycles without host alternation evolved independently in several lineages of Aphidinae (von Dohlen et al. 2006). Currently the Aphidinae are subdivided into two major subclades, Macrosiphini (144 genera) and Aphidini (33 genera). A formerly separated third group (Pterocommatini) is probably nested within Macrosiphini as sister to the genus *Cavariella* (von Dohlen et al. 2006). In some publications the name Pterocommatini was extended to include also *Cavariella* (Kim et al. 2011). Species of Pterocommatini live on Salicaceae and have simple life cycles without host alternation (about 45 species of genera *Pterocomma*, *Fullawaya*, *Paducia*, *Plocamaphis*, *Neopterocomma*). *Cavariella* has 38 species alternating between Salicaceae and Apiaceae or Araliaceae. Monophyly of Macrosiphini is also put into question by molecular phylogenetic analyses resulting in a nested position of Aphidini inside Macrosiphini (e.g., Novakowa et al. 2013). Notable genera of Macrosiphini are *Myzus* (about 50 species) and *Acyrthosiphon* (about 80 species in the Holarctic ecozone), which include the first two species for which genomic resources are available: *Acyrthosiphon pisum* (The International Pea Aphid Consortium 2010) and *Myzus persicae* (unpublished, some data available at www.

aphidbase.com). The tribe Aphidini was subject to a more detailed phylogenetic analysis recently (Kim et al. 2011). In that study a clear split between Rhopalosiphina and Aphidina was supported. Among Rhopalosiphina (e.g., genera *Rhopalosiphum*, *Schizaphis* and *Melanaphis*) many species have a complete life cycle on grasses (Poaceae, a few as well on Cyperaceae or Typhaceae), while others show host alternation between Poaceae and Rosaceae. Among Aphidina *Aphis* is the most diverse genus in all aphids, containing more than 500 species, predominantly found on shrubs and herbs of the northern hemisphere, but with a few native species in South America, New Zealand and Australia. One publication suggested that the origin of the northern species was derived from southern hemisphere ancestors (von Dohlen and Teulon 2003), but that is not supported by the phylogenetic analysis (Kim et al. 2011). Many of the species are associated with ants in a mutual relationship.

23. Lachninae with 14 Holarctic genera and about 350 species. Life cycles are without host alternation. Four tribes: (a) Cinarini, with only one species *Pseudessigella brachychaeta* from *Pinus* of Pakistan; (b) Lachnini with six genera, Holarctic on Fagaceae, Rosaceae, Aceraceae, Salicaceae and woody Rosaceae; many species associated with ants; *Longistigma* species are generalists with a very broad set of host trees; (c) Eulachnini comprises four genera (*Essigella*, *Eulachnus*, *Schizolachnus*, *Cinara*) living on Pinaceae (mainly *Pinus*) and Cupressaceae in the Holarctic region; the highly diverse genus *Cinara* contains about 200 rather large species (up to 6 mm); (d) Tramini with three Holarctic genera (*Eotrama*, *Protrama*, *Trama*), which spend their complete life cycles on the roots of Asteraceae.

Taxonomic Resources

Due to the strong economic impact of many species as tree and crop pests, there is a comparatively good taxonomic record of aphids in print and online. Major publications covering the complete diversity of aphids were the books of Roger Blackman and the late Victor Eastop (Blackman and Eastop 1994, Blackman and Eastop 2006), of which most contents are yet accessible in an online database (http://www.aphidsonworldsplants.info), also providing an updated bibliography. Other helpful online databases are the aphid species record (http://aphid.speciesfile. org) coordinated by Colin Favret and David Eades, and the extensive collection of aphid photographs and information from Bob Dransfield (http://influentialpoints.com/ Gallery/Aphid_genera.htm).

Keywords: molecular phylogeny, Aphidoidea, Aphididae, evolution, taxonomy

References

Aoki, S. 1977. *Colophina clematis*, an aphid species with "soldiers". Kontyu 45: 276–282.
Aoki, S. and U. Kurosu. 2010. A review of the biology of Cerataphidini (Hemiptera: Aphididae, Hormaphididae), focusing mainly on their life cycles, gall formation and soldiers. Psyche 2010: 380351.

Berlocher, S.H. and J.L. Feder. 2002. Sympatric speciation in phytophagous insects: moving beyond controversy? Ann. Rev. Entomol. 47: 773–815.

Blackman, R.L. and V.F. Eastop. 1994. Aphids on the World's Trees. CAB International, Wallingford.

Blackman, R.L. and V.F. Eastop. 2006. Aphids on the World's Herbaceous Plants and Shrubs. Wiley, Chichester.

Buchner, P. 1965. Endosymbiosis of Animals with Plant Microorganisms. Interscience Publishers, New York.

Chen, J., L.Y. Jiang and G.X. Qiao. 2014. A total-evidence phylogenetic analysis of Hormaphidinae (Hemiptera: Aphididae), with comments on the evolution of galls. Cladistics 30: 26–66.

Cui, Y., Q. Xie, J.M. Hua, K. Dang, J.F. Zhou, X.G. Liu, G. Wang, X. Yu and W.J. Bu. 2013. Phylogenomics of Hemiptera (Insecta: Paraneoptera) based on mitochondrial genomes. Syst. Entomol. 38: 233–245.

Grimaldi, D.A. and M.S. Engel. 2005. Evolution of the Insects. Cambridge University Press, Cambridge UK, New York.

Havill, N.P. and R.G. Foottit. 2007. Biology and evolution of Adelgidae. Ann. Rev. Entomol. 52: 325–349.

Havill, N.P., R.G. Foottit and C.D. von Dohlen. 2007. Evolution of host specialization in the Adelgidae (Insecta: Hemiptera) inferred from molecular phylogenetics. Mol. Phylogenet. Evol. 44: 357–370.

Hawthorne, D.J. and S. Via. 2001. Genetic linkage of ecological specialization and reproductive isolation in pea aphids. Nature 412: 904–907.

Heie, O. and P. Wegierek. 2009. A classification of the Aphidomorpha (Hemiptera: Sternorrhyncha) under consideration of the fossil taxa. Redia XCII: 69–77.

Huang, X.L., J.G. Xiang-Yu, S.S. Ren, R.L. Zhang, Y.P. Zhang and G.X. Qiao. 2012. Molecular phylogeny and divergence times of Hormaphidinae (Hemiptera: Aphididae) indicate Late Cretaceous tribal diversification. Zool. J. Linn. Soc. 165: 73–87.

Husnik, F., T. Chrudimsky and V. Hypsa. 2011. Multiple origins of endosymbiosis within the Enterobacteriaceae (gamma-Proteobacteria): convergence of complex phylogenetic approaches. BMC Biol. 9: 87.

Jousselin, E., A. Cruaud, G. Genson, F. Chevenet, R.G. Foottit and A. Coeur d'acier. 2013. Is ecological speciation a major trend in aphids? Insights from a molecular phylogeny of the conifer-feeding genus *Cinara*. Front. Zool. 10: 56.

Kim, H., S. Lee and Y. Jang. 2011. Macroevolutionary patterns in the Aphidini aphids (Hemiptera: Aphididae): diversification, host association, and biogeographic origins. PLoS One 6: e24749.

Li, X.Y., L.Y. Jiang and G.X. Qiao. 2014. Is the subfamily Eriosomatinae (Hemiptera: Aphididae) monophyletic? Turkish J. Zool. 38: 285–297.

Liu, Q.H., J. Chen, X.L. Huang, L.Y. Jiang and G.X. Qiao. 2015. Ancient association with Fagaceae in the aphid tribe Greenideini (Hemiptera: Aphididae: Greenideinae). Syst. Entomol. 40: 230–241.

Michalik, A., A. Golas, M. Kot, K. Wieczorek and T. Szklarzewicz. 2013. Endosymbiotic microorganisms in *Adelges (Sacchiphantes) viridis* (Insecta, Hemiptera, Adelgoidea: Adelgidae): molecular characterization, ultrastructure and transovarial transmission. Arthrop. Struct. & Dev. 42: 531–538.

Miller, D.G. and B. Crespi. 2003. The evolution of inquilinism, host-plant use and mitochondrial substitution rates in *Tamalia* gall aphids. J. Evol. Biol. 16: 731–743.

Misof, B., S. Liu, K. Meusemann, R.S. Peters, A. Donath, C. Mayer, P.B. Frandsen, J. Ware and 93 others. 2014. Phylogenomics resolves the timing and pattern of insect evolution. Science 346: 763–767.

Moran, N.A., P.H. Degnan, S.R. Santos, H.E. Dunbar and H. Ochman. 2005a. The players in a mutualistic symbiosis: insects, bacteria, viruses, and virulence genes. Proc. Natl. Acad. Sci. USA 102: 16919–16926.

Moran, N.A., J.A. Russell, R. Koga and T. Fukatsu. 2005b. Evolutionary relationships of three new species of Enterobacteriaceae living as symbionts of aphids and other insects. Appl. Env. Microbiol. 71: 3302–3310.

Nieto Nafria, J.M., M.P. Mier Durante and G. Remaudiere. 1998. Les noms des taxa du group-famille chez les Aphididae (Hemiptera). Revue fr. Ent. 19: 77–92.

Novakova, E., V. Hypsa, J. Klein, R.G. Foottit, C.D. von Dohlen and N.A. Moran. 2013. Reconstructing the phylogeny of aphids (Hemiptera: Aphididae) using DNA of the obligate symbiont *Buchnera aphidicola*. Mol. Phylogenet. Evol. 68: 42–54.

Ortiz-Rivas, B. and D. Martinez-Torres. 2010. Combination of molecular data support the existence of three main lineages in the phylogeny of aphids (Hemiptera: Aphididae) and the basal position of the subfamily Lachninae. Mol. Phylogenet. Evol. 55: 305–317.

Ouvrard, D., B.C. Campbell, T. Bourgoin and K.L. Chan. 2000. 18S rRNA secondary structure and phylogenetic position of Peloridiidae (Insecta, Hemiptera). Mol. Phylogenet. Evol. 16: 403–417.

Peccoud, J., J.C. Simon, C. von Dohlen, A. Coeur d'acier, M. Plantegenest, F. Vanlerberghe-Masutti and E. Jousselin. 2010. Evolutionary history of aphid-plant associations and their role in aphid diversification. Compt. R. Biol. 333: 474–487.

Ramsey, J.S., S.J. Macdonald, G. Jander, A. Nakabachi, G.H. Thomas and A.E. Douglas. 2010. Genomic evidence for complementary purine metabolism in the pea aphid, *Acyrthosiphon pisum*, and its symbiotic bacterium *Buchnera aphidicola*. Ins. Mol. Biol. 19: 241–248.

Remaudiere, G. and M. Remaudiere. 1997. Catalogue des Aphididae du monde. INRA, Paris.

Sano, M. and S.I. Akimoto. 2011. Morphological phylogeny of gall-forming aphids of the tribe Eriosomatini (Aphididae: Eriosomatinae). Syst. Entomol. 36: 607–627.

Song, N., A.P. Liang and C.P. Bu. 2012. A molecular phylogeny of Hemiptera inferred from mitochondrial genome sequences. Plos One 7: e48778.

Sorensen, J.T., B.C. Campell, R.J. Gill and J.D. Steffen-Campbell. 1995. Non-monophyly of Auchenorrhyncha ("Homoptera"), based upon 18S rDNA phylogeny: eco-evolutionary and cladistic implications within pre-Heteropterodea Hemiptera (*S. L.*) and a proposal for new monophyletic suborders. Pan-Pacific Entomol. 71: 31–60.

Sugimoto, S. 2011. Revision of the genus *Parachaitophorus* Takahashi (Hemiptera: Aphididae: Drepanosiphinae), with descriptions of five morphs and biology of *P. spiraeae* (Takahashi, 1924) on *Spiraea cantoniensis* (Rosaceae). Insecta Matsumurana N.S. 67: 33–40.

The International Pea Aphid Consortium. 2010. Genome sequence of the pea aphid *Acyrthosiphon pisum*. PLoS Biol. 8: e1000313.

Toenshoff, E.R., D. Gruber and M. Horn. 2012a. Co-evolution and symbiont replacement shaped the symbiosis between adelgids (Hemiptera: Adelgidae) and their bacterial symbionts. Env. Microbiol. 14: 1284–1295.

Toenshoff, E.R., T. Penz, T. Narzt, A. Collingro, S. Schmitz-Esser, S. Pfeiffer, W. Klepal, M. Wagner and others. 2012b. Bacteriocyte-associated gammaproteobacterial symbionts of the *Adelges nordmannianae/piceae* complex (Hemiptera: Adelgidae). ISME J. 6: 384–396.

Toenshoff, E.R., G. Szabo, D. Gruber and M. Horn. 2014. The pine bark adelgid, *Pineus strobi*, contains two novel bacteriocyte-associated gammaproteobacterial symbionts. Appl. Env. Microbiol. 80: 878–885.

von Dohlen, C.D. and N.A. Moran. 1995. Molecular phylogeny of the Homoptera—a paraphyletic taxon. J. Mol. Evol. 41: 211–223.

von Dohlen, C.D. and N.A. Moran. 2000. Molecular data support a rapid radiation of aphids in the Cretaceous and multiple origins of host alternation. Biol. J. Linn. Soc. 71: 689–717.

von Dohlen, C.D., U. Kurosu and S. Aoki. 2002. Phylogenetics and evolution of the eastern Asian-eastern North American disjunct aphid tribe, Hormaphidini (Hemiptera: Aphididae). Mol. Phylogenet. Evol. 23: 257–267.

von Dohlen, C.D. and D.A.J. Teulon. 2003. Phylogeny and historical biogeography of New Zealand indigenous Aphidini aphids (Hemiptera: Aphididae): an hypothesis. Ann. Entomol. Soc. Am. 96: 107–116.

von Dohlen, C.D., C.A. Rowe and O.E. Heie. 2006. A test of morphological hypotheses for tribal and subtribal relationships of Aphidinae (Insecta: Hemiptera: Aphididae) using DNA sequences. Mol. Phylogenet. Evol. 38: 316–329.

Wieczorek, K. and L. Kajtoch. 2011. Relationships within Siphini (Hemiptera, Aphidoidea: Chaitophorinae) in light of molecular and morphological research. Syst. Entomol. 36: 164–174.

Wilson, A.C.C., P.D. Ashton, F. Calevro, H. Charles, S. Colella, G. Febvay, G. Jander, P.F. Kushlan and others. 2010. Genomic insight into the amino acid relations of the pea aphid, *Acyrthosiphon pisum*, with its symbiotic bacterium *Buchnera aphidicola*. Insect Mol. Biol. 19: 249–258.

Xie, Q., Y. Tian, L. Zheng and W. Bu. 2008. 18S rRNA hyper-elongation and the phylogeny of Euhemiptera (Insecta: Hemiptera). Mol. Phylogenet. Evol. 47: 463–471.

Zhang, G. and G. Qiao. 1998. The phylogenetic analysis of Saltusaphidinae. Entomologia Sinica 5: 301–309.

Zhang, H.C. and G. Qiao. 2007. Molecular phylogeny of Fordini (Hemiptera: Aphididae: Pemphiginae) inferred from nuclear gene EF-1 alpha and mitochondrial gene COI. Bull. Entomol. Res. 97: 379–386.

Zhang, H. and G. Qiao. 2008. Molecular phylogeny of Pempheginae (Hemiptera: Aphididae) inferred from nuclear gene EF-1 alpha sequences. Bull. Entomol. Res. 98: 499–507.

2

The Ontogenesis of the Pea Aphid *Acyrthosiphon pisum*

Henrike Schmidtberg* and Andreas Vilcinskas

Introduction

Phenotypic plasticity is an adaptive response to environmental cues (DeWitt and Scheiner 2004, Ghalambor et al. 2007, Simpson et al. 2011). An extreme example is phenotypic polyphenism, in which two or more discrete phenotypes can arise from the same genotype (West-Eberhard 1989, 2003, Hall 1999, Nijhout 1999, 2003, Ogawa and Miura 2014). The main phenotypic differences are manifested in morphs and developmental processes, but also in physiology, biochemistry and behavior. All these parameters in turn influence ecology and fitness (West-Eberhard 1989, Whitman and Ananthakrishnan 2009, Simon et al. 2011a). In some cases, behavioral patterns rather than environmental signals generate phenotypic plasticity. Such patterns have probably evolved as an immediate response and therefore may influence morphology (Eberhard 1980, 1982, Bernays 1986; reviewed by West-Eberhard 1989).

Polyphenism is a major reason for the success of insects and many examples have been reported (reviewed by Simpson et al. 2011). The phenomenon has been observed in the Lepidoptera, Diptera, Coleoptera and Hymenoptera. Many authors name the resulting morphs according to the corresponding physical or biological stimuli, e.g., diet-mediated, sexually-selected or seasonal phenotypes (West-Eberhard 2003, Miura 2005, Simpson et al. 2011, Whitman and Ananthakrishnan 2009). Several recent studies have addressed the molecular and genetic basis of polyphenism in insects (reviewed by Evans and Wheeler 2001). Some environmental cues directly affect gene expression whereas others intervene between the environmental signal and target

Institute for Insect Biotechnology, Justus-Liebig University, Heinrich-Buff-Ring 26-32, 35392 Giessen, Germany.
* Corresponding author: Henrike.Schmidtberg@ime.fraunhofer.de

genes. The latter often depend on neuro-endocrine pathways and other regulators of gene expression (Simon et al. 2011).

Aphids provide ideal examples of phenotypic plasticity because they undergo cyclical parthenogenesis in which parthenogenetic generations alternate with sexual generations. Parthenogenesis in aphids evolved more than 200 million years ago and led to the evolution of telescoping generations (pedogenesis) in which embryos begin to develop in the ovarioles of their grandmothers. This enables them to achieve prodigious reproduction rates and also results in the appearance of different phenotypes in the same generation (Stevens 1904, Dixon 1985).

The phloem-sucking pea aphid *Acyrthosiphon pisum* (Harris) is an emerging model organism with a recently sequenced and annotated genome (International Aphid Genomics Consortium 2010). It switches its reproductive mode in response to seasonal changes. Here, the environmental signals are perceived during the early stages of ontogeny and thus have a significant impact on embryonic development. Because different generations are formed in each female parthenogenetic aphid, including sexual and asexual individuals in the same clone, the pea aphid exemplifies phenotypic plasticity based on the influence of genotype, phenotype and environment (Simon et al. 2011a). Parthenogenetic, viviparous females may share the same genotype as sexual, egg-laying females, but they differ greatly in terms of reproductive anatomy, external appearance, behavior and the molecular mechanisms underpinning early development (O'Neill 2012). This offers a prime example of how epigenetic processes, acting on genes and their products, can have a substantial impact on the phenotype. The pea aphid therefore provides an ideal model for the functional analysis of developmental genes in different morphs. In this review, we discuss the current state of the art in aphid developmental biology as a framework for further investigations of genetic and/or epigenetic regulatory mechanisms underlying the phenomenon of phenotypic plasticity (see Chapter 4 "Epigenetic Control of Polyphenism in Aphids" in this volume).

The Life Cycle of the Pea Aphid

Aphids adopt numerous polyphenic forms, e.g., different body-color morphs on particular host plants, and different aestivating forms depending on current environmental conditions. Others generate specialized altruistic solder nymphs or a perfected caste polyphenism (Essig and Abernathy 1952, Aoki 1977, Miyazaki 1987; reviewed by Itô 1989, Stern and Foster 1996, Fukatsu 2010, Tsuchida et al. 2010, Hattori et al. 2013). Numerous investigations of aphid life cycles have summarized the variety of phenotypic morphs (Kawada 1987, Miyazaki 1987), or investigated physiological processes such as the influence of hormones that control their development (Hardie 1987a,b, Hardie and Lees 1983).

The pea aphid is a hemimetabolous insect with incomplete metamorphosis, generating four nymphal instars before molting to the adult aphid (Fig. 1). There are two forms of seasonal polyphenism: wing polyphenism and reproductive polyphenism. As discussed below, these are facultative seasonal polymorphisms because they are represented by morphological changes triggered by environmental cues (West-Eberhard 1989).

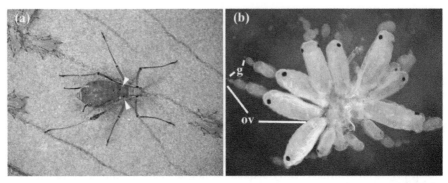

Fig. 1. (a) Adult female *A. pisum* with nymph stages. Arrowheads indicate the eyes of late embryos inside the mother. (b) Two dissected ovaries of a single mother with the germaria (g) at the tips and late developmental stages of embryos at the base of each ovariole (ov).

Reproductive Polyphenism

Pea aphids switch between sexual and parthenogenetic (asexual) modes of reproduction, a process known as reproductive polyphenism. Environmental cues directly influence the corresponding ontogenetic processes, starting with differences in oocyte determination represented by meiotic oogenesis in sexual females and mitotic division in asexual females. The changes become fixed during subsequent morphogenesis of the reproductive, sensorial and neural systems. The developmental environment also differs in the two modes of reproduction: the sexual females lay eggs that develop externally whereas asexual development continues within the female (Duncan et al. 2013a).

The pea aphid life cycle begins in the spring with a wingless foundress called the fundatrix which gives birth to nymphs (Hales et al. 1997, Miura et al. 2003, Jaquiéry et al. 2012, Ogawa and Miura 2014) (Fig. 2). Oogenesis and embryonic development of new aphid progenies are already underway in these nymphs due to the abovementioned phenomenon of pedogenesis. In long summer days with high temperatures, the pea aphid still reproduces asexually and viviparously, and the new females may be winged (alate) or wingless (apterous). Crowding and host-plant quality can induce these morphological changes during the prenatal stages. After repeated cycles of asexual reproduction, the autumnal shortening of day length suppresses juvenile hormone production in the aphid hemolymph. This leads to the development of a parthenogenetic female (sexuparae), which produces only one generation, consisting of wingless sexual oviparous females, and also winged and wingless males, the latter choice depending on genotype (see below). After mating, sexual females produce eggs for overwintering, which are then fertilized at oviposition. This initiates a frost-resistant diapause in embryonic development. The eggs are laid on herbaceous sward or grass clumps (Leather 1993) and hatch in the spring as asexual females (Moran 1992).

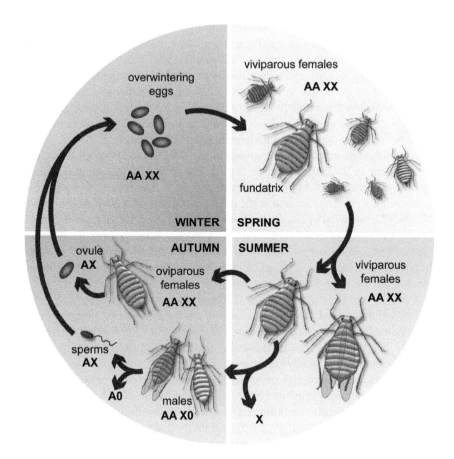

Fig. 2. Annual life cycle of the pea aphid and ploidy levels for autosomes (A) and sexual chromosomes (X) (according to Hales et al. 1997, Miura et al. 2003, Jaquiéry et al. 2012, 2013, Ogawa and Miura 2014). Oviposited eggs overwinter in diapause and hatch in spring to a wingless female (fundatrix) that gives birth parthenogenetically to clonal viviparous asexual females. In summer, the females reproduce repeatedly by parthenogenesis and produce both winged and wingless females. At shorter autumn day lengths, the asexual generations produce one sexual-producing generation, the so called gynopara. These parthenogenetic viviparous females produce wingless sexual females as well as winged and wingless males. After mating, oviparous (sexual) females produce fertilized eggs for overwintering. Aphids employ the XO sex-determination system. Therefore, viviparous and oviparous females possess two X chromosomes, while males only possess one X chromosome. Males are produced parthenogenetically by removal of one X-chromosome during the maturation division of meiosis. Only sperm cells possessing one X-chromosome are viable, those completely lacking an X chromosome (A0) degenerate. Therefore, the hatched fundatrices are entirely female (XX).

Influence of Photoperiod and Temperature

The induction of sexual generations (males and oviparous females) and the production of eggs that survive cold temperatures are short-term adaptive responses to environmental cues such as shortening day length and low temperatures, indicating the onset of winter (Hardie 1990, Simon et al. 1996, 2002, 2010, Dixon 1998, Rispe and

Pierre 1998, Rispe et al. 1998). Aphids sense the night length (scotophase) and respond in a transgenerational manner by triggering a change in embryonic development that favors a sexual phenotype, thus producing males and females (Tagu et al. 2005, Simon et al. 2011a, 2011b). The targets for photoperiod shortening are the germ cells, which are located in the germarium of the ovaries in the embryonic sexual female. Pedogenesis ensures that the adapted sexual phenotype appears two or three generations after exposure to the new scotophase, and embryos can directly sense photoperiod changes while inside the maternal abdomen (Simon et al. 2010).

Asexual aphids perceive the increase in night length via photoperiod-specific photoreceptors in the dorsal protocerebrum. The nature of the photoreceptors is unknown (Gao et al. 1999) but the photoperiod signal is transduced by two groups of five neuro-endocrine cells located in the dorsal part of the pars intercerebralis (Steel and Lees 1977; reviewed by Hardie and Lees 1985). Although the axons of such neuro-endocrine cells project to the corpora cardiaca in most insects, the projections from one group of neuro-endocrine cells in the aphid pass ventrally and follow the dorsal neuropile tracts to the thoracic ganglion mass, leaving through the median abdominal nerve, and are therefore presumed to reach the vicinity of the ovarioles where they release neurohormones (Steel 1976, 1977, Huybrechts et al. 2011).

Transcriptome analysis has been used to investigate genetic programs in the aphid regulated by photoperiod shortening and the reproductive switch (Le Trionnaire et al. 2007, 2009). Many genes are induced or repressed in the heads of mothers of the sexual individuals, with putative functions including regulation of the visual system, photoreception, and neuro-endocrine signaling. Unexpectedly, cuticular proteins were found to be strongly regulated by photoperiod shortening, such that the network between cuticular proteins and chitin might soften in the heads of aphids reared under short-day conditions (Gallot et al. 2010), and the N-β-alanyldopamine biosynthesis pathway responsible for cuticle sclerotization was suppressed (Le Trionnaire et al. 2009). The latter indicates that dopamine production is regulated by photoperiod shortening (Gallot et al. 2012).

Several pea aphid genes are specifically regulated by photoperiodic changes (Cortés et al. 2008). *ApSDI-1* is overexpressed under short-photoperiod conditions (Ramos et al. 2003), thus triggering the sexual response by initiating a cascade of processes in the sensory, nervous and endocrine systems (Hales et al. 1997). ApSDI-1 is similar to amino acid transporter proteins in GABAergic neurons, and may therefore play a role in aphid GABAergic transmission (Ramos et al. 2003). The GABAergic system regulates melatonin levels in circadian cycles and seasonal reproduction. When this hormone is fed to asexual pea aphids it induces sexual development even if the insects are reared under long-photoperiod conditions (Gao and Hardie 1997).

The pea aphid has a critical photoperiod of 13–14 h for the induction of sexual individuals (Lamb and Pointing 1972, Smith and MacKay 1989) whereas sexual reproduction in other aphid species is induced by physiological changes in the host plants (Forrest 1970, Dixon 1998). Strictly asexual species tend to be distributed in low-latitude regions in which nymphal or adult aphids can overwinter (Simon et al. 1996, 2002, 2010, Rispe and Pierre 1998, Dixon 1998). Some populations living preferentially in warmer regions with moderate winters undergo continuous cycles of asexual reproduction without a sexual component (International Aphid Genomics

Consortium 2010). Furthermore, the aphid life cycle can switch in certain species, with the sexual generation appearing in spring to lay fertilized eggs, whereas other species produce eggs by parthenogenesis (reviewed in Ogawa and Miura 2014).

Aphids producing diapausing eggs have several advantages: when food is scarce the egg benefits because there are no mothers to feed and an advanced stage of maturation is unnecessary for survival. The diapause period involves a developmental arrest, during which the insect is resistant to environmental influences (Behrendt 1963). In all holocyclic species, the sexual generation is the last in the season and produces the resistant eggs (Dixon 1985). It is presumed that the egg requires low temperatures to develop normally, which suggests that the diapause is an adaption to cold winters. When the egg enters diapause, embryonic development slows down considerably and the current stage of development is prolonged (see below). Several studies have shown that the delayed development of the embryo is temperature independent and proceeds at a slow rate even at higher temperatures. In the spring, asexual females hatch approximately 100 days after the eggs are laid (Brisson and Stern 2006).

Several studies have addressed the developmental properties of diapausing eggs in different aphid species (Lushai et al. 1996, Komazaki 1998). Behrendt (1963) described three developmental phases: a temperature-dependent phase, followed by the temperature-independent diapause and finally a further temperature-dependent phase. Although development features these stages, the embryos in the eggs do not experience complete developmental arrest. Instead they regulate the developmental rate and undergo slow, continuous growth which is not influenced by the environment (Shingelton et al. 2003).

Wing Polyphenism

Flying insects can locate distant food and mating partners so that reproduction can spread to new locations, but such dispersal comes at a considerable cost (Harrison 1980, Roff 1990, Braendle et al. 2006). In addition to the energy consumption, insects must develop sensory and reproductive systems adapted for flight. Therefore, many insect species have lost the ability to disperse and use the energy instead to maximize fecundity and longevity. Others have evolved morphological adaptations that enhance survival at ground level, including weapons for intraspecific and interspecific competition (Harrison 1980, Dixon and Howard 1986, Roff 1990, Roff and Fairbairn 1991). Wing polyphenism in aphids is widely distributed and the underlying mechanisms are discussed in several studies (reviewed by Müller et al. 2001 and by Braendle et al. 2006). As considered below, environmental conditions during embryonic development determine whether or not wings emerge from existing wing buds (Lees 1967).

Influence of Crowding, Plant Quality and Predation

In the summer season, parthenogenetic females may be winged or wingless depending on the colony density (Fig. 2). The wingless forms show adaptations to maximize fecundity when colony density is low, whereas the winged forms are more prevalent under more crowded conditions or when the host-plant quality and/or quantity is reduced, causing food stress (Sutherland 1969a,b). The perception of high colony density is probably based on chemical cues or tactile stimulation between aphids (Johnson 1965, Kunert and Weisser 2005; cited by Braendle et al. 2006). In the fall, asexual females give rise to sexual males that can be winged or wingless, and to sexual wingless females. Here the factors affecting wing formation are slightly different: the opportunity to meet new mating partners and to reach new food sources due to the depletion of host plants in autumn favors the dispersal morphs, and reduces inbreeding costs (Huang and Caillaud 2012). Interestingly, the variation of wing genesis in asexual females is linked to that of the winged male forms (Braendle et al. 2005, Brisson 2010). In both cases there is a trade-off between reproduction and flight (Zera and Denno 1997). Wing development is also induced by interactions among different aphid species on the same host plant and again appears to be dependent on increased tactile stimulation (Lamb and MacKay 1987; reviewed by Braendle et al. 2006).

Wingless pea aphids can also be induced to produce winged offspring in response to attacks by predators (Sutherland 1969a, Dixon 1985, Dixon and Agarwala 1999, Weisser et al. 1999, Kunert and Weisser 2003). In this case, behavioral and physiological changes are induced by the release of the alarm pheromone (E)-β-farnesene (Podjasek et al. 2005). This initiates an escape response in the surrounding aphids, causing them to drop off the leaves. These movements enhance the crowding effect because there the number of antennal contacts between the fleeing aphids increases, a phenomenon known as pseudo-crowding (Kunert et al. 2005). Pea aphids produce winged offspring not only in response to predators, but also their remnants, e.g., the feces and eggs of adult predatory ladybird beetles (Purandare et al. 2014).

Winged aphids are also produced as a defense mechanism when colonies are attacked by parasites such as parasitoid wasps or the plants are infected by pathogens (Müller et al. 2001, Sloggett and Weisser 2002, Leonardo and Mondor 2006). They generate comparatively more winged offspring than colonies which are attacked by common predators (Dixon 1998). In contrast, certain insect species can inhibit wing development in aphids: the presence of ants protects aphids against predators and therefore suppresses the escape response (El-Ziady and Kennedy 1956, Kleinjan and Mittler 1975; cited by Braendle et al. 2006).

Braendle et al. (2006) compared the morphological differences between wingless and winged aphid forms that reflect their divergent life styles: winged females possess ocelli and expanded thoraces to accommodate the powerful flight musculature. Their cuticles are also more heavily sclerotized and the dimensions of the legs and siphunculi differ from those of wingless females. The two morphs also exhibit different behaviors, i.e., wingless morphs have a shorter generation times and higher fecundity (Müller et al. 2001). There are also obvious differences in the context of male and female dimorphism: in males, both winged and wingless forms have ocelli and the antennae of both forms are similar to the winged female antennae. In contrast to the environmentally-determined wing polymorphism in parthenogenetic females, the development of wings in male pea aphids is determined by genotype: an X-linked polymorphism at the *aphicarus* (*api*) locus controls the development of male wing morphs (Caillaud et al. 2002, Braendle et al. 2005, 2006, Brisson 2010).

Wing Development

All embryos (winged and wingless) develop wing primordia and all first nymphal instar stages emerge from the embryonic stages with wing buds that extend from the thoracic body wall (Ishikawa and Miura 2007, Ishikawa et al. 2008, Ogawa and Miura 2013). In wingless morphs, these primordia degenerate after the second nymphal instar and there is no further development of the flight muscles (Johnson and Birks 1960, Tsuji and Kawada 1987, Ishikawa et al. 2008). In winged forms, the wing buds develop slowly until the final molt to the adult aphid (Brisson et al. 2010). The environmental cues responsible for wing development are perceived by the mother, which transmits a permissive signal to her embryos to initiate wing development (Sutherland 1969a). Only embryos that are 24–48 h from birth can respond to this signal, whereas in many other species wings can be induced up to the third nymphal instar (Müller et al. 2001, Hille Ris Lambers 1966, Lees 1966).

The genetic program controlling wing development is interrupted at the nymphal stage of future wingless morphs (Ishikawa et al. 2008). Every gene in *Drosophila melanogaster* with a reported role in wing development has a counterpart controlling wing patterning and differentiation in the pea aphid. For example, the principal genes *apterous* and *decapentaplegic* are also expressed in duplicate in pea aphids (Brisson et al. 2010). They have two *apterous* paralogs and four *decapentaplegic* paralogs. The anteroposterior patterning genes *engrailed* and *hedgehog*, the dorsal-ventral patterning genes *wingless*, *distalless*, the wing hinge development gene *homothorax*, and the homeotic gene *ultrabithorax* are also involved. These genes are differently expressed across the four post-embryonic nymph stages (Brisson et al. 2010).

Sex Determination

As stated above, the targets for photoperiod shortening are the germ cells in the germarium of the embryos in future sexual females. In parthenogenetic embryos, these germ cells do not undergo meiosis and therefore contain double the normal haploid chromosome number (Blackman 1987, Bickel et al. 2013). These oocyte-like cells directly begin embryogenesis in the absence of fertilization (Le Trionnaire et al. 2009). There is evidence of lineage-specific duplications of both mitotic regulators (e.g., *Cdk1*, *Polo* and *Wee*) and mitosis-related genes such as *SM6* (Srinivasan et al. 2010) in the pea aphid.

Aphids use the XX/XO sex-determination system: viviparous and oviparous females possess two X-chromosomes whereas males carry only one (Fig. 2). Both sexes are diploid for the autosomes (AA). Overwintering eggs (XX/AA) yield asexual females (XX/AA). After several generations of parthenogenesis, asexual females produce males and sexual females parthenogenetically. Sexual females are therefore strict clones of asexual females, hence the XX/AA karyotype (Jaquiéry et al. 2012). In oviparous sexual females, oocytes are generated by a normal meiosis. Males produce sperms containing a single X-chromosome (AX) because one X-chromosome is removed during the maturation division of meiosis and sperms lacking an X-chromosome degenerate (Wilson et al. 1997, Caillaud et al. 2002). When male and female gametes fuse, the eggs are diploid for the X-chromosomes and autosomes, and develop into asexual females (AA/XX), whereas males are diploid for the autosomes and haploid for the X-chromosome (AA/XO). The loss of one X-chromosome may be triggered by specific photoperiodic and/or temperature regimes during oogenesis (Orlando 1974, Blackman 1987) (Fig. 3). A small spermatocyte lacking an X-chromosome accordingly undergoes the first meiotic division, but always degenerates, presumably because it lacks the X-chromosome-specific factors that are essential for complete spermiogenesis (Blackman 1987). Sex determination may depend on the concentration of juvenile hormone, with high titers inducing the development of females and lower titers inducing the development of males (Hales and Mittler 1987).

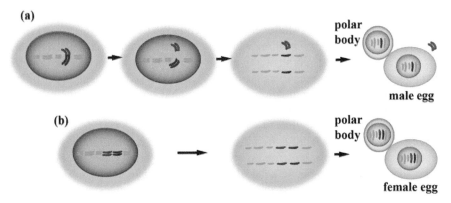

Fig. 3. Distribution of X-chromosomes during maturation divisions of parthenogenetic eggs (according to Orlando 1974 and Blackman 1987). (a) In oocytes destined to be male, sex determination is achieved by a specific meiosis only involving the X-chromosomes (red and blue). They join and form a bivalent with a c-shaped configuration. The terminal connection is then broken, and one X-chromosome (red) is excluded. The remaining X-chromosome (blue) then divides alongside the autosome (green) and one X-chromosome passes into the egg while the other moves into the polar body. (b) In oocytes destined to be female, the meiosis involves both X-chromosomes and they are evenly distributed to polar body and egg.

Influence of Hormones

In all known cases of polyphenism in insects, the switching mechanisms leading to alternative phenotypes involve the timing of hormone secretion, the timing of a hormone-sensitive period, or a threshold of hormone sensitivity (Nijhout 1999, Hardie et al. 1985). Whether the offspring become male, oviparous female or viviparous female is determined by juvenile hormone titers in the mother after photoperiod sensing (Ishikawa et al. 2012). Hormones are therefore the main factors controlling the transduction of photoperiodic signals from the brain to the ovarioles (Le Trionnaire et al. 2008). Prenatal treatment of oviparous embryos with juvenile hormone or its analogs has shown that aphids first develop a mix of viviparous and oviparous ovarioles and then follow the developmental pathway towards the viviparous morph (Mittler et al. 1976, Hardie 1980, Corbitt and Hardie 1985, Hardie and Lees 1985). The adoption of viviparity underlies the maternal regulation of polyphenism. Here, environmental cues are transferred to embryos through the mother via neuro-endocrine signals that ultimately reach the ovarioles (Huybrechts et al. 2010, Ogawa and Miura 2014).

Wing polyphenism is controlled by the endocrine system, which also regulates the development of nymphs into adults (Lees 1977). Ecdysone may also be involved in wing dimorphism (Braendle et al. 2006). The pea aphid neuropeptide hormones are encoded by numerous precursor genes (Christie 2008, Huybrechts et al. 2010, Jedlička et al. 2012). One group belongs to the adipokinetic hormone/red pigment-concentrating hormone (AKH/RPCH) family. AKHs are synthesized and released by neurosecretory cells in the corpora cardiaca and act as mobilizing factors for carbohydrates, lipids and proline (Gäde and Auerswald 2003). This neuropeptide is expressed in the brain of all female forms and in the ovaries of the winged and wingless parthenogenetic forms (Jedlička et al. 2012) (see Chapter 3 "Functional and Evolutionary Genomics in Aphids" in this volume).

Genes encoding the main enzymes that synthesize and degrade juvenile hormone have been detected in the pea aphid genome (Ishikawa et al. 2012), and some of these developmental genes are methylated (Walsh et al. 2010). This supports the hypothesis that methylation plays a role in the developmental polyphenism of aphids as it does in other insects (see Chapter 3 "Functional and Evolutionary Genomics in Aphids" and Chapter 4 "Epigenetic Control of Polyphenism in Aphids" in this volume).

Development of the Female Reproductive System

Like other aphid species, pea aphids have telotrophic meroistic ovaries with six to eight ovarioles (Fig. 1), each featuring a germarium at its anterior tip that evolves from germline precursor cells during embryogenesis. The germarium includes presumptive oocytes (oogonia) and nurse cells (trophocytes). The latter remain in the germarium while a trophic cord supports the transport of nutrients to the oocytes, eggs and embryos (Büning 1985, Blackman 1987).

The primordial germ cells are the common origins of oocytes and represent the ancestors of the germline (Chang et al. 2006). In both asexual and sexual pea aphids, germ cell specification depends on maternally inherited pre-formed germ plasm which can be detected in the posterior region of the egg chamber before the blastoderm stage (Chang et al. 2009, Lin et al. 2014). During blastulation, germline segregation occurs at the posterior of the egg. During early gastrulation, the germ cells are carried inward by the invaginating germ band (Chang et al. 2007). They form two groups and then separate into the future germaria.

A single embryonic germ cell creates 32 oogonial cells within each undifferentiated germarium (Orlando and Crema 1968). As the female embryo grows, the germaria on each side become aligned with the anteroposterior axis, and mesodermal connections arise, linking each germarium to the paired common oviducts (Blackman 1987). Then the gonads follow either the parthenogenetic or amphigonic pathways according to the maternal control center and environment.

Ovarioles of Parthenogenetic Viviparous Females

Each ovariole in a parthenogenetic female contains embryos at various stages of development slightly unsynchronized with the other ovarioles (Fig. 1, Fig. 4). Older embryos already contain early embryonic stages of the subsequent generation, so that up to three generations are represented (Brisson and Stern 2006). They undergo 20 developmental stages and hatch into first-instar nymphs one at a time (Miura et al. 2003).

(a)

(b)

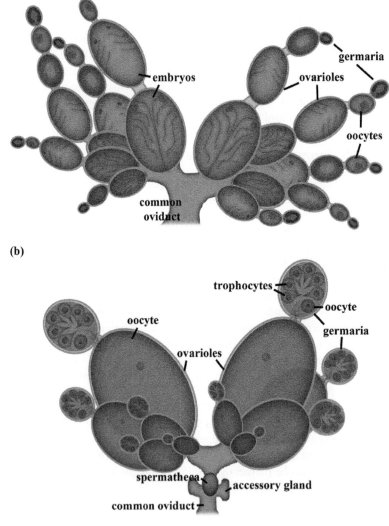

Fig. 4. Reproductive systems of the pea aphid (according to Blackman 1978). (a) Ovary of a parthenogenetic female aphid. (b) Ovary of a sexual female aphid.

The germ cells form spherical, hollow cysts and consecutively develop into trophocytes and presumptive oocytes (Büning 1985, Miura et al. 2003) (Fig. 5). As soon as the germarium is fully formed, the oogonial cell at the posterior region of the germarium increases in size and ovulates in a new follicle chamber enclosed by follicle cells. The trophic cord connects the growing oocyte with the more anterior cells in the germarium.

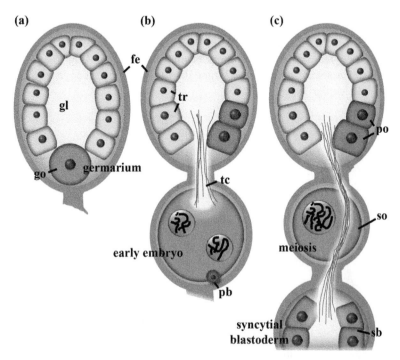

Fig. 5. Organisation of the germarium and developing oocytes in the asexual reproductive system of the pea aphid (according to Blackman 1978, Riparbelli et al. 2005). (a) The follicle epithelium (fe) surrounds the germarium. Inside, trophocytes (tr) envelope the germarial lumen (gl). The first oocyte (go) in the posterior part of germarium differentiates and enters its growth phase. (b) Germarium and first embryo at the 2-cell stage are connected via a trophic cord (tc). Further cells differentiate into new growing oocytes in the germarium. The polar body (pb) is located at the posterior area of the ovarian follicle. (c) While a second oocyte (so) is segregated from the germarium, the first embryo achieves the syncytial blastoderm (sb) stage.

Oocytes in parthenogenetic aphids do not undergo meiosis, a phenomenon known as apomixis. All maternal chromosomes and hence genes are retained (Blackman 1987). This oocyte skips the first meiotic division and undergoes a single maturation division (a modified meiosis II division) which results in a diploid clonal oocyte and a discarded polar body at the posterior end of the follicle (Blackman 1978, 1987). Mitotic cleavage divisions initiate rapid embryonic development. As soon as the current oocyte is dispensed from the germarium, the next presumptive oocyte begins to differentiate (Blackman 1987).

In viviparous oocytes, new centrosomes and microtubule-based asters are spontaneously self-organized (Riparbelli et al. 2005, Le Trionnaire et al. 2008), whereas the centrosome in oviparous oocytes is acquired from the male gamete during fertilization (Riparbelli et al. 2005, Rodrigues-Martins et al. 2007).

Ovarioles of Sexual Oviparous Females

The germarium cells of a sexual female pea aphid are derived from the same single cell that produces the 32 oogonial cells discussed above (Blackman 1978, Miura et al. 2003). Like parthenogenetic development, some of these cells are destined to become oocytes, some of the others become trophocytes (Fig. 6). Generally, the germaria in sexual morphs are larger than those of parthenogenetic females (Tsitsipis and Mittler 1976).

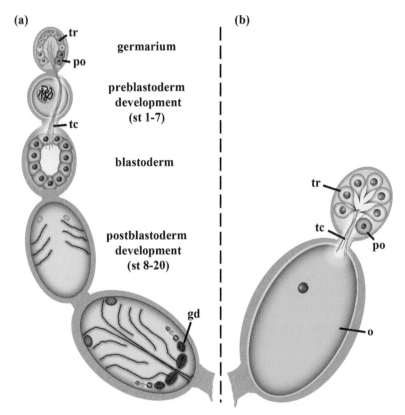

Fig. 6. Organisation of the ovarioles in viviparous and oviparous development (according to Le Trionnaire et al. 2008, Bermingham and Wilkinson 2009). (a) Ovariole of a parthenogenetic female aphid. Besides the trophocytes (tr) in the germarium, the trophic cord (tc) provides nutrition for the oocytes and the embryos. In the germarium the development starts with the differentiation of the first presumptive oocytes (po). Three major developmental stages can be summarized: (I) the preblastoderm development, (II) the blastoderm stage and (III) the postblastoderm development. Mature embryos have already formed new ovarioles with developing embryos (go, granddaughters). (b) Ovariole of a sexual female aphid with germarium and segregated oocyte (o). The trophocytes (tc) display a characteristic teardrop-shape.

The trophocytes arranged in a sphere enlarge substantially by endomitosis and adopt a teardrop shape. The nuclei of the trophocytes become polyploid (Miura et al. 2003). By accumulating large quantities of yolk, the biggest oocyte is released from its germarium and undergoes meiotic recombination. Meiosis is blocked during the first meiotic prophase until the egg is fertilized and laid (Blackman 1987). The oocyte is fully grown when it passes into the common oviduct. Spermathecae at the lower end of the oviduct release sperm to fertilize each oocyte as it passes (Fig. 4). Following spermatozoid penetration, two maturation divisions yield two polar bodies (Blackman 1987). The accessory glands secrete a protective coating and adhesive substances onto the surface of the egg just before oviposition. The fertilized oocytes commence embryogenesis once they have been deposited onto leaves.

Ovarioles in Ambiphasic Females

The arrangement of the ovarioles in the mother differs in asexual and sexual females according to the specific conditions in which they evolved. Individuals whose ovaries contain both parthenogenetic and sexual ovarioles are described as ambiphasic (Pagliai 1965, Crema 1971, 1973). The parthenogenetic ovarioles are located anteriorly in ambiphasic females that are born when the mother stops producing parthenogenetic daughters and the production of sexual daughters commences. In contrast, the parthenogenetic ovarioles are located posteriorly in ambiphasic females born when the mother ceases sexual development and initiates the development of parthenogenetic daughters (Crema 1973).

Embryogenesis

The parthenogenetic cycle is thought to have evolved more than 250 million years ago from primitive aphids with an obligatory sexual reproductive mode (Dixon 1990, Moran 1992, Tomiuk et al. 1994, Hales et al. 1997, Gallot et al. 2010). The adelgids (Adelgidae) and phylloxerids (Phylloxeridae) are closely related to the true aphids and have similar alternating parthenogenetic and sexual generations, but both the parthenogenetic and sexual females are oviparous (Granett et al. 2001, Havill and Foottit 2007, Ogawa and Miura 2014). Parthenogenesis was already firmly established before viviparity, but although the pea aphid alternates between the two facultative reproductive modes, a large number of aphid species have lost the sexual phase of their life cycles (Moran 1992, Simon et al. 2002, 2003).

Cytological observations during aphid parthenogenesis suggest that asexual oogenesis could have evolved directly from a mitotic process or as a modification of meiosis, but there is more evidence for the latter (Srinivasan et al. 2014). Furthermore, most pea aphid genes relevant to meiosis have similar expression profiles in the sexual and asexual ovaries, although members of the Argonaute family are exceptional in that they are not expressed in sexual ovaries (Lu et al. 2011).

Parthenogenetic egg development requires the mother to contribute nutrients, whereas the embryo in a sexually-produced egg is independent and can draw resources only from the yolk. Therefore, parthenogenetic embryos develop in 10–15 days, whereas the sexually-produced embryos require 80–140 days. Despite the differences in maturation time, the two reproduction modes generate apparently identical first-instar nymphs (Blackman 1978, Miura et al. 2003).

Detailed descriptions of pea aphid embryonic development have been provided by several authors (Blackman 1978, Büning 1985). Furthermore, numerous pea aphid orthologs developmental genes have been identified and excellent morphological data are often provided by whole-mount in situ hybridization (Chang et al. 2006, 2007, 2009, 2013, Lin et al. 2014). In the following overview we refer to staging schemes for parthenogenetic (Fig. 7, Fig. 8) and sexual embryos based on the morphological characteristics described by Miura et al. (2003) and complete them with data from the expression profiling of developmental genes.

Parthenogenetic Embryonic Development

Stage 0: Development begins with the formation of the oocyte in the germarium. The follicle cells form an epithelium enclosing an area devoid of cells at the posterior of the germarium. A nucleus and its associated cytoplasm become the prospective oocyte. They move into this area from the germarium and increase in size. Among genes that are preferentially expressed in germ cells, *nanos* and *vasa* are among the most conserved germline markers in animals. The pea aphid has a single *nanos* homolog (*Apnos*) and four *vasa* homologs, although only *Apvas1* is germline specific and can be used as a germline marker throughout viviparous development and also an assumed marker in oviparous embryos (Lin et al. 2014). *Apnos* and *Apvas1* (hereafter *Apvas*, because the early literature did not distinguish among the four paralogs) are found in the cytoplasm of the nurse cells and in the developing oocytes.

Stage 1: An oocyte is segregated from the germarium but still remains connected by the trophic cord until blastoderm formation. *Apnos* is preferentially expressed in granules in the anterior region and posterior to the oocyte nucleus. Weak expression is visible in the peripheral region of the oocyte and in the oocyte cytoplasm. *Apvas* is evenly distributed without distinct localization.

Stage 2: The oocyte undergoes a modified form of meiosis involving a single maturation division (without a reduction division) and produces a polar body that remains at the posterior of the oocyte. The nucleus moves towards the anterior. *Apnos* is expressed primarily in the cytoplasmic granules, around the oocyte, and in the posterior region of the oocyte.

Stage 3: Early nuclear cleavage divisions (karyokinesis) occur synchronously four times without division of the cytoplasm to produce a 16-nucleus embryonic syncytial stage. *Apnos* transcripts are uniformly distributed in the cytoplasm. *Apvas* is randomly distributed in the oocyte cytoplasm in stage 3.

Stage 4: By the time the embryo has 16 nuclei it has doubled in length and most of the nuclei migrate peripherally to form a syncytium. The uniform distribution of *Apnos* transcripts is retained.

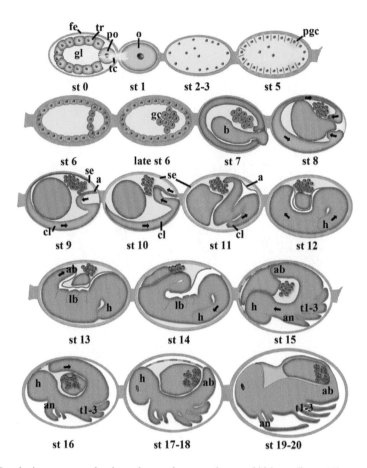

Fig. 7. Developing oocytes and embryos in a parthenogenetic pea aphid (according to Miura et al. 2003, Chang et al. 2007, 2009, 2013, Huang et al. 2010). Eggs in lateral view: anterior on the left, posterior on the right, dorsal above and ventral below. (stage (st) 0) The germarium with surrounding follicle epithelium (fe), growing presumptive oocytes (po), and trophocytes (tr) that envelop the germarial lumen (gl). (st 1) Segregated oocyte (o) with one nucleus supported by the trophic cord (tc). (st 2–4) Nuclear division after DNA replication and migration of the dividing nuclei to the periphery of the embryo. (st 5) Nuclei migrate to the periphery of the egg and cell membranes are formed around the nuclei. The primordial germ cells (pgc) are detectable as a ring around the posterior region of the egg. (st 6) Largely cellularized blastoderm and morphogenesis as well as progeny of germ cells (gc) which are localized in the posterior interior of the oocyte. (st 7) Incorporation of maternal endosymbiotic bacteria (b). (st 8) Germband with thickened cells on the ventral side is migrating posteriorly and cells at the posterior part of the embryo invaginate. (st 9) Gastrulation processes start (arrow), the two extraembryonic membranes serosa (se) and amnion (a) are differentiated and the cephalic lobe (cl) develops. (st 10) Invaginating germband induces anatrepsis. (st 11) S-shaped embryo. (st 12) Invaginating germband bends dorsally and the prospective head (h) region can be distinguished. (st 13) Limb bud (lb) formation and dorsal movement of the prospective abdomen (ab). (st 14) Induction of katatrepsis by ventral and anterior movements of the embryo (arrow). (st 15) Katatrepsis and formation of the antennae (an) and growth of thoracic (t1-3) and abdominal anlagen. (st 16) Embryo after katatrepsis with head at the anterior region of the egg chamber, germ cells in dorsolateral region. Movement of abdominal region towards the posterior part of the egg. (st 17–18) Germband retraction and pigmentation of eye spots. (st 19–20) Mature embryo prior to larviposition. Germband retraction is completed. Germ cells dorsally and bilaterally adjacent to the dorsal midline. Final eye, brain and ganglia differentiation as well as muscle formation.

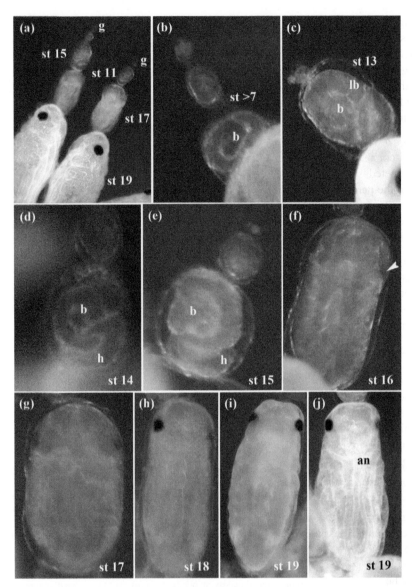

Fig. 8. Microscopic graphs of embryos in different stages of development in dissected ovarioles of parthenogenetic females (according to the staging scheme by Miura et al. 2003). (a) Arrangement of embryonic stages in two ovarioles with younger embryos located adjacent to the germaria (g). (b) Early developmental stages with incorporated maternal endosymbiotic bacteria (b) which are blue translucent. These bacteria invade at stage 7 and can be tracked throughout development. (c) Stage 13 is characterized by the limb bud (lb) formation. The bacteria move dorsally. (d) In stage 14 the embryo is curled, segmentation is completed and had formed all segments and the head (h) region is distinguishable. Bacteria move to the dorsoanterior pole. (e) In stage 15 the head of the embryo moves towards the anterior pole and after a total inversion of the embryo, the head becomes oriented towards the germarium. (f) In stage 16, the first developing eye spots become visible (arrowhead). (g) At stage 17, eye pigmentation begins. (h) Dorsal view of stage 18. The muscle segments are completed. (i) Dorsal view of an almost mature embryo at stage 19. (j) Ventral view of stage 19: the long antennae (an) and legs are attached to the body.

Stage 5: At the 32-cell stage, cell membranes begin to form. But some of the posterior-most nuclei do not cellularize and dorsal-ventral asymmetry appears. Syncytial nuclei also remain in the blastocoel and form the yolk cells (vitellophages). At this stage, the trophic cord still maintains a functional connection with the trophocytes in the germarium. Around the most posterior cellularized nuclei of the blastoderm, the formation of the primordial germ cells occurs and *Apnos* transcripts are found in their cytoplasmic granules. *Apvas* is also expressed in the presumptive germ cells.

Stage 6: The blastomeres divide further, especially at the anterior end, and the trophic cord connection is then lost. Shortly after the cellularization of the blastoderm is completed, some of the posterior primordial germ cells of the blastoderm invaginate from the periphery towards the center of the embryo. *Apnos* and *Apvas* transcripts are restricted to the cytoplasm of these morphologically identifiable germ cells. So called mycetoblasts that form a syncytial tissue in the posterior part of the yolk await the invasion of endosymbionts.

Stages 7–8: In the late blastoderm just before gastrulation, maternal endosymbiotic bacteria are incorporated through an opening at the posterior pole (described in detail later). *Apnos* transcripts cannot be identified in the primordial germ cells, which move dorsally towards the invading bacterial mass. Instead, the germ cells can be specifically stained for *Apvas* transcripts from this stage until the end of development.

Stage 9: Dorsoventral differentiation becomes prominent during gastrulation: the dorsal blastoderm becomes thinner and later differentiates into a distinct extraembryonic membrane (serosa) whereas the ventral blastoderm thickens and gives rise to the germ band. The dorsal cells of the invaginating germ band form the amnion. The serosa is necessary to protect insect eggs from infection (Jacobs and van der Zee 2013, Jacobs et al. 2013). The serosa and the amnion are involved in several key morphogenetic events (blastokinesis, including anatrepsis and katatrepsis) at specific developmental stages (Panfilio 2008).

Stage 10: Germ cell specification is completed and germline migration is driven by the invaginating germ band, which induces the process of anatrepsis. The bacterial channel through the embryo has been closed and the symbionts move towards the anterior of the embryo. The germ cells start migrating out of the posterior egg chamber and keep in contact with the bacteria and the invaginating germ band, but they remain in their dorsal position. The ventral epithelium is specified to form the cephalic lobe, and the serosa forms from the dorsal-most epithelium, which subsequently envelops the embryo and the yolk and secretes the cuticle.

Stage 11: The germ band continues to invaginate from the dorsal and ventral sides, and after reaching the anterior pole it continues to elongate, flexing ventrally so that the embryo adopts a highly-folded, S-shaped configuration. This invagination (anatrepsis) results in the inversion of the anteroposterior and dorsoventral axes of the embryo. All of the bacteria have been transferred and the channel through the follicular epithelium has presumably been closed. The symbionts are pushed towards the anterior of the embryo, apparently enclosed by a membrane with nuclei in close proximity to the package of bacteria. The germ cells remain closely associated with the bacteria and the invaginating germ band.

Stage 12: The invaginating germ band bends dorsally. The germ cells remain associated with the anterior tip of the invagination. The epithelium on the ventral side of the embryo thickens while the dorsal epithelium, the serosa, becomes thinner.

Stage 13: The germ band elongates further and the limb buds are formed. The bacteriocytes move slightly anteriorly in the direction of the germarium. The germ cells remain at the anteroposterior midline but are dorsally displaced. The cephalic lobe is positioned ventrally, the thorax in the middle and the elongating abdomen dorsally.

Stage 14: The bacteriocytes are pushed to the most anterior pole of the embryo and are then moved dorsally as the germ band elongates. The germ cells are located in an anterodorsal region, sitting above the future abdomen. The four developing segments are the three thoracic segments and the first abdominal segment.

Stage 15: As the germ band continues to elongate it twists. After the completion of segmentation, the embryo undergoes a dramatic repositioning (katatrepsis) resulting in the inversion of the dorsoventral and anteroposterior axes. The head of the embryo, which has been positioned ventrally until this point, moves anteriorly until it reaches the anterior pole. First, the bacteriocytes continue to be pushed dorsally and by the end of this stage, the germ cells are located mainly in the center of the egg and dorsal to the twisted embryo. At the end of this stage, three cephalic, three thoracic, and five abdominal segmental anlagen can be distinguished.

Stage 16: After katatrepsis, the embryo has migrated to the anterior region of the egg chamber so that the head is positioned towards the germarium. The posterior of the germ band is folded dorsally. The germ cells are dorsally beneath the future abdomen and now separated into two groups of cells located bilaterally behind the thorax. The bacteriocytes are also positioned dorsally, nestled within the folded germ band. The limbs continue to elongate.

Stage 17: The germ band begins to retract during stage 17 and the legs continue to grow. Two groups of germ cells appear as a "V" shape in the dorsal part of the embryo. In late stage 17, the germ band has largely retracted, and the posterior tip of the abdomen is retracting to the posterior dorsal region of the embryo. The germ cells are now adjacent to the posterior end of the abdomen. The bacteria are now surrounding the germ cells.

Stages 18–20: The legs and antennae are extended to the posterior tip of the body, and the germ band is almost completely retracted. The germ cells are located dorsally and bilaterally adjacent to the midline in the abdomen. The brain and thoracic ganglia are compacted, the compound eyes differentiate and individual muscles begin to form. The embryo also grows rapidly and the muscle cells fuse. The mature embryo, just prior to larviposition, has a fully formed cuticle.

Embryonic Development in the Fertilized Eggs

Development starts inside the oviparous ovariole of the sexual female mother with the formation of the oocytes. Immediately after egg deposition onto leaves, the egg passes through the stages of cleavage, invagination, segmentation, and organization of the body appendages. Further development is delayed at low winter temperatures but does not cease completely. With rising temperatures in spring, the embryos undergo blastokinesis in the egg (Böhmel and Jancke 1942). When deposited, the egg has a vitelline membrane and a cuticle which darkens due to tanning by the serosa (Miura et al. 2003).

Previous studies of pea aphid sexual embryogenesis have used a developmental timescale based on days rather than the stages used for parthenogenetic embryogenesis (Miura et al. 2003, Lin et al. 2014). The comparison of parthenogenesis and sexual development is easier if the systems are superimposed. Stages 1–4 take place in the oviparous ovariole of the sexual female, whereas subsequent stages are external (Fig. 9).

Stage 0: Development starts with the formation of the oocyte. The teardrop-shaped polyploid trophocytes produce yolk for the developing oocytes. *Apvas* is uniformly expressed in the cytoplasm of the trophocytes. The oocytes have grown to different sizes, and the first to be released from its germarium is the largest which has accumulated large quantities of yolk.

Stage 1: A previtellogenic oocyte is segregated from the germarium. The expression of *Apvas*, labeling the presumptive germ plasm, is evenly distributed in the oocyte. In larger vitellogenic oocytes, *Apvas* is detected in the anterior and posterior regions of the egg.

Stage 2: The mature oocytes undergo oviposition. The eggs are filled with yolk except for a posterior region where a package of endosymbiotic bacteria (mycetome) from the maternal hemocoel is positioned. Meiosis is blocked at the first meiotic prophase until the egg is fertilized by the sperms stored in the spermathecae.

Stage 3 (0–16 h after egg laying (AEL)): A freshly-deposited sexual egg is cream colored and begins to undergo superficial cleavage. Synchronous mitosis gives rise to numerous cleavage energids in small islands of cytoplasm. The embryos undergo 10 consecutive nuclear divisions. The mitotic nuclei have an anteroposterior mitotic gradient, and the posterior of the egg contains the bacterial mass. An *Apvasa*-positive stripe of primordial germ cells with dividing nuclei lies anterior to the bacterial mass.

Stage 4 (16–24 h AEL): Cellularization of the superficial energids takes place and the syncytial blastoderm becomes cellular with cells migrating to the periphery while some remain within the egg to become vitellophages.

Stage 5 (16–24 h AEL): The nuclei of the blastoderm at the periphery are not distributed uniformly and their allocation changes several times. At the beginning of this stage, the greatest density of nuclei is found in the anterior region and then they move towards the posterior pole. *Apvas* remains as a stripe in the posterior cortex, anterior to the bacteria.

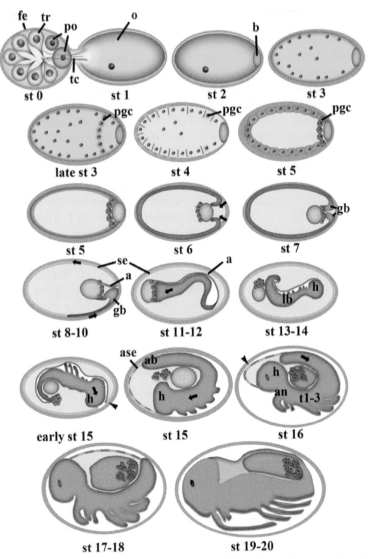

Fig. 9. Developing oocytes and embryos in oviparous pea aphid (according to Miura et al. 2003, Shingelton et al. 2003, Lin et al. 2014). Eggs on lateral view: anterior on the left, posterior on the right, dorsal above and ventral below. (st 0) Germarium with surrounding follicle epithelium (fe) comprises tear-drop like trophocytes (tr) and primordial oocytes (po). A trophic cord (tc) connects the germarium with the first two egg chambers. (st 1) Segregated oocyte in different stages of vitellogenesis (previtellogenic and vitellogenic). (st 2) Bacterial endosymbionts (b) are visible in the egg posterior. (st 3) Energids undergo synchronous divisions and numerous nuclei migrate to the periphery. The density of nuclei at the egg periphery changes several times. A stripe of cell nuclei appear at the posterior region of the egg, which can be labeled with germ line markers. These become the presumptive germ cells (pgc). (st 4) Beginning of cellularization of the syncytial blastoderm. (st 5) In the blastoderm stage, the germ cells (gc) are detectable at the posterior region of the egg anterior to the bacteria. (st 6) Late blastoderm stage at the beginning of gastrulation with bacterial symbionts that invade through an open tube in the posterior blastoderm (arrows). (st 7) Germ cells are located between bacterial mass and invading blastoderm. Dorsal and ventral invaginating blastoderms

Fig. 9. contd....

Stage 6 (24–36 h AEL): An invaginating furrow in the posterior blastoderm becomes a temporary tube which drops into the center of the egg. At the leading end of the tube the bacteria move into the interior of the egg. The primordial germ cells move to encounter the anterior border of the bacterial mass.

Stage 7 (36–48 h AEL): The bacteria migrate towards the inner region of the egg and become fully immersed within the yolk, while the peripheral germ cells remain located between the migrating bacterial mass and the dorsal and ventral invaginating posterior blastoderm, which successively evolve into the germband.

Stages 8–10 (2–3 d AEL): By the second day, the egg turns green and the invaginating blastomeres at the posterior of the egg become more columnar and constitute the germ band at the ventral side of the tube. The remaining blastomeres form two embryonal envelopes: the serosa and the amnion on the dorsal side of the tube. By the third day, a serosal cuticle is generated.

Stages 11–12 (3–6 d AEL): The germ band moves towards the anterior region of the egg. Anatrepsis begins with the inversion of the anteroposterior and dorsoventral axes of the embryo, so that the dorsal side of the embryo points towards the ventral side of the egg. In the earliest embryos, it is possible to define three thoracic segments and the first abdominal segment.

Stage 13 (6–14 d AEL): When the germ band elongates further, the segmentation of the limb buds continues, the head capsule is formed, and the thorax and abdomen become distinct. The *Apvas* stripe is transformed into a globular, extra-embryonic structure located in close proximity to the elongating abdomen.

Stage 14 (14–25 d AEL): When the embryos are fully segmented, the primordial germ cells form a U-shaped ring. The embryos enter the diapause at the end of anatrepsis and remain in this state over winter (Shingelton et al. 2003).

Stage 15 (end of diapause): In early stages, the amnion comes in contact with the serosa. During katatrapsis, the amniotic cavity opens and the embryo turns to adopt its final orientation, the head moves ventrally and anteriorly. The embryo is then orientated with its head towards the anterior of the egg and its ventral surface towards

---

Fig. 9. contd.

fuse and the germband (gb) is formed. (st 8–10) The dorsal side becomes the amnion (a). The developing serosa (se) spreads across the egg. The germband thickens and migrates towards the center of the egg. (st 11–12) The germband elongates and reaches the anterior pole. Anatrepsis starts. (st 13–14) The embryo completely sinks into the yolk. At the end of anatrepsis, the head (h) points towards the posterior of the egg and the thorax region with the first limb buds (lb) become visible. The dorsal side of the embryo is located adjacent to the ventral side of the egg. (st 15) At early stage 15 the katatrepsis process begins and the head region migrates towards the ventral side of the egg (arrow). Serosa and amnion converge (arrowhead) and form the amnioserosa (ase). During katatrepsis the head moves to the anterior pole. The abdomen (ab) is located dorsoanteriorly. (st 16) The embryo has a cap of amnioserosa at its anterior end (arrowhead). Antennae (an) and legs (t1-3) are growing. (st 17–18) The germband retracts and the amnioserosal cap is reduced at the dorsal side of the embryo. The germ cells migrate to a posterior position bilaterally adjacent to the dorsal midline. (st 19–20) Embryo is in a stage of final differentiation. The integument is formed and the dorsal closure is completed.

the ventral surface of the egg. The amnion and the serosa have fused together into the amnioserosa. The primordial germ cells and bacteria are separated and located bilaterally in the dorsal region of the abdomen.

Stage 16: The embryo has almost completed katatrapsis and has a cap of amnioserosa at its anterior end. The abdomen moves to the posterior part of the egg.

Stages 17–18: After katatrepsis, the embryo is stretched on the yolk surface, with its ventral side directed outwards. The amniosersal cap is reduced and the limbs continue to elongate. The germ cells are now adjacent to the posterior end of the abdomen.

Stages 19–20: The lateral body walls grow up to close over the dorsal surface and the embryo has formed an embryonic cuticle. The final stages of organs differentiation take place.

Embryo Nutrition

Aphid ovarioles are meroistic telotrophic structures in which a trophic cord links the germarium to the ovariole. At the beginning of development, both the resting oocytes in the germarium and the newly extruded oocytes that have already left the germarium receive nutrients from the trophocytes via these trophic cords. The trophic cord allows the first two follicular chambers to keep in contact with the trophocytes. Structurally, they are microtubule-rich, elongated intercellular bridges of the trophocytes that penetrate through the central core of the germarium to contact the oocytes (Bennet and Stebbings 1979, Couchman and King 1979, Hyams and Stebbings 1979, Huebner 1981, Büning 1985, 1998, Blackman 1987). It is presumed that nutrients are transported by peristaltic movement and electric and/or osmotic forces (Bermingham and Wilkinson 2009).

During later development from the blastula stage onwards, the connection via the trophic cord breaks and the ovariole sheath thickens and adopts a trophic function, thus adopting the most important role in the supply of nutrients from the maternal hemolymph to the oocytes (Bermingham and Wilkinson 2009). The ovariole sheath consists of a unicellular epithelial layer enclosing the germarium and the growing oocytes and embryos, with numerous lacunae and septate junctions. The nutrients that sustain the developing embryos are transferred by passive diffusion through intercellular spaces in the sheath (Couchman and King 1980, Blackman 1987, Bermingham and Wilkinson 2009). Particularly during the vitellogenic growth phase, receptor-mediated uptake of vitellogenin by the oocytes may promote oocyte growth (Büning 1998). The embryo subsequently develops the serosa and later the amnioserosal membrane, which thickens and accumulates more organelles. Together with the ovariole sheath, the serosa and amnioserosal membrane provide the embryo with nutrients (Blackman 1987, Bermingham and Wilkinson 2009).

Almost all aphids contain bacterial symbionts that contribute to their survival (see Chapter 5 "Bacterial Symbionts of Aphids (Hemiptera: Aphididae)" in this volume). The bacteria are localized in specialized cells known as bacteriocytes (Braendle et al. 2003, Baumann 2005, Moran et al. 2008, Koga et al. 2012), which cluster in the hemocoel close to the ovarioles, and are transferred from adult to sexual or parthenogenetic eggs (Douglas 1989). The bacteriocytes remain tightly associated with the germ cells throughout embryonic development (Miura et al. 2003, Bright and Bulgheresi 2010). The intracellular bacterial species *Buchnera aphidicola* is an obligate symbiont in the body cavity of the pea aphid and lives permanently within its host cells. This bacterium was probably present in the common ancestor of aphids more than 150 million years ago (Baumann et al. 1997).

The pea aphid genome has lost many genes, and the synthesis of several essential amino acids requires genes that are distributed between the genomes of the pea aphid and *B. aphidicola* (Shigenobu et al. 2000, International Aphid Genomics Consortium 2010) (see Chapter 3 "Functional and Evolutionary Genomics in Aphids" in this volume). The bacteria produce essential amino acids that the aphid cannot synthesize *de novo* (Mira and Moran 2002, Koga et al. 2012). *B. aphidicola* genes therefore particularly active in young embryos, especially those required for the production of riboflavin (vitamin B$_2$) and the flagellar apparatus, which may facilitate the transport of maternally-expressed proteins across cell boundaries (Bermingham and Wilkinson 2009). Tritium-labeled thymidine can be incorporated into embryonic nuclei 30 min after injecting the mother (Blackman 1974). *B. aphidicola* gene expression can be regulated by the aphid host in an age-dependent manner, i.e., endosymbionts in old aphids synthesize proteins that are not produced in the young and vice versa (Ishikawa 1983).

The pea aphid may also host several facultative symbionts which are discussed in more detail in Chapter 5 "Bacterial Symbionts of Aphids (Hemiptera: Aphididae)" in this volume. They are not present in all individuals and are not necessary for survival or reproduction, but they can but nevertheless strongly influence aphid resistance mechanisms against stress factors such as hear or parasites (Montllor et al. 2002, Oliver et al. 2003, Kerry et al. 2005, Leonardo and Mondor 2006). Some of them can affect the reproduction of their host and thereby enhance their transmission in the host population (Simon et al. 2011). Moreover, aphids that are infected by *Spiroplasma* spp. cannot produce male descendants because they induce a male-killing phenotype (Simon et al. 2011). The facultative symbiont *Regiella insecticola* can influence aphid dispersal by reducing the production of winged offspring and changing the timing of sexual development (Leonardo and Mondor 2006).

B. aphidicola is found in bacteriocytes consisting of large polyploid cells. These are grouped in so-called mycetomes or bacteriomes in the aphid hemocoel and remain associated with the germ cells throughout development (Douglas 1989, Miura et al. 2003, Bright and Bulgheresi 2010). Symbionts are often transmitted transovarialy through the female germ line (Buchner 1965, Bright and Bulgheresi 2010). Vertical transmission is dominant with occasional horizontal transfer within or between species or from the environment. Pea aphid females often mate with more than one male, so that the offspring can acquire the symbionts from several different males (Moran and Dunbar 2006). Therefore, biparental transfer of bacteria in the pea aphid often includes facultative symbionts such as *Serratia symbiotica*, *Hamiltonella defense* and *R. insecticola* (Moran et al. 2005, Moran and Dunbar 2006, Bright and Bulgheresi 2010, Oliver et al. 2010).

The mechanism of embryonic colonization by *B. aphidicola* varies according to its sexual or parthenogenetic origin (Fig. 10). During parthenogenetic development, maternal bacteria are deposited within a syncytium in the center of the blastoderm of young embryos through the posterior follicular epithelium (Buchner 1965, Miura et al. 2003). Several hypotheses have been proposed to explain the transfer of bacteria from mother to embryo (reviewed by Koga et al. 2012). For example, a membrane-bound maternal bacterial package may fuse with the follicular epithelium in the region of enlarged posterior follicle cells surrounding the blastula, allowing a channel to form between the follicle cells through which bacteria flow into the posterior of the embryo (Miura et al. 2003). Alternatively, *B. aphidicola* cells may be transmitted from maternal bacteriocytes, released into the extracellular space and then transferred to the cytoplasm of the adjacent syncytial blastula by endocytosis (Koga et al. 2012). It remains unclear whether the blastula induces the exocytosis of *B. aphidicola* cells from the adjacent maternal bacteriocyte or whether the maternal bacteriocyte induces endocytosis by

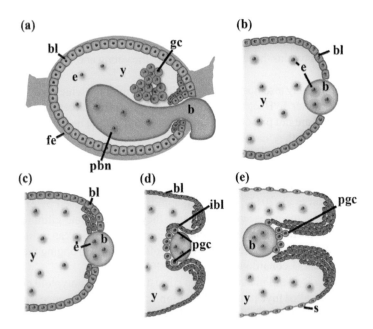

Fig. 10. Maternal transfer of endosymbiotic bacteria in parthenogenetic and fertilized eggs of the pea aphid (according to Miura et al. 2003, Lin et al. 2014). (a) The follicle epithelium (fe) surrounds an egg chamber at stage 7 during parthenogenetic development. The cellularization of the blastoderm (bl) is in progress and the germ cells (gc) lie dorsally to invading bacterial mass (b), which encloses presumptive bacteriocyte nuclei (pbn). The inner yolk (y) contains energids (e), which subsequently become vitellophages. (b–e) Incorporation of bacteria during oviparous development. (b) The endosymbiotic bacteria invade from the posterior pole of the egg. (c) Blastodermal cells invaginate adjacent to the bacterial mass. (d) The blastodermal cells and the bacterial mass migrate into the egg chamber. The primordial germ cells are peripherally located before they co-migrate with the bacteria to the inner region of the egg, pushed by the continuous invaginating blastoderm (ibl). (e) The blastoderm thickens and extends towards the interior of the egg chamber. When the bacteria are immersed in the yolk mass, the primordial germ cells migrate to the posterior cortex of the bacterial mass. At this time, the first extraembryonic membrane, the serosa (s) is differentiated.

the syncytial cytoplasm of the posterior blastula region. Freely circulating symbionts in the hemolymph could also infect the posterior region of the blastula.

The bacteriocytes descend from two different populations of nuclei in the primary embryonic syncytium, which are recruited at different times and locations during aphid development (Braendle et al. 2003). These nuclei express genes that play important roles during later embryonic development, including *distalless* (later active in limb development), which appears in the posterior blastoderm nuclei before bacterial invasion, *Ultrabithorax* and *abdominal-A* (later involved in patterning the thorax and abdomen) which are expressed after the bacteria have assembled with the nuclei, and *engrailed* (later required for segmentation) which is expressed when bacterial transfer is complete. The syncytial nuclei then develop into cellularized bacteriocytes. Later, the same genes are expressed in a second population of nuclei in the dorsal posterior embryonic region, and these migrate along the germ band, connect with the first population of bacteriocytes and take up more bacteria (Braendle et al. 2003).

Prior to fertilization in the sexual reproductive mode, posterior follicle cells lengthen along their apical-basal axis just before symbiont invasion (Bright and Bulgheresi 2010). Bacteria are released by maternal bacteriocytes and form a cup around the posterior part of the egg. The bacteria probably enter the embryo through filamentous actin-reinforced channels in the follicular epithelium (Bright and Bulgheresi 2010) or they may be taken up by endocytosis (Buchner 1965, Douglas 1998, Mira and Moran 2002). The invagination of the blastoderm then forms a blastopore and the bacteria are pulled into the center of the egg in close proximity to the prospective germ cells (Lin et al. 2014). The embryonic bacteriocyte nuclei then migrate to the bacterial mass and form the bacteriome (Mira and Moran 2002).

Gene Expression during Aphid Development

Pea aphids are host plant specialists that feed and develop on pea plants (*Pisum sativum*), they show several types of polyphenism and they require obligate bacterial symbionts (International Aphid Genomics Consortium 2010). These unusual features explain several unique genomic features, including numerous aphid-lineage-specific gene losses and gene duplications. For example, the pea aphid genome contains all the genes required for epigenetic regulation by DNA methylation (see Chapter 4 "Epigenetic Control of Polyphenism in Aphids" in this volume) whereas many genes encoding immune system components have been lost (see Chapter 6 "Aphid Immunity" in this volume). Numerous well-conserved genes that control insect embryogenesis are present in the pea aphid but others have been lost or duplicated, including those representing key signaling pathways and transcription factors (Shigenobu et al. 2010). Although pea aphid populations descending from a single mother are genetically identical, the asexual lineages mutate rapidly and some of these modifications are adaptive and transferable (reviewed by Loxdale 2008).

The genes thus far known to contribute to pea aphid embryogenesis have been compiled and discussed in detail by Cortés et al. 2008, Shigenobu et al. (2010), International Aphid Genomics Consortium (2010), Bickel et al. (2013) and Duncan et al. (2013a). The following sections focus on genes that control aspects of axis formation and germ-line specification that distinguish parthenogenetic and sexual development (see also Chapter 3 "Functional and Evolutionary Genomics in Aphids" in this volume). These include sets of genes expressed solely in one developmental mode, whereas others are required in both developmental processes (Shigenobu et al. 2010). For example, *Isd1* is thought to be responsible for yolk accumulation and is specifically expressed during sexual oogenesis (Gallot et al. 2012). Aphid-lineage-specific gene duplications that produce gene families with roles in development are likely to have acquired new functions in the later stages of each developmental mode (Shigenobu et al. 2010).

Body Axis Formation—Key Maternal and Terminal Patterning Genes

Differential gene expression during aphid development supports the divergent developmental programs in the oviparous and viviparous reproductive cycles (Gallot et al. 2012, Lin et al. 2014). Duncan et al. (2013a) presumed that differences in the developmental environment (asexual development inside the mother's body versus external sexual development) trigger differences in gene expression and that early parthenogenetic embryogenesis has evolved as a novel mode of development.

The so-called terminal patterning system is responsible for defining and patterning the anterior and posterior of the embryo (Weisbrod et al. 2013). Gallot et al. (2012) found that more than 30 axis patterning genes were differentially expressed in a comparison of synchronized asexual and sexual embryonic developmental phases. Some genes were unique to one developmental mode, and others were expressed in both modes but the spatiotemporal expression profiles and/or expression levels were distinct. Terminal patterning is initiated by the broadly-distributed tyrosine kinase receptor Torso (Sprenger et al. 1989). In the pea aphid, this system is conserved during oviparous development, but is inactive during viviparous development: oviparous ovaries express *torso-like* in somatic posterior follicle cells and activate ERK MAP kinase in the posterior of the oocyte, whereas these components are not expressed during viviparous development (summarized in Bickel et al. 2013).

In insects, the anteroposterior and dorsoventral body axes are established in the oocytes. Maternal effect genes expressed in the maternal ovaries produce mRNAs that are placed in different regions of the oocyte, and this asymmetric localization breaks the egg symmetry before fertilization (Chang et al. 2013). After fertilization, the regulatory proteins encoded by these mRNAs diffuse through the syncytial blastoderm and form so-called morphogen gradients that activate or repress the expression of specific zygotic genes in a concentration-dependent manner. Key maternal genes that pattern the embryonic axes include *bicoid* (*bcd*), *hunchback* (*hb*) and *orthodenticle* (*otd*), which regulate the production of anterior structures, as well as *caudal* (*cd*) and *nanos* (*nos*), which regulate the formation of the posterior structures (Duncan et al. 2013). The expression profiles of these genes are distinct during the early development of viviparous and oviparous embryos (Bickel et al. 2013, Chang et al. 2013, Duncan et al. 2013b, Lin et al. 2014) (Fig. 11, Fig. 12). Neither viviparous nor oviparous embryos appear to express *bcd* (Huang et al. 2010, Shigenobu et al. 2010) and instead this function is carried out by genes that otherwise act downstream of *bcd*, namely

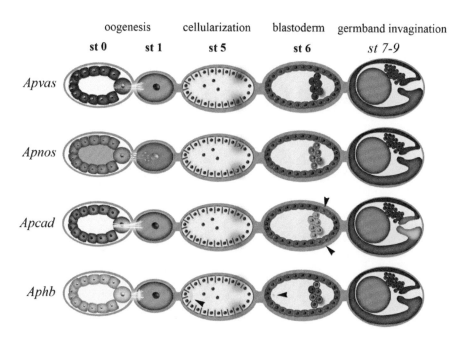

Fig. 11. Expression of key maternal and patterning genes involved in early development of oocytes and embryos of viviparous pea aphid (according to Duncan et al. 2013a,b, Chang et al. 2006, 2007, 2009, 2013, Lin et al. 2014). Anterior is to the left, posterior to the right and dorsal uppermost. *Apvas* (highlighted in red) can be detected in the cytoplasm of the trophocytes and in the primordial oocyte in the germarium (stage 0). In stage 5, in the primordial germ cells, and from stage 6 onward, *Apvas* indicates the germ cells throughout the while embryogenesis. *Apnos* (highlighted in blue) transcripts accumulate in the germarial lumen, in the cytoplasm of the trophocytes and the primordial oocytes, and in the trophic cord (stage 0). In segregated oocytes (stage 1), *Apnos* shows weak expression in the cytoplasm, but occurs in granules in the anterior region of the oocytes and posteriorly to the oocyte nucleus. In stages 3–4 (not shown) *Apnos* is evenly distributed in the cytoplasm of the oocytes, whereas in stage 5 and stage 6, *Apnos* transcripts are restricted to the cytoplasmic granules of the primordial germ cells. From stage 7 onwards, the expression of *Apnos* is unidentifiable in the germ cells. *Apcad* (highlighted in green) transcripts are first detectable in germ cells and their surrounding blastoderm (arrowheads) at stage 6. In the following development, its expression is down-regulated in the germ cells but is upregulated in the posterior blastoderm (arrowheads) and later in the invaginating germ band (stages 7–9). In later stages (not shown) *Apcad* expression becomes restricted to the posterior most region of the abdomen. *Aphb* (highlighted in yellow) transcripts are visible in the trophocytes, in the primordial oocytes, and in the germarium (stage 0). Furthermore, they are expressed in stages 1–5 at the anterior pole of the oocytes and in spots in the embryos (arrowheads), but they are undetectable from stage 6 onwards.

hb and *otd* (Huang et al. 2010). Accordingly, maternal *Aphb* mRNA is localized in the trophocytes, germarial lumen and anterior pole of oocytes in sexual embryos, but also in the trophocytes, germarial lumen, trophic cords, anterior pole of oocytes and anterior part of the embryo in early parthenogenetic embryos. This localization breaks the symmetry of the oocytes and specifies the anteroposterior axis (Chang et al. 2013, Huang et al. 2010). *Aphb* is not found in the posterior of either sexual or parthenogenetic pea aphid oocytes prior to the formation of the blastoderm (Huang et al. 2010, Chang et al. 2013, Duncan et al. 2013a). In parthenogenetic embryos,

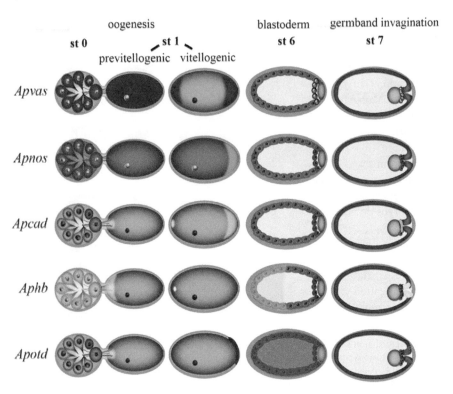

Fig. 12. Expression of key maternal and patterning genes involved in early development of oocytes and embryos of oviparous pea aphids (according to Huang et al. 2010, Duncan et al. 2013a,b, Chang et al. 2013, Lin et al. 2014). Anterior is to the left, posterior to the right and dorsal uppermost. *Apvas* (highlighted in red) is evenly distributed in the cytoplasm of the primordial oocytes and the trophocytes in the germarium (stage 0), and throughout the cytoplasm of the previtellogenic oocytes, whereas *Apvas* expression becomes obvious in anterior and posterior regions of vitellogenic oocytes. After cellularization of the syncytial blastoderm and *Apvas*-positive stripe of cells appears in the posterior region of the egg, anterior to the symbiotic bacteria. During germband invasion in stage 7, *Apvas* expressien is detectable in cells posterior to the migrating bacteria. *Apnos* (highlighted in blue) is maternally provided and detectable in the germarium (germarial lumen, trophocytes and primordial oocytes, stage 0) and troughout the cytoplasm of previtellogenic oocytes (stage 1). In vitellogenic oocytes the posterior most region of the oocyte lacks *Apnos* expression. In further development, *Apnos* is no longer identifiable as a germline marker. *Apcad* (highlighted in green) is maternally expressed to the anterior and posterior vitellogenetic oocyte and further transcripts can first be detected again at stage 7 during germband invagination. Here, *Apcad* is expressed in cells close to the bacteriocytes in the germband. *Aphb* (highlighted in yellow) is maternally expressed and visible in the cytoplasm of the trophocytes, primordial oocytes in the germarium (stage 1) and anteriorly localized in oocytes. In blastoderm stage, the signals are visible as an anterior cap (stage 6) and later throughout the invaginating germband (stages 7–9). *Apotd* (highlighted in pink) is exclusively expressed in posterior regions of vitellogenic oocytes (stage 1), in blastoderm stage throughout the oocyte (stage 6), and later in two patches of cells in the invaginating germband (stage 7).

posterior *Aphb* expression is first observed during late blastulation in newly-formed primordial cells and later in gastrulation in the posterior-most region of the germ band (McGregor et al. 2009, Chang et al. 2013, Duncan et al. 2013a). In sexually-produced embryos, *Aphb* is expressed at the anterior of the blastoderm (associated with energids) and expands towards the posterior. *Aphb* is also expressed in all cells of the invaginating germ band (Duncan et al. 2013a).

Maternal *Apotd* is not expressed during asexual oogenesis, and the embryonic gene remains inactive until gastrulation suggesting it has no role in anteroposterior axis specification (Huang et al. 2010, Duncan et al. 2013a). However, *Apotd* is expressed during sexual oogenesis in a group of follicle cells at the posterior of post-vitellogenetic oocytes, and after fertilization expression expands throughout the blastoderm (Duncan et al. 2013a) (Fig. 11, Fig. 12).

Maternal *Apcad* expression during sexual oogenesis is visible prior to blastoderm formation in the ooplasm adjacent to the trophic cord and during late oogenesis it accumulates in the posterior of the oocyte. After fertilization, embryonic *Apcad* is not expressed before germ band invagination. In contrast, asexual germaria, oocytes and embryos do not express maternal *Apcad* prior to the formation of the blastoderm (Chang et al. 2013, Duncan et al. 2013a) and embryonic expression is first detected in primordial germ cells and later in the invaginating germ band (Duncan et al. 2013a). The *Apcad* does not act as a posterior organizer in asexual pea aphid embryos (Chang et al. 2013). Because of *Apotd* and *Apcad* do not take part in anteroposterior axis specification in asexual embryos, it is like that anterior *Aphb* and posterior *Apnos* mRNAs assume this role (Chang et al. 2006, 2013).

Germ Line Specification and Germ Cell Migration

Although the two modes of pea aphid development share common genetic programs, the gene expression profiles differ substantially in germ cells and oocytes (Gallot et al. 2012). Pea aphid germ line specification relies on the germ plasm at the posterior pole of the oocyte prior to blastoderm formation (Chang et al. 2007). The *vasa* gene product is usually required for the translation of *gurken* during oogenesis, which in turn is needed to establish the posterior region of the oocyte, as well as *oskar* and *nanos*, whose products are also components of the germ plasm (Styhler et al. 1998, Tomancak et al. 1998, Tsuda et al. 2003, Chang et al. 2006). Interestingly, there is no *oskar* ortholog in the pea aphid even though all the upstream and downstream maternal-effect genes are conserved (International Aphid Genomics Consortium 2010).

In parthenogenetic pea aphids, the germ cells are already determined in the embryo and both *Apnos* and *Apvas* are expressed before germ cell formation in the germ plasm. *Apnos* is already localized around the nucleus of developing oocytes in the germarium and in segregated oocytes, whereas *Apvas* appears for the first time in the posterior of the segregated oocyte (Chang et al. 2006, 2007, 2009). After cellularization, *Apvas* and *Apnos* are expressed in the primordial germ cells, which are the first differentiated and recognizable cells in the blastoderm as it begins to invaginate at the posterior pole of the egg and forms the blastopore (Tannreuther 1908, Chang et al. 2007). *Apvas* expression declines just after blastoderm formation, but increases again when the symbiotic bacteria invade the embryo and thereafter (Chang et al. 2007). In contrast, *Apnos* is only involved in germline specification, but not in germ cell migration during the subsequent developmental stages (Chang et al. 2009, Lin et al. 2014) (Fig. 11).

In sexual aphids, the expression of maternal *Apvas* in oviparous ovarioles is evenly distributed in the cytoplasm of the nurse cells in the germaria and can also be detected in the previtellogenic and vitellogenic oocytes. *Apvas* expression declines just before oviposition and reappears in the early syncytial posterior blastoderm, marking the pre-formed germ plasm. During cellularization, *Apvas* is incorporated in the primordial germ cells before migration (Lin et al. 2014) (Fig. 12).

These studies indicate that germline specification is similar in sexual and asexual pea aphids. In both developmental modes, germline specification involves the early segregation of germ plasm and the specification and migration of the germ cells, therefore depending on preformed maternally-inherited germ plasm and not on zygotic induction (Lin et al. 2014). Thus, germ line specification is deterministic in both modes of development and environmental cues have no effect thus eliminating developmental plasticity (Lin et al. 2014).

Signal Transduction Pathways and Transcription Factors

The major signaling pathways that control developmental processes in animals are largely conserved in the pea aphid genome, although the components are also often affected by gene loss and gene duplication. For example, the Wnt signaling pathway is conserved despite the absence of several Wnt ligand genes, but many of the downstream components have undergone sequence duplication and divergence suggesting they have adapted to fulfil new functions (Shigenobu et al. 2010). Gene loss and gene duplication has also been demonstrated in the highly-conserved EGF, TGFβ, Notch, RTK, Hedgehog, JAK/STAT, and Nuclear Hormone Receptor signaling pathways (Shigenobu et al. 2010, International Aphid Genomics Consortium 2010).

The pea aphid genome contains clusters of *Hox* genes encoding homeobox transcription factors involved in pattern formation, but their sequences have diverged significantly from *Hox* genes in other insects (International Aphid Genomics Consortium 2010). As well as their well-characterized role in positional specification along the anteroposterior axis, novel members of the pea aphid *Hox* family have assumed wider roles, such as *HoxR* which may help to buffer differences between early sexual and asexual development allowing the convergence of the developmental pathways so that they yield morphologically identical offspring (O'Neill 2012).

Acknowledgements

The authors wish to thank *Richard M. Twyman* for his helpful comments and editing of the chapter and to LOEWE Center for Insect Biotechnology & Bioresources (Giessen, Germany) for valuable help.

Keywords: pea aphid, life cycle, reproductive polyphenism, wing polyphenism, sex determination, parthenogenetic embryogenesis, embryogenesis in fertilized eggs, embryo nutrition, gene expression

References

Aoki, S. 1977. *Colophina clematis* (Homoptera, Pemphigidae), an aphid species with soldiers. Kontyû. 45: 276–282.

Baumann, P. 2005. Biology of bacteriocyte-associated endosymbionts of plant sap-sucking insects. Annu. Rev. Microbiol. 59: 155–189.

Behrendt, K. 1963. Über die Eidiapuse von *Aphis fabae* Scop. (Homoptera, Aphididae). Zool. Jahrb. Physiol. 70: 309–398.

Bernays, E.A. 1986. Diet-induced head allometry among foliage-chewing insects and its importance for graminivores. Science 23: 495–497.

Bennet, C.E. and H. Stebbings. 1979. Redundant nutritive tubes in insect ovarioles: the fate of an extensive microtubule transport system. Cell Biol. Int. Rep. 3: 439–443.

Bermingham, J. and T.L. Wilkinson. 2009. Embryo nutrition in parthenogenetic viviparous aphids. Physiol. Entomol. 34: 103–109.

Bickel, R.D., H.C. Cleveland, J. Barkas, C.C. Jeschke, A.A. Raz, D.L. Stern and G.K. Davis. 2013. The pea aphid uses version of the terminal system during oviparous, but not viviparous, development. EvoDevo 4: 10.

Blackman, R.L. 1978. Early development of parthenogenetic egg in three species of aphids (Homoptera: Aphididae). Int. J. Insect Morph. Embryol. 7: 33–44.

Blackman, R.L. 1974. Incorporation of thymidine into the chromosomes of aphid (Myzus persicae) embryos. Experientia 30(10): 1136–1137.

Blackman, R.L. 1979. Stability and variation in aphid clonal lineages. Bio. J. Linn. Soc. 11: 259–277.

Blackman, R.L. 1987. Reproduction, cytogenetics and development. pp. 163–195. *In*: A.K. Minks and P. Harrewijn (eds.). Aphids, their Biology, Natural Enemies and Control. Elsevier, Amsterdam.

Böhmel, W. and O. Jancke. 1942. Beitrag zur Embryonalentwicklung der Wintereier von Aphiden. J. Appl. Entomol. 29(4): 636–658.

Braendle, C., T. Miura, R. Bickel, A.W. Shingelton, S. Kambhampati and D.L. Stern. 2003. Developmental origin and evolution of bacteriocytes in the aphid-*Buchnera* symbiosis. PloS Biol. 1: 70–76.

Braendle, C., M.C. Caillaud and D.L. Stern. 2005. Genetic mapping of *aphicarus*—a sex-linked locus controlling a wing polymorphism in the pea aphid (*Acyrthosiphon pisum*). Heredity 94: 435–442.

Braendle, C., G.K. Davis, J.A. Brisson and D.L. Stern. 2006. Wing dimorphism in aphids. Heredity 97: 192–199.

Bright, M. and S. Bulgheresi. 2010. A complex journey: transmission of microbial symbionts. Nat. Rev. Microbiol. 8: 218–230.

Brisson, J.A. 2010. Aphid wing dimorphisms: linking environmental and genetic control of trait variation. Phil. Trans. R. Soc. B Biol. Sci. 365: 605–616.

Brisson, J.A. and D.L. Stern. 2006. The pea aphid, *Acyrthosiphon pisum*: an emerging genomic model system for ecological, developmental and evolutionary studies. BioEsssays 28: 747–755.

Brisson, J.A., A. Ishikawa and T. Miura. 2010. Wing development genes of the pea aphid and differential gene expression between winged and unwinged morphs. Insect Mol. Biol. 19: 63–73.

Buchner, P. 1965. Endosymbiosis of Animals with Plant Microorganisms. Wiley and Sons, New York.

Büning, J. 1985. Morphology, ultrastructure, and germ cell cluster formation in ovarioles of aphids. J. Morph. 186: 209–221.

Büning, J. 1998. The ovariole: structure, type and phylogeny. pp. 897–932. *In*: M. Locke and H. Harrison (eds.). Microscopic Anatomy of Invertebrates, Vol. 11C. Wiley-Liss. Inc., New York.

Caillaud, M.C., M. Boutin, C. Braendle and J.-C. Simon. 2002. A sex-linked locus controls wing polymorphism in males of the pea aphid, *Acyrthosiphon pisum* (Harris). Heredity 89: 346–352.

Chang, C.-C., W.-C. Lee, C.E. Cook, G.-W. Lin and T. Chang. 2006. Germ-plasm specification and germline development in the parthenogenetic pea aphid *Acyrthosiphon pisum*: Vasa and Nanos as markers. Int. J. Dev. Biol. 50: 413–421.

Chang, C.-C., G.W. Lin, C.E. Cook, S.-B. Horng, H.-J. Lee and T.-Y. Huang. 2007. Apvasa marks germ-cell migration in the parthenogenetic pea aphid *Acyrthosiphon pisum* (Hemiptera: Aphidoidea). Dev. Genes Evol. 217: 275–287.

Chang, C.-C., T.Y. Huang, C.E. Cook, G.W. Lin, C.L. Shih and R.P. Chen. 2009. Developmental expression of *Apnanos* during oogenesis and embryogenesis in the parthenogenetic pea aphid *Acyrthosiphon pisum*. Int. J. Dev. Biol. 53: 169–176.

Chang, C.-C., Y.-M. Hsiao, T.Y. Huang, C.E. Cook, S. Shienobu and T.-H. Chang. 2013. Noncanonical expression of caudal during early embryogenesis in the pea aphid *Acyrthosiphon pisum*: maternal cad-driven posterior development is not conserved. Insect Mol. Biol. doi: 10.1111/imb.12035.

Christie, A.E. 2008. *In silico* analysis of peptide paracrines/hormones in Aphidoidea. Gen. Comp. Endocrinol. 159(1): 67–79.

Corbitt, T.S. and J. Hardie. 1985. Juvenile hormone effects on polymorphism in the pea aphid *Acyrthosiphon pisum*. Entomol. Exp. Appl. 38: 131–136.

Cortés, T., D. Tagu, J.C. Simon, A. Moya and D. Martínez-Torres. 2008. Sex versus parthenogenesis: a transcriptomic approach of photoperiod response in the model aphid *Acyrthosiphon pisum* (Hemiptera: Aphididae). Gene 408: 146–156.

Couchman, J.R. and P.E. King. 1979. Germarial structure and oogensis in *Brevicoryne brassica* (L.) (Hemiptera: Aphididae). Int. Insect Morph. Embryol. 8: 1–10.

Couchman, J.R. and P.E. King. 1980. Ovariole sheath structure and its relationship with developing embryos in a parthenogenetic viviparous aphid. Acta Zool. 61(3): 147–155.

Crema, R. 1971. Changes in the ovary of *Acyrthosiphon pisum* Harris (Homoptera: Aphididae) during transition from aphigony to parthenogenesis. Ital. J. Zool. DOI: 10.1080/00269786.1971. 10736167.

Crema, R. 1973. Structure and determination of the ambiphasic ovary of *Acyrthosiphon pisum* (Homoptera: Aphididae). Entomol. Exp. Appl. 16(4): 427–432.

DeWitt, T.J. and S.M. Scheiner. 2004. Phenotypic Plasticity: Functional and Conceptual Approaches. Oxford University Press, New York.

Dixon, A.F.G. 1985. Structure of aphid populations. Ann. Rev. Entomol. 30: 155–174.

Dixon, A.F.G. 1990. Evolutionary aspects of parthenogenetic reproduction in aphids. Acta Phytopath. Entomol. Hung. 25: 41–56.

Dixon, A.F.G. 1998. Aphid Ecology. Chapman and Hall, London, UK.

Dixon, A.F.G. and M.T. Howard. 1986. Dispersal in aphids, a problem in resource allocation. pp. 145–151. *In*: W. Danthanarayana (ed.). Insect Flight: Dispersal and Migration. Springer Verlag, Berlin.

Dixon, A.F.G. and B.K. Agarwala. 1999. Ladybird induced life history changes in the aphids. Proc. R. Soc. Lond. Biol. Sci. 266: 1549–1553.

Douglas, A.E. 1989. Mycetocyte symbiosis in insects. Biol. Rev. 69: 409–434.

Douglas, A.E. 1998. Nutritional interactions in insect-microbial symbiosis: aphids and their symbiotic bacteria *Buchnera*. Ann. Rev. Entomol. 43: 17–37.

Duncan, E.J., M.P. Leask and P.K. Dearden. 2013a. The pea aphid (*Acyrthosiphon pisum*) genome encodes two divergent early developmental programs. Dev. Biol. 377: 262–274.

Duncan, E.J., M.A. Benton and P.K. Dearden. 2013b. Canonical terminal patterning is an evolutionary novelty. Dev. Biol. 2013: 245–261.

Eberhard, W.G. 1980. Horned beetles. Sci. Am. 242(3): 166–182.

Eberhard, W.G. 1982. Beetle horn dimorphism. Making the best of a bad lot. Am. Nat. 119: 420–426.

El-Ziady, S. and J.S. Kennedy. 1956. Beneficial effects of the common garden ant, *Lasius niger* L., on the black bean aphid, *Aphis fabae* Scop. R. Entomol. Soc. Lond. 31: 61–65.

Essig, E.O. and F. Abernathy. 1952. The Aphid Genus *Periphyllus* (Family Aphidae); A Systematic, Biological, and Ecological Study. University of California Press, Berkley.

Evans, J.D. and D.E. Wheeler. 2001. Gene expression and the evolution of insect polyphenisms. BioEssays 23: 62–68.

Forrest, J.M.S. 1970. The effect of maternal and larval experience on morph determination in *Dysaphis devecta*. J. Insect Physiol. 16: 2281–1192.

Fukatsu, T. 2010. A fungal past to insect color. Science 328: 574–575.

Gäde, G. and L. Auerswald. 2003. Mode of action of neuropeptides from adipokinetic hormone family: a new take on biodiversity. Ann. N.Y. Acad. Sci. 1163: 125–136.

Gallot, A., C. Rispe, N. Letere, J.P. Gauthier, S. Jaubert-Possamai and D. Tagu. 2010. Cuticle proteins and seasonal photoperiodism in aphids. Insect Biochem. Mol. Biol. 40: 235–240.

Gallot, A., S. Shigenobu, T. Hashiyma, T. Jaubert-Possamai and D. Tagu. 2012. Sexual and asexual oogenesis require the expression of unique and shared sets of genes in the insect *Acyrthosiphon pisum*. BMC Genomics 13: 76. doi: 10.1186/1471-2164-13-76.

Gao, N. and J. Hardie. 1997. Melatonin and the pea aphid, *Acyrthosiphon pisum*. J. Insect Physiol. 43: 615–620.

Gao, N., M. von Schantz, R.G. Forster and J. Hardie. 1999. The putative brain photoperiodic photoreceptors in the vetch aphid Megoura viciae. J. Insect Physiol. 45: 1011–1019.

Ghalambor, C.K., J.K. McKay, S.P. Carroll and D.N. Reznick. 2007. Adaptive versus non-adaptive phenotypic plasticity and the potential for contemporary adaption in new environments. Funct. Ecol. 21: 394–407.

Granett, J., M.A. Walker, L. Kocsis and A.D. Omer. 2001. Biology and management of grape phylloxerae. Annu. Rev. Entomol. 46: 387–412.

Hales, D.F. and T.E. Mittler. 1987. Chromosomal sex determination in aphids controlled by juvenile-hormone. Genome 29: 28–37.

Hales, D.F., J. Tomiuk, K. Wöhrmann and P. Sunnacks. 1997. Evolutionary and genetic aspects of aphid biology: a review. Eur. J. Entomol. 94: 1–55.

Hall, B.K. 1999. Evolutionary developmental biology. Dordrecht: Kluwer. doi: 10.1007/978-94-011-3961-8.

Hardie, J. 1980. Juvenile hormone mimics the photoperiodic apterization of the alate gynopara of the aphid *Aphis fabae*. Nature 602–604.

Hardie, J. 1987a. The photoperiodic control of wing development in the black bean aphid, *Aphis fabae*. J. Insect Phys. 33: 543–549.

Hardie, J. 1987b. The corpus allatum neurosecretion and photoperiodically controlled polymorphism in an aphid. J. Insect Physiol. 33: 201–205.

Hardie, J. and A.D. Lees. 1983. Photoperiodic regulation of the development of winged gynoparae in the aphid, *Aphis fabae*. Physiol. Entomol. 8: 385–391.

Hardie, J. and A.D. Lees. 1985. Endocrine control of polymorphism and polyphenism. pp. 441–490. *In*: G.A. Kerhut and L.I. Gilbert (eds.). Comprehensive Insect Physiology, Biochemistry and Pharmacology. Pergamon Press, Oxford.

Hardie, J., F.C. Baker, G.C. Jamieson, A.D. Lees and D.A. Schooley. 1985. The identification of an aphid juvenile-hormone and its titer in relation to photoperiod. Physiol. Entomol. 10: 297–302.

Hardie, J. 1990. The photoperiodic counter, quantitive day-length effects and scotophase timing in the vetch aphid Megoura viciae. J. Insect Physiol. 36: 939–949.

Harrison, R.G. 1980. Dispersal polyphenisms in insects. Ann. Rev. Ecol. Syst. 11: 95–118.

Hattori, M., O. Kishida and T. Itino. 2013. Soldiers with large weapons in predator-abundant midsummer: reproductive plasticity in eusocial aphid. Evol. Ecol. 27: 847–862.

Havill, N.P. and R.G. Foottit. 2007. Biology and evolution of Adelgidae. Annu. Rev. Entomol. 52: 325–349.

Hille Ris Lambers, L. 1966. Polymorphism in the Aphididae. Ann. Rev. Entomol. 11: 47–78.

Huang, M.H. and M.C. Caillaud. 2012. Inbreeding avoidance by recognition of close kin in the pea aphid, *Acyrthosiphon pisum*. J. Insect Sci. 12: 39.

Huang, T.Y., C.E. Cook, G.K. Davis, S. Shigenobu, R.P. Chen and C.-C. Chang. 2010. Anterior development in the parthenogenetic and viviparous form of the pea aphid, *Acyrthosiphon pisum*: *hunchback* and *orthodenticle* expression. Insect Mol. Biol. 19(2): 75–85.

Huebner, E. 1981. Nurse cell-oocyte interaction in the telotrophic ovarioles of an insect, *Rhodnius prolixus*. Tiss. Cell 13: 105–125.

Huybrechts, J., J. Bonhomme, S. Minoli, N. Prunier-Leterme, A. Dombrovsky, M. Abdel-Latief, A. Robichon, J.A. Veenstra and D. Tagu. 2010. Neuropeptide and neurohormone precursors in the pea aphid, *Acyrthosiphon pisum*. Insect Mol. Biol. 2: 87–95.

Huybrechts, J., J. Bonhomme, S. Minoli, N. Prunier-Leterme, A. Dombrovsky, M. Abdel-Latief, A. Robichon, J.A. Veenstra and D. Tagu. 2011. Neuropeptide and neurohormone precursors in the pea aphid Acyrthosiphon pisum. Insect Mol. Biol. 19: 87–95.

Hyams, J.S. and H. Stebbings. 1979. The mechanisms of microtubule associated cytoplasmic transport. Cell Tissue Rev. 196: 103–116.

International Aphid Genomics Consortium. 2010. Genome sequence of the pea aphid *Acyrthosiphon pisum*. PLoS Biol. 8: e1000313.

Ishikawa, A. and T. Miura. 2007. Morphological differences between wing morphs of two Macrosiphini aphid species, *Acyrthosiphon pisum* and *Megoura crassicauda* (Hemiptera, Aphididae). Sociobiology 50: 881–893.

Ishikawa, A., S. Hongo and T. Miura. 2008. Morphological and histological examination of polyphenic wing formation in the pea aphid *Acyrthosiphon pisum* (Hemiptera, Hexaposa). Zoomorphology 127: 121–133.

Ishikawa, A., K. Ogawa, H. Gotoh, T.K. Walsh, D. Tagu, J.A. Brisson, C. Rispe, S. Jaubert-Possamai, T. Kanbe, T. Tsubota, T. Shiotsuki and M. Toru. 2012. Juvenile hormone titer and related gene expression during the change of reproductive modes in the pea aphid. Insect Mol. Biol. 12: 49–60.

Itô, Y. 1989. The evolutionary biology of sterile soldiers in aphids. Trends Ecol. Evol. 4: 69–73.

Jacobs, C.G.C. and M. van der Zee. 2013. Immune competence in insect eggs depends on the extraembryonic serosa. Dev. Comp. Immunol. 41: 263–269.

Jacobs, C.G.C., G.L. Rezende, G.E.M. Lamers and M. van der Zee. 2013. The extraembryonic serosa protects the insect egg against desiccation. Proc. R. Soc. B 280: 20131082. doi: 10.1098/rspb.2013.1082.

Jaquiéry, J., S. Stoeckel, C. Rispe, L. Mieuzet, F. Legeai and J.-C. Simon. 2012. Accelerated evolution of sex chromosomes in aphids, an XO system. Mol. Biol. Evol. 29(2): 837–847.

Jaquiéry, J., C. Rispe, D. Roze, F. Legeai, G. Le Trionnaire, S. Stoeckel, L. Mieuzet, C. Da Silva, J. Poulain, N. Prunier-Leterme, B. Ségurens, D. Tagu and J.-C. Simon. 2013. Masculinization of X chromosome in the pea aphid. PloS Genetics 9(8): e1003690.

Jedlička, P., V. Steinbauerová, P. Šimek and H. Zahradníčková. 2012. Functional characterization of the adipokinetic hormone in the pea aphid, *Acyrthosiphon pisum*. Comp. Biochem. Physiol. A Mol. Integr. Physiol. 162: 51–58.

Johnson, B. 1965. Wing polymorphism in aphids. II. Interaction between aphids. Entomol. Exp. Appl. 8: 49–64.

Johnson, B. and P.R. Birks. 1960. Studies on wing polymorphism in aphids. I. The developmental process involved in the production of different forms. Entomol. Exp. Appl. 3: 327–329.

Kawada, K. 1987. Polymorphism and morph determination. See Ref. 117(2A): 255–266.

Kerry, O.M., N.A. Moran and M.S. Hunter. 2005. Variation in resistance to parasitism in aphids is due to symbiosis not host genotype. Proc. Nat. Acad. Sci. USA 102: 12795–12800.

Kleinjan, J.E. and T.E. Mittler. 1975. A chemical influence of ants on wing development in aphids. Entomol. Exp. Appl. 18: 384–388.

Koga, R., X.-Y. Meng, T. Tsuchida and T. Fukatsu. 2012. Cellular mechanism for selective vertical transmission of an obligate insect symbiont at the bacteriocyte-embryo interface. PNAS. doi: 10.1073/pnas.11192212109.

Komazaki, S. 1998. Difference of egg diapause in two host races of the spirea aphid, *Aphis spiraecola*. Entomol. Exp. Appl. 89: 201–205.

Kunert, G. and W.W. Weisser. 2003. The interplay between density- and trait-mediate effects in predator-prey interactions: a case study in aphid wing polyphenism. Oecologia 135: 304–312.

Kunert, G. and W.W. Weisser. 2005. The importance of antennae for pea aphid wing induction in the presence of natural enemies. Bull. Entomol. Res. 95: 125–131.

Kunert, G., S. Otto, U.S.R. Rose, J. Gershenzon and W.W. Weisser. 2005. Alarm pheromone mediates production of winged dispersal morphs in aphids. Ecol. Lett. 8: 596–603.

Lamb, R.J. and P.J. Pointing. 1972. Sexual morph determination in the aphid, *Acyrthosiphon pisum*. J. Insect Physiol. 18: 2029–2042.

Lamb, R.J. and P.A. MacKay. 1987. Acyrthosiphon kondoi influences alata production by the pea aphid A. pisum. Entomol. Exp. Appl. 45: 195–204.

Le Trionnaire, G., S. Jaubert, B. Sabater-Munoz, A. Benedetto, J. Bonhomme, N. Prunier-Leterme, D. Martinez-Torres, J.C. Simon and D. Tagu. 2007. Seasonal photoperiodism regulates the expression of cuticular and signaling protein genes in the pea aphid. Insect Biochem. Mol. Biol. 37: 1094–1102.

Le Trionnaire, G., J. Hardie, S. Jaubert-Possamai, J.C. Simon and D. Tagu. 2008. Shifting from clonal to sexual reproduction in aphids: physiological and developmental aspects. Biol. Cell 100: 441–451.

Le Trionnaire, G., F. Francis, S. Jaubert, J. Bonhomme, E. De Pauw, J.-P. Gauthier, E. Haubruge, F. Legeai, N. Prunier-Leterme, J.C. Simon, S. Tanguy and D. Tagu. 2009. Transcriptomic and proteomic analysis of seasonal photoperiodism in the pea aphid. BMC Genomics 10(456): 14.

Leather, S.R. 1993. Overwintering in six arable aphid pests: a review with particular relevance to pest management. J. Appl. Ent. 116: 217–233.

Lees, A.D. 1966. The control of polymorphism in aphids. Adv. Ins. Phys. 3: 207–277.

Lees, A.D. 1967. The production of apterous and alate forms in the pea aphid *Megoura viciae* Buckton, with special references to the role of crowding. J. Insect Physiol. 13: 289–318.

Lees, A.D. 1977. Action of juvenile hormone mimics on the regulation of larval-adult and alary polymorphismus in aphids. Nature 267: 46–48.

Leonardo, T.E. and E.B. Mondor. 2006. Symbiont modifies host life history traits that affect gene flow. Proc. R. Soc. Lond. Ser. B 273(1590): 1079–1084.

Lin, G.-W., C.E. Cook, T. Miura and C.-C. Chang. 2014. Posterior localization of ApVas1 positions the preformed germ plasm in the sexual oviparous pea aphid *Acyrthosiphon pisum*. EvoDevo 5: 18. doi: 10.1186/2041-9139-5-18.

Loxdale, H.D. 2008. The nature and reality of the aphid clone: genetic variation, adaption and evolution. Agr. Forest Entomol. 10: 81–90.

Lushai, G., J. Hardie and R. Harrington. 1996. Diapause termination and egg hatching in the bird cherry aphid, *Rhopalosiphum padi*. Entomol. Exp. Appl. 81: 113–115.

McGregor, A.P., M. Pechmann, E.E. Schwager and W.G. Damen. 2009. An ancestral regulatory network for posterior development in arthropods. Commun. Integr. Biol. 2: 174–176.

Mira, A. and N.A. Moran. 2002. Estimating population size and transmission bottlenecks in maternally transmitted endosymbiotic bacteria. Microb. Ecol. 44: 137–143.

Mittler, T.E., S.G. Nassar and G.B. Staal. 1976. Wing development and parthenogenesis induced in progenies of kinoprene-treated gynoparae of *Aphis fabae* and *Myzus persicae*. J. Insect Physiol. 22: 1717–1725.

Miura, T. 2005. Developmental regulation of caste-specific characters in social-insect polyphenism. EvoDevo 7(2): 122–129.

Miura, T., C. Braendle, A. Shingleton, G. Sisk, S. Kambhampati and D. Stern. 2003. A comparison of the parthenogenetic and sexual embryogenesis of the pea aphid *Acyrthosiphon pisum* (Hemiptera: Aphidoidea). J. Exp. Zool. B Mol. Dev. Evol. 295: 59–81.

Miyazaki, M. 1987. Forms and morphs of aphids. pp. 27–50. *In*: A.K. Minks and P. Harrewijin (eds.). Aphids, their Biology, Natural Enemies, and Control. Elsevier, Amsterdam.

Montllor, C.B., A. Maxmen and A.H. Purcell. 2002. Facultative bacterial endosymbionts benefit pea aphids *Acyrthosiphon pisum*, under heat stress. Ecol. Ent. 27: 189–195.

Moran, N.A. 1992. The evolution of aphid life cycles. Ann. Rev. Entomol. 37: 321–348.

Moran, N.A. and H.E. Dunbar. 2006. Sexual aquisition of beneficial symbionts in aphids. Proc. Natl. Acad. Sci. USA 103: 12803–12806.

Moran, N.A., J.A. Russel, R. Koga and T. Fukatsu. 2005. Evolutionary relationships of three new species of Enterobacteriaceae living as symbionts of aphids and other insects. Appl. Environ. Microbiol. 71: 3302–3310.

Moran, N.A., J.P. McCutcheon and A. Nakabachi. 2008. Genomics and evolution of heritable bacterial symbionts. Annu. Rev. Genet. 42: 165–190.

Müller, C.B., I.S. Williams and J. Hardie. 2001. The role of nutrition, crowding, and interspecific interactions in the development of the winged aphids. Ecol. Ent. 26: 330–340.

Nijhout, H.F. 1999. Control mechanisms of polyphenic development in insects—in polyphenetic development, environmental factors alter same aspects of development in an orderly and predictable way. Bioscience 49: 181–192.

Nijhout, H.F. 2003. Development and evolution of adaptive polyphenism. Evol. Dev. 5: 9–18.

Oliver, K.M., J.A. Russel, N.A. Moran and M.S. Hunter. 2003. Facultative bacterial symbionts in aphids confer resistance to parasitic wasps. Proc. Nat. Acad. Sci. USA 100: 1803–1807.

Oliver, K.M., P.H. Degnan, G.R. Burke and N.A. Moran. 2010. Facultative symbionts in aphids and the horizontal transfer of ecologically important traits. Ann. Rev. Entomol. 55: 247–266.

O'Neill, M.P. 2012. A sub-set of the Hox genes in the pea aphid, *Acyrthosiphon pisum*. M.S. Thesis. University of Otago, Dunedin, NZ.

Ogawa, K. and T. Miura. 2013. Two developmental switch points for the wing polymorphisms in the pea aphid *Acyrthosiphon pisum*. EvoDevo 4: 30.

Ogawa, K. and T. Miura. 2014. Aphid polyphenisms: trans-generational developmental regulation through viviparity. Frontiers Zool. 5: 1–11.

Orlando, E. and R. Crema. 1968. Growth of follicular cells in aphids. A cytophotometric and autoradiographic study. Experientia 24(10): 1038–1039.

Orlando, E. 1974. Sex determination in *Megoura viciae* Buckton (Homoptera, Aphididae). Italian J. Zool. 8: 61–70.

Pagliai, A.M. 1965. A new category of females in the life cycle of *Brevicoryne brassicae* (L.): the ambiphasic females. Experientia 21: 283.

Panfilio, K.A. 2008. Extraembryonic development in insects and the acrobatics of blastokinesis. Dev. Biol. 313: 471–491.

Podjasek, J.O., L.M. Bosnjak, D.J. Booker and E.B. Mondor. 2005. Alarm pheromone induces a transgenerational wing polyphenism in the pea aphid, *Acyrthosiphon pisum*. Can. J. Zool. 83: 1138–1141.

Purandare, S.R., B. Tenhumberg and J.A. Brisson. 2014. Comparison of the wing polyphenic response of pea aphids (*Acyrthosiphon pisum*) to crowding and predator cues. Ecol. Entomol. 39: 263–266.

Ramos, S., A. Moya and D. Martinez-Torres. 2003. Identification of a gene overexpressed in aphids reared under short photoperiod. Insect Biochem. Mol. 33(3): 289–298.

Riparbelli, M.G., D. Tagu, J. Bonhomme and G. Callaini. 2005. Aster self-organization at meiosis: a conserved mechanism in insect parthenogenesis? Dev. Biol. 278: 220–230.

Rispe, C., J.-S. Simon and P.-H. Gouyon. 1998. Models of sexual and asexual coexistance in aphids based on constraints. J. Evol. Biol. 11: 685–701.

Rispe, C. and J.S. Pierre. 1998. Coexistence between cyclical parthenogens, obligate parthenogens, and intermediates in a fluctuating environment. J. Theor. Biol. 195: 97–110.

Rodrigues-Martins, A., M. Riparbelli, G. Callaini, D.M. Glover and M. Bettencourt-Dias. 2007. Revisiting the role of mother centriole in centriole biogenesis. Science 316: 1046–1050.

Roff, D.A. 1990. The evolution of flightlessness in insects. Ecol. Monogr. 60: 389–421.

Roff, D.A. and D.J. Fairbairn. 1991. Wing dimorphisms and the evolution of migratory polymorphisms among Insecta. Am. Zool. 31: 243–251.

Shigenobu, S., H. Watanabe, M. Hattori, Y. Sakaki and H. Ishikawa. 2000. Genome sequence of the endocellular bacterial symbiont of aphids *Buchnera* sp. APS. Nature 407: 81–86.

Shigenobu, S., R.D. Bickel, J.A. Brisson, T. Butts, C.C. Chang, O. Christiaens, G.K. Davis, E.J. Duncan, D.E. Ferrier, M. Iga, R. Janssen, G.W. Lin, H.L. Lu, A.P. McGregor, T. Miura, G. Smagghe, J.M. Smith, M. van der Zee, R.A. Velarde, M.J. Wilson, P.K. Dearden and D.L. Stern. 2010. Comprehensive survey of developmental genes in the pea aphid, Acyrthosiphon pisum: frequent lineage-specific duplications and losses of developmental genes. Insect Mol. Biol. 2: 47–62.

Shingelton, A.W., G.C. Sisk and D.L. Stern. 2003. Diapause in the pea aphid (*Acyrthosiphon pisum*) is a slowing but not a cessation of development. BMC Dev. Biol. doi: 10.1186/1471-213X-3-7.

Simon, J.C., D. Martinez-Torres, A. Latorre, A. Moya and P.D. Hebert. 1996. Molecular characterization of cyclic and obligate parthenogens in the aphid *Rhopalosiphum padi* (L.). Proc. Biol. Sci. 263: 481–486.

Simon, J.C., C. Rispe and P. Sunnucks. 2002. Ecology and evolution of sex in aphids. Trends Ecol. Evol. 17: 34–39.

Simon, J.C., F. Delmotte, C. Rispe and T. Crease. 2003. Phylogenetic relationships between parthenogens and their sexual relatives: the possible routes to parthenogenesis in animals. Biol. J. Linn. Soc. 79: 151–163.

Simon, J.C., S. Stoeckel and D. Tagu. 2010. Evolutionary and functional insides into reproductive strategies of aphids. C. R. Biol. 333: 488–496.

Simon, J.C., M.E. Pfrender, R. Tollrian, D. Tagu and J.K. Colbourne. 2011a. Genomics of environmentally induced phenotypes in two extremely plastic arthropods. J. Hered. doi: 10.1093/jhered/esr1020.

Simon, J.-C., S. Boutin, T. Tsuchida, R. Koga, J.-F. Le Gallic, A. Frantz, Y. Outreman and T. Fukatsu. 2011b. Facultative symbiont infections affect aphid reproduction. PLoS ONE 6(7): e21831. doi:10.1371/journal.pone.0021831.

Simpson, S.J., G.A. Sword and N. Lo. 2011. Polyphenism in insects. Curr. Biol. 21: R738–R749.

Sloggett, J.J. and W.W. Weisser. 2002. Parasitoids induce production of the dispersal morphs of pea aphid, *Acyrthosiphon pisum*. Oikos 98: 323–333.

Smith, M.A.H. and P.A. MacKay. 1989. Genetic variation in male alary dimorphism in populations of pea aphid, *Acyrthosiphon pisum*. Entomol. Exp. Appl. 51: 125–132.

Sprenger, F., L.M. Stevens and C. Nusslein-Volhard. 1989. The *Drosophila* gene torso encodes putative receptor tyrosine kinase. Nature 338: 478–483.

Srinivasan, D.G., B. Fenton, S. Jaubert-Possamai and M. Jaouannet. 2010. Analysis of meiosis and cell cycle genes of the facultative asexual pea aphid *Acyrthosiphon pisum* (Hemiptera: Aphididae). Insect Mol. Biol. 19: 229–239.

Srinivasan, D.G., A. Abdelhady and D.L. Stern. 2014. Gene expression analysis of parthenogenetic embryonic development of the pea aphid, *Acyrthosiphon pisum*, suggests that aphid parthenogenesis evolved from meiotic oogenesis. PloS One 9(12): e115099. doi: 10.137/journal.pone.0115099.

Steel, C.G.H. 1976. Neurosecretory control of polymorphism in aphids. pp. 117–130. *In*: M. Lüscher (ed.). Phase and Caste Determination in Insects. Pergamon Press, Oxford and New York.

Steel, C.G.H. 1977. The neurosecretory system in the aphid *Megoura viciae*, with reference to unusual features associated with long distance transport of neurosecretion. Gen. Comp. Endocrinol. 31: 307–322.

Steel, C.G.H. and A.D. Lees. 1977. The role of neurosecretion in the photoperiodic control of polymorphism in the aphid *Megoura viciae*. J. Exp. Biol. 67: 117–135.

Stern, D.L. and W.A. Foster. 1996. The evolution of soldiers in aphids. Biol. Rev. Cam. Philos. Soc. 71: 27–79.

Stevens, N.M. 1904. A study of the germ cells of *Aphis rosae* and *Aphis oenotherae*. J. Exp. Zool. 2: 313–333.

Styhler, S., A. Nakamura, A. Swan, B. Suter and P. Lasko. 1998. *Vasa* is required for Gurken accumulation in the oocyte and is involved in oocyte differentiation and germline cyst development. Development 125: 1569–1578.

Sutherland, O.R.W. 1969a. The role of crowding in the production of winged forms by two strains of the pea aphid, *Acyrthosiphon pisum*. J. Insect. Phys. 15: 1385–1410.

Sutherland, O.R.W. 1969b. The role of host plant in the production of winged forms by two strains of the pea aphid, *Acyrthosiphon pisum*. J. Insect Phys. 15: 2179–2200.

Tagu, D., B. Sabater-Munoz and J.C. Simon. 2005. Deciphering reproductive polyphenism in aphids. Invertebr. Rep. Dev. 48: 71–80.

Tannreuther, G.W. 1908. History of the germ cells and early embryology of certain aphids. Ph.D. Thesis. University of Chicago, Department of Zoology, Chicago.

Tomancak, P., A. Guichet, P. Zavorszky and A. Ephrussi. 1998. Oocyte polarity depends on regulation of *gurken* by Vasa. Development 125: 1723–1732.

Tomiuk, J., V. Loeschke and M. Schneider. 1994. On the origin of polyploidy parthenogenetic races in the weevil *Polydrusus mollis* (Coleoptera: Curculionidae). J. Theor. Biol. 167: 89–92.

Tsitsipis, J.A. and T.E. Mittler. 1976. Embryogenesis in parthenogenetic and sexual females of *Aphis fabae*. Entomol. Exp. Appl. 19: 263–270.

Tsuchida, T., R. Koga, M. Horikawa, T. Tsunoda, T. Maoka, S. Matsumoto, J.C. Simon and T. Fukatsu. 2010. Symbiotic bacterium modifies aphid body color. Science 330: 1102–1104.

Tsuda, M., Y. Sasaoka, M. Kiso, K. Abe, S. Haraguch, S. Kobayashi and Y. Saga. 2003. Conserved role of Nanos protein in germ cell development. Science 301: 1239–1241.

Tsuji, H. and K. Kawada. 1987. Development and degeneration of wing buds and indirect flight muscles in the pea aphid *Acyrthosiphon pisum* (Harris). Jpn. J. Appl. Entomol. Z. 31: 247–252.

Walsh, T.K., J.A. Brisson, H.M. Robertson, K.H. Gordon, S. Jaubert-Possamai, D. Tagu and O.R. Edwards. 2010. A functional DNA methylation system in the pea aphid *Acyrthosiphon pisum*. Insect Mol. Biol. 19(2): 215–228.

Weisbrod, A., M. Cohen and A.D. Chipman. 2013. Evolution of the insect terminal patterning system—insights from the milkweed bug, *Oncopeltus fasciatus*. Dev. Biol. 380: 125–131.

Weisser, W.W., C. Braendle and N. Minoretti. 1999. Predator induced morphological shift in the pea aphid. Proc. R. Soc. Lond. B 266: 1175–1181.

West-Eberhard, M.J. 1989. Phenotypic plasticity and the origins of diversity. Ann. Rev. Syst. 20: 249–278.

West-Eberhard, M.J. 2003. Developmental Plasticity and Evolution. Oxford University Press, Oxford.

Whitman, D.W. and T.N. Ananthakrishnan. 2009. Phenotypic plasticity in insects: mechanisms and consequences. Science Publishers, Enfield, NH.

Wilson, A.C.C., P. Sunnucks and D.F. Hales. 1997. Random loss of an X chromosome at male determination in an aphid, *Sitobion near fragariae*, detected using an X-linked polymorphic microsatellite marker. Gen. Res. 69: 233–236.

Zera, A.J. and R.F. Denno. 1997. Physiology and ecology of dispersal polymorphism in insects. Ann. Rev. Ent. 42: 207–230.

3

Functional and Evolutionary Genomics in Aphids

Denis Tagu,[1,] Federica Calevro,[2] Stefano Colella,[2]*
Toni Gabaldón[3,4,5] and Akiko Sugio[1]

Introduction

Genomic resources refer to different types of data containing information on the organization, structure, evolution and function of elements present in a genome. These resources integrate "gene" expression regulatory network data, from epigenetics to metabolites. This includes also the physical and functional interactions these molecules might have within a cell/tissue/organ/organism. The central genomic resource for an organism, on which most of the others resources refer to, is the complete sequence of its genome. To generate it, the nuclear DNA (talking inclusively here about eukaryotes) is extracted, sequenced, assembled and annotated. This so-called "reference genome" has to be seen in most cases as a consensus genome of the concerned species, since it is very rare that such a genome sequence comes from only one individual. The Human reference genome has been obtained about 15 years ago, just at the entry in the 21st century (The International Human Genome Sequencing Consortium 2000). Getting a reference genome is essential, but that is only a starting point: a genome is a potential atlas of all the possible functions of a cell, and it contains the footprints

---

[1] INRA, UMR1349, Institute of Genetics, Environment and Plant Protection, Domaine de la Motte, BP35327, 35653 Le Rheu cedex, France.
[2] UMR203 BF2I, Biologie Fonctionnelle Insectes et Interaction, INRA, INSA-Lyon, Université de Lyon, F-69621 Villeurbanne, France.
[3] Bioinformatics and Genomics Programme. Centre for Genomic Regulation (CRG). Dr. Aiguader, 88. 08003 Barcelona, Spain.
[4] Universitat Pompeu Fabra (UPF), 08003 Barcelona, Spain.
[5] Institució Catalana de Recerca i Estudis Avançats (ICREA), Pg. Lluís Companys 23, 08010 Barcelona, Spain.
* Corresponding author: denis.tagu@rennes.inra.fr

of past evolutionary paths walked by the species. A genome sequence in itself does not say so much: that is why—for the Human Genome Project—several consortia or initiatives were organized to exploit these genome data for (i) the reconstruction the evolutionary history of the human genome compared to other primate, mammal or eukaryotic genomes (phylogenomics) (Gabaldon et al. 2009), (ii) the characterization of the genetic diversity within the human species for biomedical research (HapMap) (1000 Genomes Project Consortium 2010), (iii) the characterization of the diversity and role of associated microorganisms able to strongly influence Human phenotypes (Human Microbiome Project),[1] (iv) the systematic characterization of all functional DNA elements controlling a cell and circumstances in which a gene is active (ENCODE) (ENCODE Project Consortium 2012), and (v) the definition of the epigenetic regulations (complementary to ENCODE) affecting the phenotype without affecting the genome sequence (Bradbury 2003, Roadmap Epigenomics Consortium 2015). All these projects have not been completed yet, and it is even difficult to anticipate whether such actions will be achieved or not. But it is already clear that the next effort will be the integration of these different data to understand the organism's complexity. In fact, it is important to know, for instance, how the microbiome can affect gene function, or how the genetic diversity influences the microbiome composition and gene function, and so on and so forth. Although the efforts around the human species are naturally leading current research, a plethora of genomic resources is being generated for other model organisms and, increasingly, for species outside the traditional sets of model species. The task at hand is thus extremely ambitious!

Despite the challenging nature of this task we can be optimistic since the prospect of using all these integrated information is more than interesting as it will allow a better understanding of biological systems. Furthermore technologies are evolving so fast that we don't know what we will be able to do within 10 years. Ten years back, for instance, it was hard to imagine the upcoming sequencing revolution enabled by next generation sequencing technologies.

What is the situation of aphids within this general framework? The human genome project and the post-genomics human projects are examples for us, working on aphids, since they represent a sort of roadmap that our communities need to follow in order to decipher the fascinating and specific original traits that make aphids able to adapt to their environment. The global aim is to link genotypes to phenotypes (or in other words, to understand the Genetic X Environment interaction equation that sculpts a phenotype) in a dynamic way. So the challenge is to develop genomic resources for aphids, and to introduce their use into other disciplines studying aphids, including evolutionary biology, genetics (for diversity studies), physiology, biochemistry and cell biology (for functional study), microbiology (for microbiomes), and even ecology (for environmental genomics) (Fig. 1).

Studying aphid biology is—in our opinion—driven by two main reasons. First, many aphids are crop pests, responsible for much damage, and forcing the farmers to use polluting and hazardous pesticides. Second, aphids have fascinating and original

---

[1] http://hmpdacc.org/overview/about.php

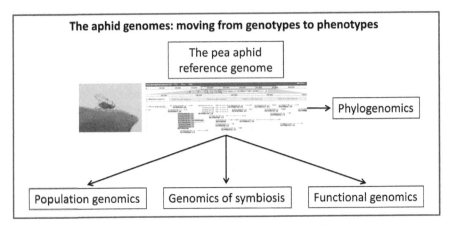

Fig. 1. Post-genomics as a future prospective for aphid biology. We propose that for each relevant aphid species (here, example given for the pea aphid), a reference genome is used as an atlas required to evolutionary approaches (phylogenomics), population studies (genome variations), metagenomics strategies (microbiome) and functional descriptions (genome functioning). This is required to link for any trait genotypes and phenotypes. All these actions require to organize the community to share efforts and knowledge.

traits that appeared 300 million years ago and that deserve deep investigation in the mechanisms underlying the diversification of life forms. We can cite the most important traits on which investigations have been, and still are, concentrated: aphids exploit a restricted ecological niche (the phloem sap) as other Hemipteran insects; aphids express several plastic traits (reproductive mode, dispersal, gall formation) as other Arthropods; aphids reproduce by parthenogenesis; aphids interact with host-plants (as other insect pests); aphids developed an obligate symbiosis, as well as facultative symbioses (as other sap-feeding insects). One can pinpoint the case of the grape phylloxera, an ancient aphid that already has some of the traits of the modern aphids (parthenogenesis for instance) and which caused the loss of most of the French vineyards in the 19th century: a species part of world heritage, from French aphidologists at least.

In this chapter, we will present an overview of the main developments in aphid genomics in the last 10 years, and try to show how these advancements pave the road for the future research in aphid biology. First, we present the available genomics resources, and then we give some examples of their use for functional and evolutionary biological studies on aphids.

Genome Resources

Pea Aphid Genomic Resources

Today, most of the available aphid genomics resources are from the pea aphid *Acyrthosiphon pisum*. This species was selected in June 2003, during the first meeting of the International Aphid Genomics Consortium (IAGC, Paris, France), as the aphid species on which the international community would focus its efforts to get the complete genome sequence. At that time, genome sequencing was still a long and very expensive endeavour, since the Next Generation Sequencing (NGS) technologies were not available by then. Some of the reasons why the pea aphid was selected are its relatively large body size (reaching several millimetres for adult forms), the relatively easy laboratory rearing (the pea aphid does not alternate plant hosts between seasons, and its complete reproductive cycle, including parthenogenesis and sexual reproduction, can be easily completed under laboratory conditions, furthermore it can be reared on artificial media), the knowledge of its physiology as well as the extensive studies of its symbiotic biology with *Buchnera aphidicola* (*Buchnera* sp. APS from *A. pisum* was the first symbiotic organism to be sequenced in 2000 (Shigenobu et al. 2000) (see Chapter 5 "Bacterial Symbionts of Aphids (Hemiptera: Aphididae)" in this volume). The history of this genome project and its interest has been reviewed elsewhere (Tagu et al. 2010); in summary, IAGC had to wait to 2006 to get the genome project funded, and to 2010 for the release and publication of the first annotated version of the pea aphid genome (IAGC 2010). The pea aphid genome is the first hemimetabolous insect genome to have been sequenced and published. This genome revealed four main peculiarities shortly summarized as follows. (i) It contains a large set of predicted genes (> 34,000, for a genome of approximately 525 megabases and organized in 2n = 8 chromosomes) including half of them with strong evidence of expression (so-called "ref-Seq" catalog made by the NCBI) and the other half with only little information on their putative expression and function. (ii) From the sequence it is apparent that many gene families have undergone extensive duplication(s) during evolution. Whether all these duplicated copies—kept as functional DNA elements in the genome—have evolved or not new expression profiles and/or functions remains to be studied. (iii) In parallel to gene duplications and expansion, some pathways are missing or lack key elements. This is the case for the immune system that lacks the IMD pathway (Gerardo et al. 2010), amino-acid transporters (see below), as well as for some genes involved in embryonic development such as hairless (Notch signaling pathway) or Wnt8 (activator of the Wnt signaling pathway) (Shigenobu et al. 2010) (see Chapter 6 "Aphid Immunity" in this volume). (iv) An extensive computational screening of the entire *A. pisum* genome, followed by phylogenetic and experimental analyses, provided strong support for the transfer of 12 genes or gene fragments from bacteria to the pea aphid genome: three LD–carboxypeptidases (*LdcA1, LdcA2,ψLdcA*), five rare lipoproteins A (*RlpA1-5*), an *N*-acetylmuramoyl-L-alanine amidase (*AmiD*), a 1,4-beta-*N*-acetyl muramidase (*bLys*), a DNA polymerase III alpha chain (*ψDnaE*), and an ATP synthase delta chain (*ψAtpH*) (Nikoh et al. 2010). Although these genes are used in the context of the symbiosis with *Buchnera*, this work excluded the hypothesis that gene transfer to the host nuclear genome has been linked to genome reduction in *Buchnera*. In parallel, other horizontal gene transfers between fungi and the pea aphid genome have been identified: this concerns the carotenoid pathway, which confers to the pea aphid the capability to alternate green and rose in its body color (Moran and Jarvik 2010). This body color is also influenced by the presence of secondary symbionts and this phenotypic trait might be related to trade-off between predatory and parasitoid attacks (Tsuchida et al. 2010) (see Chapter 5 "Bacterial Symbionts of Aphids (Hemiptera: Aphididae)" in this volume).

In parallel to the genome resource, several transcriptomes have been obtained and described, from different tissues, organs and conditions. We can mention collections of transcripts from whole aphid colonies (containing a mixture of developmental stages and morphs); from whole bodies of parthenogenetic wingless or winged individual as well as sexual males and females, embryos, heads, guts, bacteriocytes, ovaries... (Sabater-Munoz et al. 2006). These data are very helpful for the identification and structural annotation of genes (e.g., to define the intron/exon junctions, the putative transcription start site, putative alternative splicing variants, putative functions by homology) and their study provide interesting leads to identify genetic programs differentially regulated to express important life traits in the pea aphid (see below). Furthermore, the reference genome can be used to identify other genomic features such as repeats and transposable elements, non-coding RNAs, genetic polymorphisms among populations or individuals, or epigenetic marks differentially present among morphs or conditions.

There are still limitations in the *A. pisum* reference genome. First, the assembly is not perfect since the genome is scattered in approximately 30,000 scaffolds. However, some of these scaffolds are large since half of the genome is represented in approximately 1000 scaffolds. Efforts are still required to improve the assembly and in consequence, its annotation. The large number of orphan genes (nearly 40% of the predicted genes) hampers sometimes the interpretation of data. However, it is intriguing that a large set of orphan genes has also been described in the *Daphnia pulex* genome, another arthropod with strong capacity of phenotypic plasticity (Colbourne et al. 2011). Most of these orphan genes in *D. pulex* are regulated in specific environmental conditions such as the response to stressors. Whether or not this is also the case for the pea aphid remains to be verified (Simon et al. 2011).

Genomic Resources for other Aphid Species

In the fall 2014, the *A. pisum* genome and in 2015, the genome of *Diuraphis noxia* has been published (Nicholson et al. 2015). But several other genome projects are ongoing and we can expect that many of them will be available within the next few years. This expectation is fuelled by international initiatives such as that of the i5k consortium that aims to provide tools, best practices and proof of principle for the sequencing and annotation of arthropod genomes (i5k Consortium 2013). The *Myzus persicae* genome is currently being sequenced and annotated: *M. persicae* is a polyphagous species—while *A. pisum* is specialized on Fabaciae—causing a lot of damages on crops, it is able to transmit a lot of viruses and it is present worldwide. Its genome is smaller than the one of *A. pisum* (~ 350 Mb vs. ~ 520 Mb) and the two species diverged several million years ago. The comparison between these two aphid genomes is not possible as yet, but it will be important to determine whether the general characteristics of the pea aphid genome are also present in *M. persicae*.

In terms of evolutionary biology and history of the aphid group, it is essential to have access to the genomes of ancient species or species basal to the phylogenetic tree of modern aphids (see Chapter 1 "Phylogeny of the Aphids" in this volume). This is the case for the cedar's aphid, *Cinara cedri*. This aphid species belongs to the clade Lachninae one of the many aphid groups that belong to the subfamily Aphididae that radiated during the Tertiary (Miocene), coinciding with the proliferation of herbaceous angiosperms. Thus, as compared to the diversification of *Myzus* and *Acyrthosiphon* (both belonging to the clade Macrosiphini), *Cinara* constitutes an early-diverging aphid lineage, which will certainly help tracing the evolution of aphid genomes. *Cinara cedri* differs from the other sequenced aphids in that is not considered a commercially relevant pest, as it does not grow on a crop but on a conifer. This fact may reveal interesting differences in the recent evolution of aphid populations from the different species, where those leaving on crops have certainly exposed to important pressures such as the use of pesticides but also to opportunities such as the global expansion associated to the cultivation and trading of certain crops. Finally, *C. cedri* is an interesting organism from the perspective of the study of the association of aphids to symbiotic bacteria. Indeed *C. cedri* hosts not one but two obligate endosymbionts in a prime example of metabolic inter-dependence resulting from co-evolution. Besides the general primary endosymbiont of aphids, *Buchnera aphidicola*, which plays a role in providing nutrients, such as essential amino acids and vitamins, which are deficient in the insect diet, *C. cedri* harbors a second obligate endosymbiont *Candidatus Serratia symbiotica*, which is only a facultative symbiont in other aphid species and which provides the consortium with some metabolic steps that have been irreversibly lost in *B. aphidicola* from *C. cedri*. The sequence of the aphid species in this tripartite living consortium will help to clarify the metabolic and evolutionary inter-dependence of these lineages as well as seed light on evolutionary processes such as symbiont establishment and replacement (see Chapter 5 "Bacterial Symbionts of Aphids (Hemiptera: Aphididae)" in this volume).

Phylloxeridae are also a basal family of aphids, represented by one species: *Daktulosphaira vitifoliae* which genome is being sequenced. Grape phylloxera is a historical pest of grapevine with worldwide economic and ecological importance. Phylloxera is native to the eastern United States of America where its natural hosts are American *Vitis* species. These natural hosts show varying levels of resistance to the insect and a few accessions of *Vitis* spp. have been used to produce rootstocks for use in commercial viticulture worldwide. The use of such rootstocks is, to date, the only means by which economically viable grape production can be reliably maintained from *Vitis vinifera* in a phylloxera infested vineyard. Phylloxera provides an interesting model for comparative genomics between aphids (in the broad sense) as well for life traits differing from modern aphids with the ability to form galls on leaves and roots or the absence of viviparous parthenogenesis. The availability of the phylloxera genome is expected to give the opportunity to study and shed light on the mechanisms underlying these processes.

It is also important to keep in mind that aphidologists will need to fill the gap between aphids and related taxa such as bugs, cicadas and leafhoppers, psyllids and whiteflies that all have specific and intricate interactions with their host plant (and their symbionts) (see also Chapter 8 "The Effect of Plant Within-Species Variation on Aphid Ecology" in this volume). But for specific phylogenomics issues to be solved, a complete genome sequence is not always required: phylogenies can be analyzed using a large set of conserved genes/proteins sequenced. Misof et al. (2014), using a set of 1478 protein-coding genes obtained from partial transcriptome reconstruction of 103 insect species, were able to date the origin of Insects, as well as their flight for instance.

Databases and Online Research Tools Supporting these Resources

Over the years, research groups interested in the study of aphid's biology using genomics approaches have directly developed dedicated databases and/or collaborated with experts to include aphid's data in their specific database or online tools. All these databases are key repositories for all genomics resources and they offer tools for online analysis. Data stored and organized in these databases are available to the aphid research community at large. It is worth mentioning that this database development and maintenance work, albeit paramount for genomics research, is very difficult to fund and publish. Furthermore, even if most of the time resources available for the human and/or other genome model can be used for the aphids genomics development, quite a lot of *ad hoc* developments are often needed to optimize these resources for the specific community.

AphidBase

In parallel to the development of genomic resources and the publication of the pea aphid genome, the reference aphid genome database, AphidBase,[2] has been set up. This information system is based at INRA (France) and contains today mainly data from the pea aphid. AphidBase is dedicated to become the repository of all aphid genomes when available. AphidBase was constructed using open-access tools based on the GMOD Chado system (Gauthier et al. 2007, Legeai et al. 2010). It contains mainly a browser that allows users to navigate in the different scaffolds, annotated with different features such as mRNAs and proteins. For each feature, a report is available indicating the number and size of introns/exons, the predicted coding sequence, the occurrence of transcriptomic evidences as well as the putative annotation. A "Blast" tool is also installed, to allow any direct similarity search on different databanks of AphidBase. And a short "quick search" is made available on the home page to more conveniently performed using keywords and gene numbers. In terms of annotation, users can use the GeneOntology description terms associated with the pea aphid genes that have orthologs in *Drosophila* and WebApollo, a tool that allows this time the users to correct, propose annotations for their favorite gene. AphidBase includes also the Galaxy platform link: users can load, display and analyse their data (such as RNA-Seq expression data) directly from this genomic database.

Very shortly, AphidBase should be upgraded to include other aphid genomes such as *Myzus persicae* and *Daktulosphaira vitifoliae*, with the same tools.

---

[2] http://www.aphidbase.com

SymbAphidBase

Symbiosis is central to aphid's biology (see following section in this chapter) as aphids harbour an obligate primary endosymbiont, *Buchnera aphidicola*, and several facultative secondary symbionts (see Chapter 5 "Bacterial Symbionts of Aphids (Hemiptera: Aphididae)" in this volume).

Recently a genomic database for aphid's symbionts has been developed using the same tools used for AphidBase (i.e., the GMOD Chado system). SymbAphidBase[3] is an *ad hoc* genome database to store and analyse aphid symbionts' genome sequences. A specific annotation pipeline was developed for these genomes and this will facilitate comparative genomic studies.

AcypiCyc

The AcypiCyc[4] is a BioCyc Pathway/Genome Database (PGDBs)[5] (Karp et al. 2005) dedicated to aphid and it constitutes a key resource for metabolism analyses. A collection of BioCyc databases generated using the Pathway Tools software (Karp et al. 2010) is developed and maintained by a large community of scientists (Caspi et al. 2014). The AcypiCyc database was developed during the genome annotation for the pea aphid (*Acyrthosiphon pisum*) using an *ad hoc* developed automated annotation management system called CycADS (Cyc Annotation Database System) (Vellozo et al. 2011)[6] in combination with Pathway Tools. CycADS allows the seamless integration of the latest sequence information/annotation into Pathway Tools for metabolic networks reconstruction. In CycADS, specific genomic data, as well as their functional annotations obtained using different methods (such as KAAS, PRIAM, PhylomeDB, Blast2GO, MetaPhOrs, Interproscan), are collected into a SQL database and later extracted, with the possibility to apply different quality filters. CycADS was recently used to create ArthropodaCyc,[7] a collection of arthropods metabolic network databases, which contains 23 organisms to date (January 2015) including a duplicate of the pea aphid database for comparisons with other species. These databases collection allows metabolic pathways visualization, and each protein page includes information about the annotation methods used, as well as hyperlinks to genome specific resources. Comparisons using interactive web functionalities, as well as user "omics" data mapping or information extraction, are also available in the BioCyc interface of ArthropodaCyc and AcypiCyc.

PhylomeDB

PhylomeDB is a repository of complete collections of evolutionary histories of genes encoded in complete genomes (Huerta-Cepas et al. 2014). Besides precomputed gene trees representing the evolutionary relationships of genes and their homologues in

---

[3] http://symbaphidbase.cycadsys.org/
[4] http://acypicyc.cycadsys.org/
[5] http://biocyc.org/
[6] http://www6.inra.fr/cycads_eng
[7] http://arthropodacyc.cycadsys.org/

other species, PhylomeDB provides sequence alignments, information on domain composition of proteins, predictions of orthology and paralogy relationships, functional annotations, and links to relevant databases such as UniProt, Ensembl, or AphidBase. Gene trees can be navigated through a potent webbrowser and complete gene tree collections and alignments can be downloaded for large-scale analyses. With over 4.5 million gene trees in its current version, PhylomeDB is the largest repository of genes evolutionary histories covering the evolution of hundreds of genomes, from bacteria to higher eukaryotes. Trees in PhylomeDB are built following a highly sophisticated pipeline (Huerta-Cepas et al. 2011). It includes finding homologues for each query sequence, building consensus alignments from six different multiple sequence alignment strategies and trimming them using a consistency-based approach. Then a suitable evolutionary model is selected using a statistical model testing approach and a full maximum likelihood reconstruction is performed. Resulting trees include computed branch lengths and supports as well as automated detection of duplication and speciation nodes, which are indicated with different colors on the tree. The trees are further decorated with taxonomic information and the domain composition of the encoded proteins, using the powerful display capabilities of the ETE toolkit (Huerta-Cepas et al. 2010a). The tree can be downloaded or manipulated on-line, and alternative leaf lables can be displayed (e.g., alternative naming or ID systems). Orthology and paralogy information is provided in the form of a table with links to the relevant sequence databases. Importantly, through a constant link to the MetaPhOrs database (Pryszcz et al. 2011) PhylomeDB extends orthology predictions to over a thousand organisms and provides a consistency-based confidence score for each prediction. PhylomeDB was used in the initial annotation of the pea aphid genome (IAGC 2010) and provides regular updates linked to newer versions of the annotations of pea aphid. PhylomeDB is currently collaborating with recent aphid genome sequencing efforts such as those of *Cinara* and *Phylloxera*, which will be included in the database. The community of aphid researchers regularly uses this resource to (1) find orthology and paralogy relationships across insects or metazoans, (2) obtain a first view of the evolution of the protein family of interest and retrieve the relevant sequences, (3) trace recent an old duplications within a gene family, (4) expand the tree available in PhylomeDB by adding additional genes from species sequenced in-house. An increasing number of publications cite PhylomeDB as an important resource for the described research.

E-RNAi

The inactivation by RNA interference is a key approach to functional studies through gene inactivation (see below in this chapter) and a good design is a very important starting step towards success. The pea aphid genome is available in the E-RNAi tool (Horn and Boutros 2010)[8] that allows the design and evaluation of RNAi reagents. This specialized resource is available to researchers and shows the importance of collaborations in developing further research tools in the genomics of aphids.

---

[8] http://www.e-rnai.org/

Functional Approaches: Gene Function Studies

The interest in developing genomics resources in aphids is not only related to the fact that these insects are pests, but also because of their very original unique traits that differ from other insects and that explain partly their ability to adapt to their environment. In this section, we will briefly summarize how genomics resources have been used to begin to decipher gene regulations and pathways responsible for different biological traits such as phenotypic plasticity, interactions with symbionts or plants.

Dispersal and Phenotypic Plasticity

Parthenogenetic females, in aphid populations, can be winged or wingless. This is a plastic trait since it is dependent on environmental conditions: an adult parthenogenetic female will sense local changes in environment and produce clonal progeny winged or wingless (Ogawa and Miura 2014) (see Chapter 2 "The Ontogenesis of the Pea Aphid *Acyrthosiphon pisum*" in this volume). There are different triggering factors of this shift of phenotype in the progeny, related to different kind of environmental conditions for which getting fly capacity is an advantage to escape. The first category of stress condition is "danger": the presence of predators or parasitoids in a colony is sensed and the information transmitted between individuals by the emission of an alarm hormone (the (*E*)-ß-farnesene) (see Chapter 9 "Chemical Ecology of Aphids (Hemiptera: Aphididae)" in this volume). This alarm system, even if it has not been clearly demonstrated, probably triggers changes in the embryogenesis programs of the future progeny of viviparous females, such as the progeny after birth and final molting will be constituted by winged parthenogenetic females able, if necessary, to escape the environment. The second category of stress producing winged progeny concerns food quality and quantity. When plants are senescing or dying, aphid clonal colonies feeding on them will contain more winged individuals in their progeny. As well, when the colony increases too much and individuals are competing on a same plant for space and food, the progeny will be mostly winged. In this case, we know that physical contacts between legs can trigger for female aphids to shift to the production of winged progeny. The molecular signaling molecules and mechanisms leading to this shift are not yet known. But a parallel can be drawn with recent studies on locusts that shift from sedentary to migrating morphs by leg contact. Neurosignalisation including serotonin and dopamine is involved in the regulation of this plasticity (Wang and Kang 2014) (as well as a specific regulation of the dopamine pathway by microRNAs. These represent strong hypotheses that need to be tested in aphids. As well insulin signalization might play a key role in wong morphs development, as suggested by a work on planthoppers indicating that two insulin receptors regulate in opposite direction the signalling pathway of Foxo (Xu et al. 2015). Wings allow insects to explore new horizons and disperse, in a foraging behaviour, even though aphids can also move walking from plant to plant. In *Drosophila*, bees and other insects, some groups of genes involved in foraging have been identified. They mainly correspond to genes/proteins expressed in the brain, and regulating phosphorylation pathways and transduction signals (Tares et al. 2013) characterized (by homology search) four different genes encoding the for genes in the pea aphid. The expression of one of these for gene is higher in wingless adults reared in crowded conditions, compared to solitary adults. This might indicate that for signaling could be active in the mothers of the future winged embryos. In parallel, the cGMP-dependent protein kinase associated with the FOR protein has its activity increased in this crowded aphid females, in correlation with the for expression. This pathway could represent one of the key mediator for this plastic trait.

In insects, most of the neuronal regulations are driven by hormone subsequent transduction. It is known that juvenile hormones (JH) play roles in aphid dispersal polyphenisms (reviewed in Ogawa and Miura 2014) even if the underlying mechanisms are unclear. We thus might expect a JH differential concentrations in females harboring future winged or wingless progeny (Schwartzberg et al. 2008) used a LC-MC approach to dose the JHIII in hemolymph extracts of such pea aphid females, but they were not able to find differences in its concentration. Either JHIII variation does not occur during the shift of embryogenesis, or variations are so localised in time and space that this dosage in hemolymph could not detect them. In the *Megoura crassicauda* aphid, this concentration of JHIII does differ between individuals experiencing crowding and those living in non-crowded conditions, but this difference is evident only post-maternaly, in third instar nymphs (Ishikawa and Miura 2013), the presumptive wingless having a higher JHIII concentration than the presumptive winged ones. In this aphid, topical application of JHIII in presumptive winged nymphs inhibited wing development, strongly indicating a role of JH in dispersal polyphenisms regulation.

Several transcriptomics approaches have been performed, that compare mainly the adult winged and wingless morphs (Brisson 2010, Brisson et al. 2007). Some genes are specifically regulated in winged morphs, and some of them have putative functions related to what is known in wing development in *D. melanogaster*. However, in order to really understand the bases of dispersal polyphenism in aphids, it should be more appropriate to compare gene functions by dissecting embryos from mothers under non-stress conditions (wingless future progeny) and under stress conditions (winged future progeny): this is where the molecular mechanisms of the shift differ, and this thus represents the stage to identify the major genetic programs involved in dispersal polyphenism.

To conclude this part, the fascinating complexity of aphid biology still remains to be understood! Not only clonal viviparous parthenogenetic females can be winged. Males of several aphid species are winged or wingless, and this has been studied in the pea aphid. These studies have shown that this trait is genetically driven, and it is not a plastic trait (see Brisson 2010). Genetic variations among pea aphid populations allowed quantitative genetics approaches that led to the identification of a locus (*aphicarus*) located on the X chromosome of the pea aphid genome (Braendle et al. 2005a, 2005b); the nature of the genes responsible for this genetically-driven traits is still unknown (see Chapter 2 "The Ontogenesis of the Pea Aphid *Acyrthosiphon pisum*" in this volume).

Reproductive Mode and Phenotypic Plasticity

Aphids can reproduce by viviparous parthenogenesis, and this was observed for the first time in 1745 by Charles Bonnet. More than one century later, it has been observed that sexual reproduction between males and oviparous females also occurred (see Chapter 2 "The Ontogenesis of the Pea Aphid *Acyrthosiphon pisum*" in this volume). Since then, physiological and molecular studies have been developed in order to describe and compare these two modes of reproduction. This is a plastic trait in aphids since during an annual life cycle, an aphid colony can alternate between sexual and asexual reproduction, depending on the local environment. Here, the triggering event is the season: the increase in autumn in the length of the scotophase of the circadian cycle produces changes in the developmental processes of the embryos inside their clonal mothers: embryos that normally (under short scotophase) develop in future viviparous parthenogenetic females now develop in future males or oviparous sexual females. Thus, in fall, the mating of sexual females with males allows, through regular meiotic reproduction, the production of sexual eggs that are laid on plants before winter. As these eggs are overwintering, they prevent the loss of a pure clonal viviparous colony in hard winter conditions, as aphids are not able to regulate their internal body temperature to compensate low external temperatures. In the following spring, the eggs hatch larvae that will develop in viviparous parthenogenetic females. This seasonal alternation between parthenogenesis and sexual reproduction is called "cyclical parthenogenesis" (see Chapter 2 "The Ontogenesis of the Pea Aphid *Acyrthosiphon pisum*" in this volume). As for the abovementioned dispersal polyphenism, this reproductive polyphenism is complementary to a variation of this trait within populations: some populations have lost the capacity to perform sexual reproduction and remain parthenogenetic all the year. These populations can survive only if winters are mild. Using quantitative genetics approaches, a 10 cM-locus located in the X chromosome on the pea aphid have been identified: it contains genetic elements required for the parthenogenetic state (Jaquiery et al. 2014).

Different molecular processes involved in reproductive polyphenism have been described using genomic resources, such as cDNA or oligonucleotide arrays or *in situ* hybridization and comparing gene expression during the shift of phenotype from sexual to asexual reproduction. This has been largely already reviewed, and the readers can refer to (Davis 2012, Le Trionnaire et al. 2008, 2012, Ogawa and Miura 2014) (see Chapter 2 "The Ontogenesis of the Pea Aphid *Acyrthosiphon pisum*" in this volume).

First (Fig. 2), sensing of photoperiod is essential to the shift of the reproductive mode. Several transcriptomic studies have been performed—mainly on the pea aphid. They used different methods such as quantitative PCR and cDNA arrays, to compare mRNA accumulations between individuals reared under short (the progeny will be sexual) or long (the progeny will be parthenogenetic) photoperiods. This shift is complex because of viviparity and the so-called "telescoping of generation" in parthenogenetic females; the mother contains embryos that already contain the first developing early stages of embryos. Thus, three generations are embedded. It has been showed that the photoperiod sensing has no major effect on the mRNA accumulation of the first generation (the grand-mother) (Le Trionnaire et al. 2009). However, different genetic programs are expressed in the second generation (the mother) (Cortes et al. 2010, Ramos et al. 2003) and more particularly in their head, where the transcriptomes have been compared (heads are easy to collect and contain the main morphological structures required for neuro-sensing and neuro-regulation).

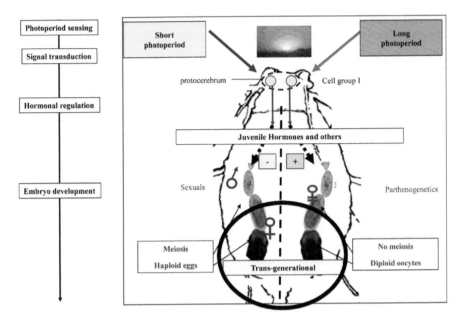

Fig. 2. The different molecular regulation steps occurring during the reproductive mode plasticity in aphids. See the text for following the role of photoperiod sensing, neuro-endocrine signal transduction and morphogenesis of the two alternative sexual and asexual morphs.

The main conclusions are that unexpectedly: (i) cuticular proteins represent a large set of regulated RNAs, (ii) there are correlations between these cuticular proteins regulation and neurological pathways, such as the dopamine pathway (Gallot et al. 2010), and (iii) insulins could also represent key mediators of the photoperiod signal (Huybrechts et al. 2010). Some of these conclusions have been confirmed for another species, the cotton aphid (Liu et al. 2014).

Second, the sensed photoperiodic signal has to be transduced from the head to the ovarioles, where embryos will develop. As already mentioned above, several neuro-endocrine regulators have already been suggested as potential candidates for this transduction: dopamine and insulin mRNA levels are down-regulated in short days, indicating that these molecules could be part of the complex process of neuro-regulation of the reproductive polyphenism. In the late 80's, melatonin (a hormone involved for regulating biological rhythms in vertebrates) has been identified in the pea aphid, but its modulation between short- and long-day reared insects has not been detected. However, topical application of melatonine to parthenogenetic females reared under long-day periodicity modifies their progeny, with the production of sexual embryos (Gao and Hardie 1997). Melatonine is synthesized by an arylalkylamine N-acetyltransferase (AANAT): four genes in the pea aphid genome have been identified for this protein, and two of them are over-expressed in individuals reared on short photoperiod (Barbera et al. 2013). This indicates that regulation by circadian rhythms of the reproductive polyphenism could occur, but deserves further study. And the fact that the genome of the pea aphid contains all the genes required for the regulation of circadian rhythm (Cortes et al. 2010) will help us deciphering this putative role.

Third, hormonal regulation of reproductive polyphenism in aphids has been demonstrated more than 30 years ago: juvenile hormones (JH) can reverse the effect of the photoperiod (Corbitt and Hardie 1985). JHIII topically applied to adult parthenogenetic females reared under short-days (programmed to produced a sexual progeny) provokes the production by these females of a parthenogenetic progeny. Thus, JHs are regulators, or transducers, of the photoperiodic signals to the ovaries. However, we still don't know which specific cells are targeted and which pathways are triggered by JHs. The concentration of JHIII has been measured and compared between parthenogenetic females producing sexual or asexual embryos (Ishikawa et al. 2012); the lower levels of JHIII in sexual-producing females is coherent with previous hypotheses (Lees 1966) that JH concentration is lower during the shortening of the photoperiod. And this lower concentration corresponds also to a lower mRNA expression level of a gene encoding a JH esterase responsible for JH degradation. These data thus confirm that, at least for the pea aphid, JHs are important for the regulation of the reproductive polyphenism, and that the production of sexual individuals is correlated with a lower concentration of these hormones.

Fourth, as a last step, the sensing and transduction of the photoperiod shortening acts on parthenogenetic embryos developing in the ovaries of the mothers. And the final fate of these embryos is either to develop meiotic (for future sexual individuals) or mitotic (future asexual individuals) gametes. This is why recent studies have focused on the regulation of genetic programs that differ in these two types of embryos. There are 20 different developmental stages during the pea aphid embryogenesis (Miura et al. 2003) (see Chapter 2 "The Ontogenesis of the Pea Aphid *Acyrthosiphon pisum*" in this volume) (Gallot et al. 2012) identified that (at least in a dedicated lab conditions and for the pea aphid LSR1 clone), Stage 17 was the last development stage that responds to the effect of the photoperiod: this suggests that germlines are not yet differentiated at this stage and that the effect of the photoperiod can still drive the developmental process of the stem cells for the sexual or asexual production of gametes. Then, between stages 17 and 18, the process becomes insensitive to the change of the photoperiod; embryos reared under short-days conditions produce sexual gametes and embryos reared under long-days produce asexual gametes. Thus, a window of time as to when and where the phenotypic plasticity occurs was defined. With this type of defined material (Gallot et al. 2012), compared the transcriptomes of such embryos from stage 17 to stage 20, and for the two types of photoperiods. The genes whose mRNAs are differentially expressed belong to four main putative functional categories: unknown functions, cell cycle, epigenetic regulation and mRNA stability. Those three last categories correspond to early stages of oogenesis in *Drosophila*: and *in situ* hybridizations performed on most of these transcripts confirmed their expression in oocyte, or early stages of development (Duncan et al. 2013) also showed differences between asexual and sexual early developmental programs in the pea aphid: they hypothesize that the parthenogenetic developmental program is not only derived from the sexual one, but rather represents a highly evolved new program required to produce asexual embryos. And the development of the posterior structures of the embryos (labelled by torso-like gens and activated ERK proteins) seems to be different in oviparous and viviparous oocytes or embryos, suggesting the viviparous development uses different mechanisms than oviparous development to specify posterior fate (Bickel et al. 2013). As well, the expression profiles of some genes known to be involved in canonical meiosis differ in sexual and asexual ovaries of the pea aphid. This is the case for Argonaute-family, as well as the homolog of Spo11 which harbours differential spliced forms in sexual and asexual aphid tissues. This suggests again that parthenogenesis in the pea aphid evolved from regular meiotic processes (Srinivasan et al. 2014).

PIWI proteins are also known to be expressed during early embryogenesis: they interact with piRNAs that are regulators of the chromatin and prevent the expression of transposable elements. Unexpectedly, the pea aphid encodes a large number of PIWI proteins, as an expanded family more particularly for the sub-clade of Argonaute proteins (Lu et al. 2011). Some of these copies are differentially expressed; for instance Api-piwi2 is over-expressed in sexual females, and Api-piwi8 is over-expressed in parthenogenetic females. These observations correlate thus the PIWI protein family with the reproductive polyphenism in the pea aphid.

Beside this comparison of asexual and sexual development in the pea aphid, the embryogenesis of the asexual fate has been more studied into details since it differs from the canonical sexual development by different means (see Bickel et al. 2013, Le Trionnaire et al. 2008) such as the size of the oocyte or reduced vitellogenesis during the viviparous development. The gene repertoire required for embryogenesis has thus been explored using the annotation of the pea aphid genome (Shigenobu et al. 2010). More precisely, the vasa and nanos mRNAs and proteins have been localised in viviparous embryos, as markers of germ-cell migration and specification in asexual pea aphids (Chang et al. 2006, 2007, 2009, Huang et al. 2010). Interestingly, during the posterior embryonic development, the expression of *caudal* differs in the pea aphid from *Drosophila* (Chang et al. 2013): its antero/posterior localisation in oviparous and viviparous pea aphid embryos differs (see Chapter 2 "The Ontogenesis of the Pea Aphid *Acyrthosiphon pisum*" in this volume).

During the early release of the asexual oocyte, as a diploid gamete, Riparbelli et al. (2005) reported that the oocyte of the parthenogenetic viviparous pea aphid is able to self-organize microtubule-based asters, independently of male material, which in turn interact with the female chromatin to form the first mitotic spindle, assembling new centrosomes. Recruitment of material along the microtubules might contribute to the accumulation of pericentriolar material and centriole precursors at the focus of the asters, thus leading to the formation of true centrosomes (see Chapter 2 "The Ontogenesis of the Pea Aphid *Acyrthosiphon pisum*" in this volume).

During the life cycle of aphids, there is another limiting step in reproductive polyphenism. In the early spring, eggs hatch and young parthenogenetic colonies develop and settle. But at that date (before mid-march) the photoperiod length (which is under 12 h) is an inducible photoperiod for the reproductive switch. It is thus remarkable that these newborn parthenogenetic colonies are unable to respond to this short photoperiod: at least several weeks (that corresponds to several parthenogenetic generations) are needed for the recovery of this sensitivity. This is the so-called "timer" or "fundatrix" effect, from the name given to the morph hatching from an egg that will fund a parthenogenetic colony (Lees 1960). This effect prevents a non-appropriate shift of the reproductive mode since there would be no advantage of producing sexual and oviparous females in spring. This process deserves further molecular studies since one can hypothesize that the molecular regulations underlying the autumn polyphenism might be similar to the one acting in early spring, but blocking the switch.

The pea aphid, during the last years, has represented an adapted model to study reproductive polyphenism for at least three main reasons: its body size (considered as "large" for an aphid), the availability of a sequenced genome since 2010, and a relatively simple annual life cycle. The pea aphid is monoecic: it performs the two different reproduction types on a same plant species. However, many aphid species alternate host plants between the sexual and asexual reproductive mode: sexual reproduction is usually performed on the primary host that is often a woody perennial plant that survives during winter. The parthenogenetic reproduction is performed on annual plants (such as cereals) during spring and summer. Thus, there are migrations from the secondary host to the primary host (in autumn) and from the primary to the secondary host (in spring) after egg hatching. The molecular bases of this complex life cycle has not been yet analysed in details. This represents however a very good system to study the interplay between the dispersal and reproductive polyphenisms described in this chapter.

Aphids in Interaction with Host Plants

Aphids are herbivorous insects and feed on plant phloem sap using a needle like mouthpart called stylet. The phloem sap contains sugars, amino acids and other molecules such as RNAs and secondary metabolites (Dinant et al. 2010). Aphids spend most of their lifetime feeding on host plants. The interactions of aphids with host plants vary depending on the aphid species (see Chapter 8 "The Effect of Plant Within-Species Variation on Aphid Ecology" and Chapter 11 "Function of Aphid Saliva in Aphid-Plant Interaction" in this volume). Similar to other herbivorous insects, most of aphids are specialized to one or a few host plant species and cannot feed and reproduce efficiently on other plant species. Fewer aphid species, known as "generalists", are able to feed on a wide range of plant species. Furthermore, as we mentioned above, some aphids colonize two taxonomically different plants and alternate their hosts depending on the season. As aphids cause damage on crop plants by removing nutrients, producing toxic saliva and transmitting viruses, understanding the mechanisms of interactions between aphids and plants is a key to select or construct aphid resistant/tolerant crops and develop better agricultural practice. Therefore, significant amount of studies have been conducted to understand the interactions between aphids and plants. Recent developments in genomic resources of plants and aphids have been contributing to the progress of this field of research and allowed us to ask fundamental questions on aphid-plant interactions. The pea aphid is the only aphid whose genome sequence has been completed so far. In addition, it forms at least 11 biotypes each of which is specialized to one or a few legume plant species (Peccoud et al. 2009). Therefore, the pea aphid complex offers great opportunities to identify the factors that are involved in the specialization of aphids to host plant (see Chapter 8 "The Effect of Plant Within-Species Variation on Aphid Ecology" in this volume). On the other hand, the green peach aphid, *M. persicae*, is a generalist and feed on two model plants, *Arabidopsis thaliana* and *Nicotiana benthamiana*. Thus, increasing number of researchers are using

M. persicae as a model aphid to study plant-aphid interactions at a molecular level, and its genome sequencing is underway.

The interactions between aphids and plants take several steps (Fig. 3). Accumulating studies indicate that plant sensing, establishment of phloem feeding and detoxification of plant secondary metabolites are the key steps to determine the compatibility between plants and aphids. Here, we would like to review researches on these three major steps, and see how genomic resources have helped the progress of this field of research.

First, particularly in case of winged forms, aphids sense the presence of a suitable host plant and land on it. In this early process of host selection, aphids are thought to use chemosensing (see also Chapter 9 "Chemical Ecology of Aphids (Hemiptera: Aphididae" in this volume). For example, to avoid plants that are already infested by aphids, *Aphis gossypii* senses volatile organic compounds (VOC) emitted by

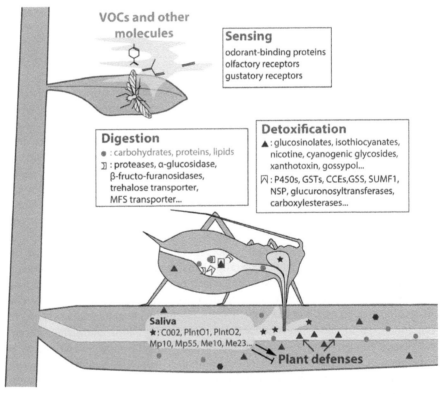

Fig. 3. An overview of plant-aphid interactions. Aphids sense VOC (volatile organic compounds) and other molecules to select suitable host plant species. Odorant-binding proteins, olfactory receptors and gustatory receptors are known to be involved in this process. If aphids are not repelled, they will insert stylets into plant and try to establish phloem feeding. The molecules secreted into aphid saliva, such as C002 protein, play important roles in the establishment of feeding. While feeding on plant sap, aphids have to cope with plant derived toxic compounds such as nicotine and glucosinolates. P450 monooxygenases, glutathione S-transferases (GSTs) and carboxy/cholinesterases (CCEs) genes are known to be involved in the detoxification step. Some plants carrying resistance genes (e.g., *Mi1-2*, *Vat*) can sense aphid attack and trigger strong defense reactions. The figure was modified from Simon et al. (2015).

infested cotton plant and is repelled by them (Hegde et al. 2011, 2012). Aphids also sense the pheromones emitted by fellow aphids. One example is an alarm pheromone (*E*)-ß-farnesene, which is released in the presence of danger and repels other aphids (Bowers et al. 1972). In aphid, the antenna is thought to be the most important organ to perceive odours. Sensing of volatiles and sapid molecules relies on a set of proteins in which OBPs (odorant-binding proteins), ORs (olfactory receptors) and GRs (gustatory receptors) have key roles. The molecules aphids sense are often highly hydrophobic. However, the binding with OBPs solubilizes them in the sensilla lymph and transports them to the receptors, which triggers, in turn, the cascade of responses (Sachse and Krieger 2011). ORs and GRs have typical seven transmembrane domain ligand gated ion channels and can be identified based on sequence homology with previously characterized proteins. By using the full genome sequence of the pea aphid, *A. pisum*, Smadja et al. identified 79 Or and 77 Gr genes (Smadja et al. 2009). Furthermore, they showed that most recently duplicated Or and Gr loci have been under positive selection and hypothesize that this could be due to the high degree of ecological specialization (e.g., host plants) of this aphid species (Smadja et al. 2009). On the contrary, only 13 OBPs have been identified in the pea aphid (Zhou et al. 2010). Interestingly, *in vitro* studies have shown that strong repellents, including (*E*)-ß-farnesene, bind OBP3 and/or OBP7 of *A. pisum* and *M. persicae*, while non-repellents showed different spectra of binding (Qiao et al. 2009, Sun et al. 2012). Genome wide studies on the pea aphid biotypes also suggested that Or and Gr genes are involved in *A. pisum* biotype formation (Jaquiery et al. 2012, Smadja et al. 2012). Furthermore, differences in copy number of certain genes of the GR and OR families were detected within *A. pisum* biotypes, highlighting the potential importance of copy number variation in the host specialization process and the importance of the sensing step in plant-aphid interactions (Duvaux et al. 2015).

If volatiles don't repel aphid, then, the aphid inserts its stylet into the plant. When the aphid is well adapted to the plant, the aphid inserts stylet into the intracellular spaces and samples the contents of epidermal and mesophyll cells (see Chapter 11 "Function of Aphid Saliva in Aphid-Plant Interaction" in this volume). The aphid stylet finally reaches phloem sieve cells and the aphid may feed on the plant for several hours (Prado and Tjallingii 2007, Tjallingii 2006). In this physical interaction between aphid and host plant, aphid saliva seems to play important roles. Aphids secrete two types of saliva: gelling saliva is solidified when it is secreted and oxidised in plant intracellular space and thought to protect the stylet, and watery saliva is secreted into plant cells (Miles 1999). In some cases, brief puncture of plant cells (epidermal level) is enough for the aphid to recognize unwanted plant, and the aphid leaves the plant (Powell and Hardie 2000). In other cases, aphids try to feed on the plant by penetrating their stylet but cannot establish phloem feeding and give up on feeding (Caillaud and Via 2000, Schwarzkopf et al. 2013). These studies demonstrated that, in some cases, aphid needs to have direct interactions with the plant cells to recognize their suitability as a food source. Aphid gustatory sensors in the stylet (Wensler and Filshie 1969) and/or aphid saliva may be involved in this recognition process (Hogenhout and Bos 2011). Aphid saliva contains proteins that are essential for the aphids to establish feeding. A pioneering work by Will et al. (2007) showed that the calcium binding proteins in aphid saliva prevents the dispersion of forisomes, which causes phloem occlusion and thought to interfere with continuous aphid feeding. A protein called C002, first identified in the pea aphid genome, is conserved in many aphid species and has an indispensable role in aphid feeding: *C002* knockout aphids cannot establish phloem feeding and die (Mutti et al. 2006, 2008, Pitino et al. 2011, Zhang et al. 2013). A large majority of the proteins secreted into aphid saliva is thought to be produced in the organ called salivary glands although some proteins produced in other organs than salivary glands can be also secreted with saliva (Rao et al. 2013). To understand the composition of proteins secreted into aphid saliva, proteomics of saliva and proteomics and transcriptomics of salivary glands were conducted in various aphid species (Atamian et al. 2013, Bos et al. 2010, Carolan et al. 2009, 2011, Elzinga and Jander 2013, Harmel et al. 2008, Rao et al. 2013, Vandermoten et al. 2014). These studies are largely benefitted by the completion of the pea aphid genome sequence and new sequencing technologies. Interestingly, overlaps of salivary gene libraries between different studies are low, indicating that each salivary gene library is not complete. Following the identification of salivary genes, functional characterization of these salivary genes have been conducted in various aphid species (Atamian et al. 2013, Bos et al. 2010, Elzinga and Jander 2013, Pitino and Hogenhout 2013). Silencing of the salivary genes by dsRNA microinjection is possible in large aphids, but technically difficult to conduct in small aphids. Thus, plant mediated aphid gene silencing and/or expression of saliva proteins *in planta* was used to assess the functions of the salivary genes on plant responses and aphid fitness. These studies identified several salivary genes that trigger or suppress plant defence reactions against aphid feeding and decrease or enhance aphid fitness (Atamian et al. 2013, Bos et al. 2010, Elzinga and Jander 2013, Pitino and Hogenhout 2013). Although many salivary genes are conserved in various aphid species, some genes have species-specific sequences. Pitino and Hogenhout demonstrated that salivary proteins from *M. persicae* enhances aphid performance when it is expressed in *A. thaliana* while the orthologous gene from legume specialized aphid, *A. pisum*, does not exhibit such function in *A. thaliana* (Pitino and Hogenhout 2013). The study revealed that aphid salivary proteins have the functions specific for their host plant species. The same study also presented that some aphid salivary genes are under positive selection, suggesting a key role of salivary genes in host plant adaptation and specialization. Previously mentioned genome wide studies population genomics studies of the pea aphid biotypes also suggested the involvement of aphid salivary proteins in biotype formation (Jaquiery et al. 2012).

In specific plant-aphid interactions, plants resistance against aphid is conferred by resistance genes that have characteristic structural feature (Nucleotide Binding Site-Leucine Rich-Repeats) of plant resistance genes against microbial pathogens: *Mi1-2* of tomato provides resistance to certain aphids, psyllids and whiteflies (Casteel et al. 2006, Kaloshian and Walling 2005, Rossi et al. 1998). The *Vat* gene confers resistance to the cotton-melon aphid *Aphis gossypii* on melon (Dogimont et al. 2010, 2014). Several miRNAs are differentially expressed in the aphid fed on *Vat*[+] and *Vat*[-] plants, indicating the involvement of miRNA in aphid responses to *Vat* mediated plant resistance reactions (Sattar et al. 2012). It has been hypothesized that the proteins secreted in aphid saliva are recognized by the plant resistance genes and trigger defence responses (Hogenhout and Bos 2011), however, the defence eliciting aphid proteins have not been identified yet.

Finally, when aphids start sucking plant phloem contents, they have to deal with toxic plant derived compounds. Whether the aphid can tolerate the toxic compounds or not can be an important question, the answer of which determines the compatibility between plant and aphid. Comparative analysis of P450 monooxygenases, glutathione S-transferases (GSTs) and carboxy/cholinesterases (CCEs) genes, which are key detoxification enzymes, between a legume specialist *A. pisum* and a generalist *M. persicae* revealed larger expansion of P450 gene family in *M. persicae* compared to *A. pisum* (Ramsey et al. 2010). Furthermore, a study of *M. persicae* races that are recently adapted to tobacco crops showed that the adapted races constitutively overexpress cytochrome P450 (*CYP6CY3*) to detoxify nicotine, highlighting the importance of detoxification step in host plant adaptation. The overexpression of *CYP6CY3* was realized by gene amplification and polymorphism in the promoter region (Bass et al. 2013). The amplification of *CYP6CY3* was also reported to be the cause of increased insecticide resistance of *M. persicae* races (Puinean et al. 2010).

Instead of detoxifying plant compounds, the cabbage aphid, *Brevicoryne brassicae*, store plant compounds, glucosinolates, which are secondary metabolites specifically produced by *Brassicaceae*, and use it to damage the natural enemies that attack the aphid. The aphid sequesters the glucosinolates in hemolymph and an enzyme myrosinase in crystalline microbodies (Bridges et al. 2002, Kazana et al. 2007). When the aphid is wounded, the myrosinase is released to catalyse the hydrolysis of glucosinolates and produces secondary compounds toxic to the natural enemies (Kazana et al. 2007).

Recently, more and more examples of microbial symbionts that interfere with plant-insect interactions have been reported (Sugio et al. 2015). In case of *A. pisum*, each aphid biotype, which is determined by the feeding plants, is associated with specific composition of facultative symbionts (Ferrari et al. 2012, Henry et al. 2013) (see Chapter 5 "Bacterial Symbionts of Aphids (Hemiptera: Aphididae)" in this volume). The strong association between the aphid biotype and the specific facultative symbiont composition can be explained by the positive impact of the symbiont on aphid performance on specific host plant. For example a facultative symbiont, *Regiella insecticola*, is reported to increase *A. pisum* fecundity on white clover (Tsuchida et al. 2004). However, such positive and plant specific effect of facultative symbionts on aphid performance is not always observed (Ferrari et al. 2007, Leonardo 2004), and combination of aphid, plant and symbiont genomes possibly determines the outcome of the multitrophic interactions. Although genome sequences of some of *A. pisum* facultative symbionts are already available (Burke and Moran 2011a, Degnan et al. 2010, Hansen et al. 2012), more genome sequences of different races of facultative symbionts will probably contribute to the bacterial genes that influence the outcome of plant-aphid interactions.

Aphids in Interaction with Bacteria

Aphids, as well as other herbivorous insects, are susceptible to interact with different types of microorganisms: intracellular symbionts (obligatory and facultative), and gut and environmental microorganisms. It is largely admitted that aphid-associated microbes can allow the acquisition of traits facilitating insect dispersion, plant use and adaptation to sub-optimal environments (Hansen and Moran 2014). Up to now, most of the progresses on the role of these bacteria have been done on the obligatory symbiosis between the pea aphid and *Buchnera aphidicola*, the first genomic insect association for which both partners' genomes have been sequenced and annotated. The recent availability of the genome sequences of secondary symbionts in the pea aphid is at present opening up several new research perspectives for this genomic model of multipartner symbiosis (see Chapter 5 "Bacterial Symbionts of Aphids (Hemiptera: Aphididae). The availability of new molecular methods and genomic resources for gut and environmental microbes are expected, in the next decades, to supply a complete picture of the interaction of this complex microbiome with insect physiology.

The use of genomic resources and functional genomic tools to decipher the molecular bases of the long-term nutritional association of aphids with *Buchnera aphidicola*

Nutritional symbioses play a central role in insects' adaptation to specialized diets and in their evolutionary success. The obligatory symbiosis between the pea aphid, and the bacterium, *Buchnera aphidicola*, is no exception as it enables this important agricultural pest insect to develop on a diet exclusively based on plant phloem sap. As a diet, phloem sap is low in nitrogen and especially low in essential amino acids, those amino acids that contribute to proteins and have a carbon skeleton that metazoans cannot synthesize *de novo* and must obtain them from their food (Dadd 1985). Even before the confirmation provided by genome sequencing projects, there was persuasive experimental evidence that *B. aphidicola* provides the insect with essential amino acids (Akman Gündüz and Douglas 2009, Douglas and Prosser 1992, Liadouze et al. 1995) and it is an obligate symbiont for aphid growth and reproduction.

Consistent with these experimental data, the genome of *B. aphidicola* from *A. pisum* (*Buchnera* APS) includes most genes required for essential amino acid biosynthesis, but lacks the genes for the amino acid degradation and the synthesis of most nonessential amino acids (Prickett et al. 2006, Shigenobu et al. 2000), suggesting that aphid metabolism supports a substantial flux of nonessential amino acids from insect to *Buchnera* and of essential amino acids in the reverse direction. But the knowledge of the genome of *Buchnera* also pointed out a paradox. *Buchnera* APS genome lacks some genes encoding essential amino acid biosynthesis enzymes, namely isoleucine, leucine, valine, phenylalanine, tyrosine and methionine (Shigenobu et al. 2000, Zientz et al. 2004), while there is strong experimental evidence that all essential amino acids are produced in the pea aphid (Douglas 1988, Febvay et al. 1999). The initial explanation for this paradox was that other *Buchnera* APS enzymes are promiscuous, mediating the 'missing' reactions (Shigenobu et al. 2000).

The availability of the aphid genome and the annotation of its metabolic network performed using the AcypiCyc database and expert follow-up annotation (Vellozo et al. 2011, Wilson et al. 2010) provided two important information about essential amino acids biosynthesis in the *Buchnera*/aphid symbiosis: (i) the pea aphid genome has the capability to complete with one or two reactions the biosynthetic capabilities of *Buchnera* and (ii) each of the aphid genes responsible for these enzymatic activities have orthologs in other insects with completely sequenced genomes, suggesting that the presence of these genes should not be interpreted as aphid adaptations to the symbiosis, but rather that the pre-existing capability of the aphid to compensate for genes lost from the *Buchnera* has facilitated genomic degradation in its genome (Wilson et al. 2010).

These genomic annotations have been recently completed by transcriptional and metabolic studies. A recent RNAseq transcriptome study of bacteriocytes (the specialized aphid cells hosting *Buchnera*), compared with other tissues, supports the integrated nature of the host-symbiont metabolic network (Hansen and Moran 2011). This work demonstrated that all six aphid genes involved in essential amino acids biosynthesis and all seven aphid nonessential amino acid pathways (except for pathways for biosynthesis of proline and asparagine) missing from *Buchnera* are up-regulated in the bacteriocyte and suggested that aphid-encoded enzymes within the bacteriocyte serve to adjust the profile of nonessential amino acids to fit the needs of both *Buchnera* and its host. The combination of transcriptome and metabolic studies has pointed out the essential role of the tyrosine biosynthesis pathway during the viviparous parthenogenetic development of the pea aphid, showing that aphid-encoded genes involved in the biosynthesis of phenylalanine, tyrosine, DOPA and dopamine are up-regulated in late embryonic and early nymphal stages, consistently with an accumulation of tyrosine in these same developmental stages (Rabatel et al. 2013) (Fig. 4). This work demonstrates the existence of an insect specific gene transcription regulation of tyrosine synthesis and complements previous observations of a lack of significant gene expression regulation of the aromatic amino acid pathways in the symbiotic bacteria during embryonic development (Bermingham et al. 2009).

Other characteristics of the pea aphid genome relative to symbiosis

Duplication of genes important for symbiosis. The high gene count in the pea aphid genome reflects both extensive gene duplications and the presence of aphid-specific orphan genes, which display no significant similarity with genes identified in other sequenced organisms and constitute about 20 per cent of the total number of genes. The functions of these duplicated and orphan genes are unknown, but the evolution of some of these genes seems to be related to the evolution of lineage-specific traits in aphids, such as symbiosis with *Buchnera*.

Using a *de novo* approach, Price et al. have characterized the complement of 40 amino acid polyamine organocation (APC) superfamily member amino acid transporters (AATs) encoded in the genome of the pea aphid (Price et al. 2011). These authors found that this APC superfamily is characterized by extensive gene duplications (*A. pisum* has more APC superfamily transporters than other fully sequenced insects), with a functional specificity that is a 10 paralog aphid-specific expansion of the *D. melanogaster* slimfast transporter (Attardo et al. 2006, Colombani et al. 2003). Expression analysis by qRT-PCR of these transporters demonstrated that bacteriocyte-biased gene expression is associated with genes having underwent aphid-specific expansion, whereas transporters highly expressed but not enriched in bacteriocytes retain one-to-one orthology with transporters in other genomes, then suggesting that these new transporters have been recruited for nutrient transport in bacteriocyte cells at the symbiotic interface. Using deep sequencing of

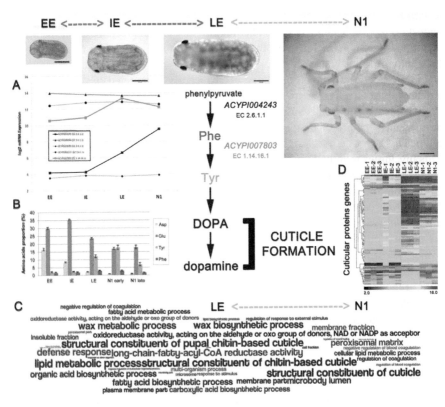

Fig. 4. Pea aphid regulation of the tyrosine pathway for cuticle formation. Summary results of integrated functional genomics analysis during parthenogenetic development of the pea aphid (from Rabatel et al. 2013). (A) mRNA expression patterns of tyrosine pathway genes encoded in the pea aphid genome (phenylpyruvate is synthesized by *B. aphidicola*) showing the induction of *ACYPI004243* and *ACYPI007803*. (B) HPLC analysis of free aminoacids linked to the tyrosine synthesis showing accumulation of Tyr in the late embryos and first nymphal stages. (C) Results of the Gene Ontology (GO) analysis of differentially expressed genes between late embryos and first nymphal stages realized with BinGO: GO were ordered based on statistical significance and binned (in 10 groups of equal size); in this representation generated in Wordle the text size is proportional to statistical test significance. (D) Hierarchical cluster analysis of cuticular genes during development showing strong induction in the late embryos and first nymphal stages.

bacteriocytes mRNA followed by whole mount *in situ* hybridizations of overrepresented transcripts encoding aphid-specific orphan proteins Shigenobu and Stern identified a novel class of genes that encode small proteins having a signal peptides, often cysteine-rich, and are overrepresented in bacteriocytes (Shigenobu and Stern 2013). These genes are first expressed at a developmental time point coincident with the incorporation of symbionts strictly in the cells that contribute to the bacteriocyte and this bacteriocyte-specific expression is maintained throughout the aphid's life. These genes show no significant similarity outside the aphid lineage supporting the hypothesis that they have evolved to mediate the symbiosis with *Buchnera*.

Functional and evolutionary studies on these and other orphan/duplicated genes are needed in the pea aphid to determine their possible regulation and their possible aphid lineage-specificity in relation to symbiosis.

Modulation of immune response in aphids. Aphids thus provide an excellent opportunity to study the immune system of an organism that is dependent on microbial symbionts but is hampered by parasites and pathogens (see Chapter 6 "Aphid Immunity" in this volume). By coupling gene annotation with functional assays, Gerardo et al. (2010) have shown that, although aphids share with other arthropods defence systems (for example, the Toll and JAK/STAT signaling pathways, HSPs, ProPO), surprisingly, several of the genes thought central to arthropod innate immunity are missing in their genome (for example, PGRPs, the IMD signaling pathway, defensins, c-type lysozymes). The authors stress the fact that they could not find aphid homologs to many insect immune genes because of the large evolutionary distance between aphids and the taxa (flies, mosquitoes and bees) from which these genes are known, the split between the ancestors of aphids and these taxa having occurred approximately 350 million years ago (Gaunt and Miles 2002). Large-scale transcriptional and proteomic studies of aphid body compartments (especially hemolymph and fat body) are expected to allow a characterization of the humoral component of immunity in aphids. The absence of genes suspected to be essential for the insect immune response suggests that the traditional view of insect immunity may not be as broadly applicable as once thought. It is noteworthy that immune cell response in aphids, that just starts to be characterized, seems to be complex and based on an important immune cell diversity (Laughton et al. 2011, Schmitz et al. 2012). In particular, Schmitz et al. (2012) have shown that some of these cells (e.g., plasmocytes and granulocytes) are able to actively phagocytize foreign bodies, but also primary and secondary symbionts when present in the hemolymph and that there is a cell-specificity in the recognition of different symbionts.

Studies on humoral and cell immunity should have to be extended to other non-model insects, to understand whether the peculiarities of the aphid immune system are representative of a broad range of insects, or is aphid specific.

Interaction of aphids with secondary symbionts. In parallel to primary obligatory symbionts, aphids also carry secondary facultative bacterial symbionts, which are predominantly vertically inherited (Russell et al. 2003) (see Chapter 5 "Bacterial Symbionts of Aphids (Hemiptera: Aphididae)" in this volume). The three best-known aphid secondary symbionts are the γ-Proteobacteria *Hamiltonella defensa*, *Regiella insecticola*, and *Serratia symbiotica*, which are common and widespread across aphid species (Russell et al. 2003). A series of other taxa have also been recorded but are less well characterized. These include a *Rickettsia* and a *Spiroplasma*, and two further γ-Proteobacteria: a *Rickettsiella* and a bacterium referred to as PAXS or the X-type (Chen et al. 1996, Fukatsu 2001, Guay et al. 2009, Moran et al. 2005, Sandström et al. 2001, Tsuchida et al. 2010).

The phenotypic effects of facultative symbionts upon hosts have been studied in some detail. Secondary symbionts affect different aspects of aphid biology including defence against natural enemies (Oliver et al. 2003, 2005, Scarborough et al. 2005), resistance to heat shock (Montllor et al. 2002), and the propensity to form winged morphs (Leonardo and Mondor 2006). However, little is known about the molecular and genomic basis of these interactions.

Recently, metabolomic profiling experiments showed that facultative symbionts in pea aphids (including *S. symbiotica*) have a large effect upon whole body metabolic pools (Burke et al. 2010). The effect of *S. symbiotica* infection was of comparable magnitude to the metabolic changes that occur in pea aphids upon heat-shock, which is remarkable given that *S. symbiotica* infect aphids at low density. More than 70% of the metabolites that contributed to the difference between infected and uninfected aphids could not be identified, making the origin (bacterial or pea aphid) of these metabolites difficult to discern.

The only whole-genome transcriptional analysis published up to now compared an uninfected laboratory colony of aphids to a genetically identical colony artificially infected with *S. symbiotica* (Burke and Moran 2011b). This study, demonstrates that only 28 genes showed changes in expression with *S. symbiotica* infection, and these changes were of small magnitude. No expression differences in genes involved in innate immunity in other insects were observed. Therefore, the large metabolic impact of *S. symbiotica* is most likely a result of metabolism of the symbiont itself, or of post-transcriptional modification of host gene expression. Thus, although *S. symbiotica* has a major influence on its host's metabolome and resistance to heat, it induces little change in gene expression in its host.

Small Non-Coding RNAs and Epigenetics Regulations

Most of efforts and genomic resources and their use for functional studies have been put on mRNAs coding proteins. However, there are many other actors of gene regulation, such as non-coding RNAs and epigenetic factors (see Chapter 4 "Epigenetic Control of Polyphenism in Aphids" in this volume).

Non-coding RNAs is a very vast world, and made of many different kinds of molecules. Some have been known for decades (rRNAs, tRNAs, soRNAs...). In the late 90's, small non-coding RNAs have been discovered as regulators of gene expression. Among them, microRNAs have been identified in aphids: microRNAs are 20 base pairs long, coded by the genome as precursors and matured by specific nucleoproteic complexes. They recognize by base complementarity mRNA targets, and the mRNA/microRNA duplex formation is generally related to down-regulation of mRNAs translation. MicroRNAs are thus post-transcriptional regulators of gene expression. More than 200 genes encoding microRNAs have been annotated in the pea aphid (Legeai et al. 2010) and unpublished data indicate that probably more than 400 microRNA genes might be present. The expression of these microRNAs have been compared and three miRNAs (Ap-mir-34, Ap-mir-X47 and Ap-mir-X103) and two mir* (Ap-mir307* and Ap-mirX-52*) showed different expression between the two parthenogenetic morphs (giving birth to sexual or parthenogenetic embryos). Mir-34 has been described in *D. melanogaster* as regulated by juvenile hormones that could make sense in the hypothesis of the role of the JH pathway in the regulation of the reproductive mode in aphids.

The state of the chromatin is a key player of the regulation of gene expression. The compaction of the DNA in the cell (required by space limitation) is badly compatible with accessibility of the different machineries involved in gene transcription. Thus, chromatin can be in compacted (closed) states, or non-compacted (open) states: and some of these states are reversible either during development or in the course of the adult life when environmental parameters change. The chromatin state is cell specific. Thus, in order to understand deeply the functioning of aphid genomes, we need to include in our studies the effect of chromatin accessibility on the expression of the genes of interest for the related phenotypes. The pea aphid genome contains mostly all the genes encoding the proteins required for the formation and modulation of the chromatin, for the two main biochemical processes known as "DNA methylation", and "histone code" (see also Chapter 4 "Epigenetic Control of Polyphenism in Aphids" in this volume). Methylation of cytosines in the DNA is associated (at least in vertebrates) with a modification of the chromatin and with an effect on gene expression. The pea aphid genome contains five DNA methyltransferase genes suggesting a functional machinery for DNA methylation (Walsh et al. 2010). As an indirect proof, it has been shown that approximately 0.7% of the cytosines in the pea aphid genome are methylated, which is much higher than in *Drosophila melanogaster* (who does not have a functional DNA methylation machinery) and about the same range as *Apis mellifera* (who has a functional DNA methylation machinery) (Kucharski et al. 2008). Cytogenetic analyses also demonstrated a methylation of DNA in the pea aphids (see Manicardi et al. 2015 for a review). First analyses of specific gene methylation indicate that methylation occurs mainly within the coding regions, as for *A. mellifera*. In particular, Walsh et al. (2010) showed a methylation in three genes involved in the JH pathway (one JH binding protein, one JH epoxide hydrolase and one JH esterase binding protein), and the methylation of the JH binding protein gene is higher in heads of unwinged adult females compared with winged individuals. In parallel, the mRNA level of Dmt2 (a DNA methyltransferase involved in the maintenance of methylation) is higher in the winged individuals (Dombrovsky et al. 2009) used inhibitors of DNA methyltransferase enzymes on unwinged adult parthenogenetic females and showed that their progeny was less winged—after crowding—than the control. All together, these data indicate a functional methylation system in the pea aphid, and an effect of this chromatin state in the expression of the dispersal polyphenism.

As well, histones and enzymes required for chromatin structure and remodelling have been searched and genes annotated in the pea aphid genome (Rider et al. 2010). Some of these families have duplications or even large expansion of members: the mof and enok histone acetyltransferases are duplicated in the pea aphid. In *Drosophila*, these proteins are involved in cell proliferation, oogenesis, morphogenesis and dosage compensation. As well, SET proteins (histone methyltransferases) are represented by 25 different genes in the pea aphid genome, which is a very large expansion in an arthropod genome. Today, it is clear that the dynamic of chromatin modelling has to be more studies in details in the frame of phenotypic plasticity in aphids, since we can easily make the hypothesis that some of the aphid-specific plastic traits are regulated by epigenetic mechanisms.

Phylogenomics and Comparative Biology

Phylogenomics can be regarded as the intersection of the fields of genomics and evolution (Eisen and Fraser 2003). In other words it means to look at genomes under the prism of evolution. Questions that are addressed by this emerging field are: (i) how the current genome content and structure came to be, and (ii) how much of it has been shaped through selective forces related to the corresponding phenotypes. Conversely one could use evolutionary and comparative analyses to uncover the genomic basis of important phenotypes that are thought to have emerged through selection. As mentioned above aphids have important traits that have varied through evolution; some are specific to aphids and some are shared by distantly related arthropods; some have singularities in evolution that have appeared only once, and some have emerged or have been lost independently in different lineages. All this is a very interesting substrate for phylogenomic and comparative research and entails a great potential to contribute to our understanding of aphids, their evolution, and their biology. There are two complementary, and somewhat overlaping aspects in phylogenomics: species evolution and gene evolution. The first relates to the establishment of evolutionary relationships among species, while the second relates to the reconstruction of evolutionary histories of gene families, shaped by duplications and losses, but also species diversifications. Regarding the reconstruction of species relationships, the aphid taxonomy and phylogeny is far from being resolved. Rapid radiations of the main groups coupled with shifts in morphologies related to adaptation to different host plants and environments complicate the phylogenetic reconstruction within this group. The availability of complete genomes carries the promise of increasing the available sequence information from which to reconstruct a robust phylogenetic framework depicting the evolution of aphids (Delsuc et al. 2005). The availability of complete genome sequences for at least a representative of each main clade, coupled with target-gene sequencing or transcriptomics from an additional set of species will certainly increase the resolution of the aphid tree of life and perhaps uncover some previously unrecognised relationships.

Regarding the second aspect of phylogenomics, the evolution of individual gene families, there has been already great progress. As mentioned above, thanks to the participation of PhylomeDB into the IAGC, the complete collection of evolutionary histories for all the aphid genes was generated alongside the first annotation of the genome, being the pea aphid the first species for which such sophisticated approach was used in the annotation. The association of PhylomeDB with the Aphid genomics initiative dates back to 2007, when the efforts of annotating the newly-sequenced *A. pisum* genome were ongoing. At that time, PhylomeDB had only been used on publicly available annotated genomes, but its potential in aiding the genome annotation efforts had been realized. Thus PhylomeDB joined the genome annotation efforts and build the pea aphid phylome that is the complete collection of evolutionary histories of all genes encoded in the *A. pisum* genome and their relatives in other sequenced arthropods. This analysis served to obtain a global overview of the *A. pisum* genome evolution, from the perspective of each one of its genes (Huerta-Cepas et al. 2010b). In addition, it revealed a dynamic gene content, with recent waves of gene duplications. Some of these large duplications comprised protein-coding transposable elements, which had escaped the masking with standard tools due to their high divergence from known transposable elements. Thus, the PhylomeDB approach served to refine the gene dataset, removing a large fraction of transposable elements which were undetected by other means. Besides transposable elements, some large expansions involved bona-fide protein-coding genes. Interestingly, some of them could be related to particular phenotypes in aphids, such as the expansion of amino acid transporters, probably related to the necessity to take the most of the available nutrients from a poorly diverse phloem sap, or of a protein involved in ovary formation, perhaps related to the morphological diversity in oviparous and viviparous females. Such examples illustrate how evolutionary analyses can point to potential new avenues of research, as it is able to establish new hypotheses, which subsequently need experimental work to refute or confirm those. The availability of new aphid genomes will definitely have an impact in the analysis of gene families as it will enable to add resolution to them. From the perspective of a single aphid genome, a lineage-specific duplication is a long evolutionary period and it is difficult to assess whether it was related to some early or late innovation within the aphid lineage. The availability of additional aphid species will help categorizing duplications into ancient and recent, as well as to discover parallel events of duplications and gene losses. Interestingly, transcriptomic and genomic analysis can be coupled into a phylogenetic framework that considers the particularities of transcriptome-based data, an approach that has already been used in early-dipterans (Jiménez-Guri et al. 2013). In aphids, the reference genome of the pea aphid was combined with transcriptomic data from additional sexual and non-sexual species to test whether asexual species were accumulating more deleterious mutations (Ollivier et al. 2012). The results did not show strong differences, indicating that all asexual species analyzed had been asexual for only a relatively short evolutionary period. We expect that phylogenomics will be a central part in the analysis of newly sequenced aphid species and that large-scale analyses of gene family evolution will help establishing new hypotheses of how important traits have appeared and evolved during the evolution of aphids.

Conclusions: towards Integration

This chapter aimed at scanning the state-of-the-art of aphid genomics, after about 10 years since the beginning of the development and use of genomic tools by aphidologists. Progresses are true and real. But they are probably nothing compared with what could be done in the near future. Today, nobody can say what the future of biology will be in the prism of innovation in genomics, informatics, automatism and nanotechnologies. But in five years, the sequencing technologies have moved every year, with "VIP" technologies that appear as quickly as they disappear (such as 454 yesterday, and Illumina probably tomorrow). New very fast, deep and cheap technologies will be in our labs soon, even if we don't know how soon this will be. What is sure in that we will still generate lots of data, hoping that informatics infrastructures and bioinformatics development will allow us to analyse those data.

And then? A key expertise in the future will probably be our capacity to integrate the different knowledge, and at different levels. The first level will be the integration of the so-called 'omics data. Tools exist already, but many things remain to be done. A second level of integration should be the interactions with the microbiomes. The third level will be "vertical", meaning getting out of the level of the organisms to integrate data from populations, either from the same species or from other species.

This integration will be more challenging since this implies good connections in our labs and community between physiologist (to be short) and population geneticists, ecologists, and evolutionary biologists. This is not yet the case even for human biology: links between ENCODE and HapMap projects for instance are still difficult. And the fourth level of integration will be between 'omics data and phenotypical and ecological data, to really analyse the link between genotype and phenotype and understand the complex processes of adaptation of individuals to its environment, a kind of grail for any biologist.

Acknowledgement

DT was funded by the French ANR "MiRNAdapt" and "GW_Aphid" program, as well as INRA SPE and Région Bretagne. FC and SC were funded by the French ANR "SPECIAPHID 2010–2014 (ANR-11-BSV7-0005)" and "IMetSym 2014–2017 (ANR-13-BSV7-00-16-03" programs. TG group research is funded in part by a grant from the Spanish ministry of Economy and Competitiveness (BIO2012-37161), a Grant from the Qatar National Research Fund grant (NPRP 5-298-3-086), and a grant from the European Research Council under the European Union's Seventh Framework Programme (FP/2007–2013)/ERC (Grant Agreement n. ERC-2012-StG-310325). AS is funded by FP7-PEOPLE-2012-CIG n° 333937 APHISPIT.

Keywords: aphids, genome resources, gene function studies, phenotypic plasticity, reproductive modes, host-plant interaction, bacterial symbiosis, epigenetic regulations

References

Akman Gündüz, E. and A.E. Douglas. 2009. Symbiotic bacteria enable insect to use a nutritionally inadequate diet. Proc. R. Soc. B 276: 987–991.

Atamian, H.S., R. Chaudhary, V. Dal Cin, E. Bao, T. Girke and I. Kaloshian. 2013. In planta expression or delivery of potato aphid *Macrosiphum euphorbiae* effectors Me10 and Me23 enhances aphid fecundity. Mol. Plant Microbe Interact. 26: 67–74.

Attardo, G.M., I.A. Hansen, S.-H. Shiao and A.S. Raikhel. 2006. Identification of two cationic amino acid transporters required for nutritional signaling during mosquito reproduction. J. Exp. Biol. 209: 3071–3078.

Barbera, M., B. Mengual, J.M. Collantes-Alegre, T. Cortes, A. Gonzalez and D. Martinez-Torres. 2013. Identification, characterization and analysis of expression of genes encoding arylalkylamine N-acetyltransferases in the pea aphid *Acyrthosiphon pisum*. Insect Mol. Biol. 22: 623–634.

Bass, C., C.T. Zimmer, J.M. Riveron, C.S. Wilding, C.S. Wondji, M. Kaussmann, L.M. Field, M.S. Williamson and R. Nauen. 2013. Gene amplification and microsatellite polymorphism underlie a recent insect host shift. Proc. Natl. Acad. Sci. USA 110: 19460–19465.

Bermingham, J., A. Rabatel, F. Calevro, J. Viñuelas, G. Febvay, H. Charles, A. Douglas and T. Wilkinson. 2009. Impact of host developmental age on the transcriptome of the symbiotic bacterium *Buchnera aphidicola* in the pea aphid (*Acyrthosiphon pisum*). Appl. Environ. Microbiol. 75: 7294–7297.

Bickel, R.D., J.P. Dunham and J.A. Brisson. 2013. Widespread selection across coding and noncoding DNA in the pea aphid genome. G3 3: 993–1001.

Bos, J.I., D. Prince, M. Pitino, M.E. Maffei, J. Win and S.A. Hogenhout. 2010. A functional genomics approach identifies candidate effectors from the aphid species *Myzus persicae* (green peach aphid). PLoS Genet. 6: e1001216.

Bowers, W.S., L.R. Nault, R.E. Webb and S.R. Dutky. 1972. Aphid alarm pheromone: isolation, identification, synthesis. Science 177: 1121–1122.

Bradbury, J. 2003. Human epigenome project-up and running. PLoS Biol. 1: E82.

Braendle, C., M.C. Caillaud and D.L. Stern. 2005a. Genetic mapping of aphicarus—a sex-linked locus controlling a wing polymorphism in the pea aphid (*Acyrthosiphon pisum*). Heredity 94: 435–442.

Braendle, C., I. Friebe, M.C. Caillaud and D.L. Stern. 2005b. Genetic variation for an aphid wing polyphenism is genetically linked to a naturally occurring wing polymorphism. Proc. Biol. Sci. 272: 657–664.

Bridges, M., A.M.E. Jones, A.M. Bones, C. Hodgson, R. Cole, E. Bartlet, R. Wallsgrove, V.K. Karapapa, N. Watts and J.T. Rossiter. 2002. Spatial organization of the glucosinolate-myrosinase system in brassica specialist aphids is similar to that of the host plant. Proc. R. Soc. B 269: 187–191.

Brisson, J.A. 2010. Aphid wing dimorphisms: linking environmental and genetic control of trait variation. Philosoph. Transact. B 365: 605–616.

Brisson, J.A., G.K. Davis and D.L. Stern. 2007. Common genome-wide patterns of transcript accumulation underlying the wing polyphenism and polymorphism in the pea aphid (*Acyrthosiphon pisum*). Evol. Dev. 9: 338–346.

Burke, G., O. Fiehn and N. Moran. 2010. Effects of facultative symbionts and heat stress on the metabolome of pea aphids. ISME J. 4: 242–252.

Burke, G.R. and N.A. Moran. 2011a. Massive genomic decay in *Serratia symbiotica*, a recently evolved symbiont of aphids. Genome Biol. Evol. 3: 195–208.

Burke, G.R. and N.A. Moran. 2011b. Responses of the pea aphid transcriptome to infection by facultative symbionts. Insect Mol. Biol. 20: 357–365.

Caillaud, M.C. and S. Via. 2000. Specialized feeding behavior influences both ecological specialization and assortative mating in sympatric host races of pea aphids. Amer. Nat. 156: 606–621.

Carolan, J.C., C.I.J. Fitzroy, P.D. Ashton, A.E. Douglas and T.L. Wilkinson. 2009. The secreted salivary proteome of the pea aphid *Acyrthosiphon pisum* characterised by mass spectrometry. Proteomics 9: 2457–2467.

Carolan, J.C., D. Caragea, K.T. Reardon, N.S. Mutti, N. Dittmer, K. Pappan, F. Cui, M. Castaneto, J. Poulain, C. Dossat, D. Tagu, J.C. Reese, G.R. Reeck, T.L. Wilkinson and O.R. Edwards. 2011. Predicted effector molecules in the salivary secretome of the pea aphid (*Acyrthosiphon pisum*): a dual transcriptomic/ proteomic approach. J. Proteome Res. 10: 1505–1518.

Caspi, R., T. Altman, R. Billington, K. Dreher, H. Foerster, C.A. Fulcher, T.A. Holland, I.M. Keseler, A. Kothari, A. Kubo, M. Krummenacker, M. Latendresse, L.A. Mueller, Q. Ong, S. Paley, P. Subhraveti, D.S. Weaver, D. Weerasinghe, P. Zhang and P.D. Karp. 2014. The MetaCyc database of metabolic pathways and enzymes and the BioCyc collection of Pathway/Genome Databases. Nucleic Acids Res. 42: D459–471.

Casteel, C.L., L.L. Walling and T.D. Paine. 2006. Behavior and biology of the tomato psyllid, *Bactericerca cockerelli*, in response to the Mi-1.2 gene. Entomol. Exp. Appl. 121: 67–72.

Chang, C.C., W.C. Lee, C.E. Cook, G.W. Lin and T. Chang. 2006. Germ-plasm specification and germline development in the parthenogenetic pea aphid *Acyrthosiphon pisum*: Vasa and Nanos as markers. Int. J. Dev. Biol. 50: 413–421.

Chang, C.C., G.W. Lin, C.E. Cook, S.B. Horng, H.J. Lee and T.Y. Huang. 2007. Apvasa marks germ-cell migration in the parthenogenetic pea aphid *Acyrthosiphon pisum* (Hemiptera: Aphidoidea). Dev. Genes Evol. 217: 275–287.

Chang, C.C., T.Y. Huang, C.E. Cook, G.W. Lin, C.L. Shih and R.P. Chen. 2009. Developmental expression of Apnanos during oogenesis and embryogenesis in the parthenogenetic pea aphid *Acyrthosiphon pisum*. Int. J. Dev. Biol. 53: 169–176.

Chang, C.C., Y.M. Hsiao, T.Y. Huang, C.E. Cook, S. Shigenobu and T.H. Chang. 2013. Noncanonical expression of caudal during early embryogenesis in the pea aphid *Acyrthosiphon pisum*: maternal cad-driven posterior development is not conserved. Insect Mol. Biol. 22: 442–455.

Chen, D.Q., B.C. Campbell and A.H. Purcell. 1996. A new rickettsia from a herbivorous insect, the pea aphid *Acyrthosiphon pisum* (Harris). Curr. Microbiol. 33: 123–128.

Colbourne, J.K., M.E. Pfrender, D. Gilbert, W.K. Thomas, A. Tucker, T.H. Oakley, S. Tokishita, A. Aerts, G.J. Arnold, M.K. Basu, D.J. Bauer, C.E. Caceres, L. Carmel, C. Casola, J.H. Choi, J.C. Detter, Q. Dong, S. Dusheyko, B.D. Eads, T. Frohlich, K.A. Geiler-Samerotte, D. Gerlach, P. Hatcher, S. Jogdeo, J. Krijgsveld, E.V. Kriventseva, D. Kultz, C. Laforsch, E. Lindquist, J. Lopez, J.R. Manak, J. Muller, J. Pangilinan, R.P. Patwardhan, S. Pitluck, E.J. Pritham, A. Rechtsteiner, M. Rho, I.B. Rogozin, O. Sakarya, A. Salamov, S. Schaack, H. Shapiro, Y. Shiga, C. Skalitzky, Z. Smith, A. Souvorov, W. Sung, Z. Tang, D. Tsuchiya, H. Tu, H. Vos, M. Wang, Y.I. Wolf, H. Yamagata, T. Yamada, Y. Ye, J.R. Shaw, J. Andrews, T.J. Crease, H. Tang, S.M. Lucas, H.M. Robertson, P. Bork,

E.V. Koonin, E.M. Zdobnov, I.V. Grigoriev, M. Lynch and J.L. Boore. 2011. The ecoresponsive genome of *Daphnia pulex*. Science 331: 555–561.

Colombani, J., S. Raisin, S. Pantalacci, T. Radimerski, J. Montagne and P. Léopold. 2003. A nutrient sensor mechanism controls *Drosophila* growth. Cell 114: 739–749.

Corbitt, T. and J. Hardie. 1985. Juvenile hormone effects on polymorphism in the pea aphid, *Acyrthosiphon pisum*. Entomol. Exp. Appl. 38: 131–135.

Cortes, T., B. Ortiz-Rivas and D. Martinez-Torres. 2010. Identification and characterization of circadian clock genes in the pea aphid *Acyrthosiphon pisum*. Insect Mol. Biol. 19: 123–139.

Dadd, R.H. 1985. Nutrition: organisms. pp. 313–390. *In*: G.A. Kerkut and L.I. Gilbert (eds.). Comprehensive Insect Physiology, Biochemistry and Pharmacology, Vol. 4. Pergamon Press, Oxford, UK.

Davis, G.K. 2012. Cyclical parthenogenesis and viviparity in aphids as evolutionary novelties. J. Exp. Zool. B Mol. Dev. Evol. 318: 448–459.

Degnan, P.H., T.E. Leonardo, B.N. Cass, B. Hurwitz, D. Stern, R.A. Gibbs, S. Richards and N.A. Moran. 2010. Dynamics of genome evolution in facultative symbionts of aphids. Environ. Microbiol. 12: 2060–2069.

Delsuc, F., H. Brinkmann and H. Philippe. 2005. Phylogenomics and the reconstruction of the tree of life. Nature Rev. Genet. 6: 361–375.

Dinant, S., J.L. Bonnemain, C. Girousse and J. Kehr. 2010. Phloem sap intricacy and interplay with aphid feeding. C. R. Biol. 333: 504–515.

Dogimont, C., A. Bendahmane, V. Chovelon and N. Boissot. 2010. Host plant resistance to aphids in cultivated crops: genetic and molecular bases, and interactions with aphid populations. C. R. Biol. 333: 566–573.

Dogimont, C., V. Chovelon, J. Pauquet, A. Boualem and A. Bendahmane. 2014. The Vat locus encodes for a CC-NBS-LRR protein that confers resistance to *Aphis gossypii* infestation and *A. gossypii*-mediated virus resistance. Plant J. 80(6): 993–1004.

Dombrovsky, A., L. Arthaud, T.N. Ledger, S. Tares and A. Robichon. 2009. Profiling the repertoire of phenotypes influenced by environmental cues that occur during asexual reproduction. Genome Res. 19: 2052–2063.

Douglas, A. 1988. Sulphate utilization in an aphid symbiosis. Insect Biochem. 18: 599–605.

Douglas, A.E. and W.A. Prosser. 1992. Synthesis of the essential amino-acid tryptophan in the pea aphid (*Acyrthosiphon pisum*) symbiosis. Journal of Insect Physiology 38(8): 565–568.

Duncan, E.J., M.P. Leask and P.K. Dearden. 2013. The pea aphid (*Acyrthosiphon pisum*) genome encodes two divergent early developmental programs. Dev. Biol. 377: 262–274.

Duvaux, L., Q. Geissmann, K. Gharbi, J.J. Zhou, J. Ferrari, C.M. Smadja and R.K. Butlin. 2015. Dynamics of copy number variation in host races of the pea aphid. Mol. Biol. Evol. 32: 63–80.

Eisen, J.A. and C.M. Fraser. 2003. Phylogenomics: intersection of evolution and genomics. Science 300: 1706–1707.

Elzinga, D.A. and G. Jander. 2013. The role of protein effectors in plant-aphid interactions. Curr. Opin. Plant Biol. 16: 451–456.

ENCODE Project Consortium. 2012. An integrated encyclopedia of DNA elements in the human genome. Nature 489: 57–74.

Febvay, G., Y. Rahbe, M. Rynkiewicz, J. Guillaud and G. Bonnot. 1999. Fate of dietary sucrose and neosynthesis of amino acids in the pea aphid, *Acyrthosiphon pisum*, reared on different diets. J. Exp. Biol. 202: 2639–2652.

Ferrari, J., C.L. Scarborough and H.C. Godfray. 2007. Genetic variation in the effect of a facultative symbiont on host-plant use by pea aphids. Oecol. 153: 323–329.

Ferrari, J., J.A. West, S. Via and H.C. Godfray. 2012. Population genetic structure and secondary symbionts in host-associated populations of the pea aphid complex. Evolution 66: 375–390.

Fukatsu, T. 2001. Secondary intracellular symbiotic bacteria in aphids of the genus *Yamatocallis* (Homoptera: Aphididae: Drepanosiphinae). Appl. Environ. Microbiol. 67: 5315–5320.

Gabaldon, T., C. Dessimoz, J. Huxley-Jones, A.J. Vilella, E.L. Sonnhammer and S. Lewis. 2009. Joining forces in the quest for orthologs. Genome Biol. 10: 403.

Gallot, A., C. Rispe, N. Leterme, J.P. Gauthier, S. Jaubert-Possamai and D. Tagu. 2010. Cuticular proteins and seasonal photoperiodism in aphids. Insect Biochem. Mol. Biol. 40: 235–240.

Gallot, A., S. Shigenobu, T. Hashiyama, S. Jaubert-Possamai and D. Tagu. 2012. Sexual and asexual oogenesis require the expression of unique and shared sets of genes in the insect *Acyrthosiphon pisum*. BMC Genomics 13: 76.

Gao, N. and J. Hardie. 1997. Melatonin and the pea aphid, *Acyrthosiphon pisum*. J. Insect Physiol. 43: 615–620.

Gaunt, M.W. and M.A. Miles. 2002. An insect molecular clock dates the origin of the insects and accords with palaeontological and biogeographic landmarks. Mol. Biol. Evol. 19: 748–761.

Gauthier, J.-P., F. Legeai, A. Zasadzinski, C. Rispe and D. Tagu. 2007. AphidBase: a database for aphid genomic resources. Bioinformatics 23: 783–784.

Genomes Project Consortium, G.R. Abecasis, D. Altshuler, A. Auton, L.D. Brooks, R.M. Durbin, R.A. Gibbs, M.E. Hurles and G.A. McVean. 2010. A map of human genome variation from population-scale sequencing. Nature 467: 1061–1073.

Gerardo, N.M., B. Altincicek, C. Anselme, H. Atamian, S.M. Barribeau, M. de Vos, E.J. Duncan, J.D. Evans, T. Gabaldón, M. Ghanim, A. Heddi, I. Kaloshian, A. Latorre, A. Moya, A. Nakabachi, B.J. Parker, V. Pérez-Brocal, M. Pignatelli, Y. Rahbe, J.S. Ramsey, C.J. Spragg, J. Tamames, D. Tamarit, C. Tamborindeguy, C. Vincent-Monegat and A. Vilcinskas. 2010. Immunity and other defenses in pea aphids, *Acyrthosiphon pisum*. Genome Biol. 11: R21.

Guay, J.-F., S. Boudreault, D. Michaud and C. Cloutier. 2009. Impact of environmental stress on aphid clonal resistance to parasitoids: role of *Hamiltonella defensa* bacterial symbiosis in association with a new facultative symbiont of the pea aphid. J. Insect Physiol. 55: 919–926.

Hansen, A.K. and N.A. Moran. 2011. Aphid genome expression reveals host-symbiont cooperation in the production of amino acids. Proc. Natl. Acad. Sci. USA 108: 2849–2854.

Hansen, A.K. and N.A. Moran. 2014. The impact of microbial symbionts on host plant utilization by herbivorous insects. Mol. Ecol. 23: 1473–1496.

Hansen, A.K., C. Vorburger and N.A. Moran. 2012. Genomic basis of endosymbiont-conferred protection against an insect parasitoid. Genome Res. 22: 106–114.

Harmel, N., E. Letocart, A. Cherqui, P. Giordanengo, G. Mazzucchelli, F. Guillonneau, E. De Pauw, E. Haubruge and F. Francis. 2008. Identification of aphid salivary proteins: a proteomic investigation of *Myzus persicae*. Insect Mol. Biol. 17: 165–174.

Hegde, M., J.N. Oliveira, J.G. da Costa, E. Bleicher, A.E.G. Santana, T.J.A. Bruce, J. Caulfield, S.Y. Dewhirst, C.M. Woodcock, J.A. Pickett and M.A. Birkett. 2011. Identification of semiochemicals celeased by cotton, *Gossypium hirsutum*, upon infestation by the cotton aphid, *Aphis gossypii*. J. Chem. Ecol. 37: 741–750.

Hegde, M., J.N. Oliveira, J.G. da Costa, E. Loza-Reyes, E. Bleicher, A.E.G. Santana, J.C. Caulfield, P. Mayon, S.Y. Dewhirst, T.J.A. Bruce, J.A. Pickett and M.A. Birkett. 2012. Aphid antixenosis in cotton is activated by the natural plant defence elicitor cis-jasmone. Phytochemistry 78: 81–88.

Henry, L.M., J. Peccoud, J.C. Simon, J.D. Hadfield, M.J.C. Maiden, J. Ferrari and H.C.J. Godfray. 2013. Horizontally transmitted symbionts and host colonization of ecological niches. Curr. Biol. 23: 1713–1717.

Hogenhout, S.A. and J.I. Bos. 2011. Effector proteins that modulate plant-insect interactions. Curr. Opin. Plant Biol. 14: 422–428.

Horn, T. and M. Boutros. 2010. E-RNAi: a web application for the multi-species design of RNAi reagents-2010 update. Nucleic Acids Res. 38: W332–339.

Huang, T.Y., C.E. Cook, G.K. Davis, S. Shigenobu, R.P. Chen and C.C. Chang. 2010. Anterior development in the parthenogenetic and viviparous form of the pea aphid, *Acyrthosiphon pisum*: hunchback and orthodenticle expression. Insect Mol. Biol. 19: 75–85.

Huerta-Cepas, J., J. Dopazo and T. Gabaldón. 2010a. ETE: a python environment for tree exploration. BMC Bioinformatics 11: 24.

Huerta-Cepas, J., M. Marcet-Houben, M. Pignatelli, A. Moya and T. Gabaldón. 2010b. The pea aphid phylome: a complete catalogue of evolutionary histories and arthropod orthology and paralogy relationships for *Acyrthosiphon pisum* genes. Insect Mol. Biol. 19: 13–21.

Huerta-Cepas, J., S. Capella-Gutierrez, L.P. Pryszcz, I. Denisov, D. Kormes, M. Marcet-Houben and T. Gabaldón. 2011. PhylomeDB v3.0: an expanding repository of genome-wide collections of trees, alignments and phylogeny-based orthology and paralogy predictions. Nucleic Acids Res. 39: D556–560.

Huerta-Cepas, J., S. Capella-Gutierrez, L.P. Pryszcz, M. Marcet-Houben and T. Gabaldón. 2014. PhylomeDB v4: zooming into the plurality of evolutionary histories of a genome. Nucleic Acids Res. 42: D897–902.

Huybrechts, J., J. Bonhomme, S. Minoli, N. Prunier-Leterme, A. Dombrovsky, M. Abdel-Latief, A. Robichon, J.A. Veenstra and D. Tagu. 2010. Neuropeptide and neurohormone precursors in the pea aphid, *Acyrthosiphon pisum*. Insect Mol. Biol. 19: 87–95.

International Aphid Genomics Consortium (IAGC). 2010. Genome sequence of the pea aphid *Acyrthosiphon pisum*. PLoS Biol. 8: e1000313.

i5k Consortium. 2013. The i5K initiative: advancing arthropod genomics for knowledge, human health, agriculture, and the environment. J. Hered. 104: 595–600.

Ishikawa, A. and T. Miura. 2013. Transduction of high-density signals across generations in aphid wing polyphenism. Physiol. Entomol. 38: 150–156.

Ishikawa, A., K. Ogawa, H. Gotoh, T.K. Walsh, D. Tagu, J.A. Brisson, C. Rispe, S. Jaubert-Possamai, T. Kanbe, T. Tsubota, T. Shiotsuki and T. Miura. 2012. Juvenile hormone titre and related gene expression during the change of reproductive modes in the pea aphid. Insect Mol. Biol. 21: 49–60.

Jaquiery, J., S. Stoeckel, P. Nouhaud, L. Mieuzet, F. Maheo, F. Legeai, N. Bernard, A. Bonvoisin, R. Vitalis and J.C. Simon. 2012. Genome scans reveal candidate regions involved in the adaptation to host plant in the pea aphid complex. Mol. Ecol. 21: 5251–5264.

Jaquiery, J., S. Stoeckel, C. Larose, P. Nouhaud, C. Rispe, L. Mieuzet, J. Bonhomme, F. Maheo, F. Legeai, J.P. Gauthier, N. Prunier-Leterme, D. Tagu and J.C. Simon. 2014. Genetic control of contagious asexuality in the pea aphid. PLoS Genet. 10: e1004838.

Jiménez-Guri, E., J. Huerta-Cepas, L. Cozzuto, K.R. Wotton, H. Kang, H. Himmelbauer, G. Roma, T. Gabaldón and J. Jaeger. 2013. Comparative transcriptomics of early dipteran development. BMC Genomics 14: 123.

Kaloshian, I. and L.L. Walling. 2005. Hemipterans as plant pathogens. Annu. Rev. Phytopathol. 43: 491–521.

Karp, P.D., C.A. Ouzounis, C. Moore-Kochlacs, L. Goldovsky, P. Kaipa, D. Ahrén, S. Tsoka, N. Darzentas, V. Kunin and N. López-Bigas. 2005. Expansion of the BioCyc collection of pathway/genome databases to 160 genomes. Nucl. Acids Res. 33: 6083–6089.

Karp, P.D., S.M. Paley, M. Krummenacker, M. Latendresse, J.M. Dale, T.J. Lee, P. Kaipa, F. Gilham, A. Spaulding, L. Popescu, T. Altman, I. Paulsen, I.M. Keseler and R. Caspi. 2010. Pathway Tools version 13.0: integrated software for pathway/genome informatics and systems biology. Brief. Bioinform. 11: 40–79.

Kazana, E., T.W. Pope, L. Tibbles, M. Bridges, J.A. Pickett, A.M. Bones, G. Powell and J.T. Rossiter. 2007. The cabbage aphid: a walking mustard oil bomb. Proc. R. Soc. B 274: 2271–2277.

Kucharski, R., J. Maleszka, S. Foret and R. Maleszka. 2008. Nutritional control of reproductive status in honeybees via DNA methylation. Science 319: 1827–1830.

Laughton, A.M., J.R. Garcia, B. Altincicek, M.R. Strand and N.M. Gerardo. 2011. Characterisation of immune responses in the pea aphid, *Acyrthosiphon pisum*. J. Insect Physiol. 57: 830–839.

Le Trionnaire, G., J. Hardie, S. Jaubert-Possamai, J.C. Simon and D. Tagu. 2008. Shifting from clonal to sexual reproduction in aphids: physiological and developmental aspects. Biol. Cell 100: 441–451.

Le Trionnaire, G., F. Francis, S. Jaubert-Possamai, J. Bonhomme, E. De Pauw, J.P. Gauthier, E. Haubruge, F. Legeai, N. Prunier-Leterme, J.C. Simon, S. Tanguy and D. Tagu. 2009. Transcriptomic and proteomic analyses of seasonal photoperiodism in the pea aphid. BMC Genomics 10: 456.

Le Trionnaire, G., S. Jaubert-Possamai, J. Bonhomme, J.P. Gauthier, G. Guernec, A. Le Cam, F. Legeai, J. Monfort and D. Tagu. 2012. Transcriptomic profiling of the reproductive mode switch in the pea aphid in response to natural autumnal photoperiod. J. Insect Physiol. 58: 1517–1524.

Lees, A. 1960. The role of photoperiod and temperature in the determination of parthenogenetic and sexual forms in the aphid *Megoura viciae* Buckton—II. The operation of the 'interval timer' in young clones. J. Insect Physiol. 4: 154–175.

Lees, A. 1966. The control of polymorphism in aphids. Adv. Insect Physiol. 3: 207–277.

Legeai, F., S. Shigenobu, J.P. Gauthier, J. Colbourne, C. Rispe, O. Collin, S. Richards, A.C. Wilson, T. Murphy and D. Tagu. 2010. AphidBase: a centralized bioinformatic resource for annotation of the pea aphid genome. Insect Mol. Biol. 19: 5–12.

Leonardo, T.E. 2004. Removal of a specialization-associated symbiont does not affect aphid fitness. Ecol. Lett. 7: 461–468.

Leonardo, T.E. and E.B. Mondor. 2006. Symbiont modifies host life-history traits that affect gene flow. Proc. R. Soc. B 273: 1079–1084.

Liadouze, I., G. Febvay, J. Guillaud and G. Bonnot. 1995. Effect of diet on the free amino acid pools of symbiotic and aposymbiotic pea aphids, *Acyrthosiphon pisum*. J. Insect Physiol. 41: 33–40.

Liu, L.J., H.Y. Zheng, F. Jiang, W. Guo and S.T. Zhou. 2014. Comparative transcriptional analysis of asexual and sexual morphs reveals possible mechanisms in reproductive polyphenism of the cotton aphid. PLoS One 9: e99506.

Lu, H.L., S. Tanguy, C. Rispe, J.P. Gauthier, T. Walsh, K. Gordon, O. Edwards, D. Tagu, C.C. Chang and S. Jaubert-Possamai. 2011. Expansion of genes encoding piRNA-associated argonaute proteins in the pea aphid: diversification of expression profiles in different plastic morphs. PLoS One 6: e28051.

Manicardi, G.C., M. Mandrioli and R.L. Blackman. 2015. The cytogenetic architecture of the aphid genome. Biol. Rev. 90: 112–125.

Miles, P.W. 1999. Aphid saliva. Biol. Rev. 74: 41–85.

Misof, B., S. Liu, K. Meusemann, R.S. Peters, A. Donath, C. Mayer, P.B. Frandsen, J. Ware, T. Flouri, R.G. Beutel, O. Niehuis, M. Petersen, F. Izquierdo-Carrasco, T. Wappler, J. Rust, A.J. Aberer, U. Aspock, H. Aspock, D. Bartel, A. Blanke, S. Berger, A. Bohm, T.R. Buckley, B. Calcott, J. Chen, F. Friedrich, M. Fukui, M. Fujita, C. Greve, P. Grobe, S. Gu, Y. Huang, L.S. Jermiin, A.Y. Kawahara, L. Krogmann, M. Kubiak, R. Lanfear, H. Letsch, Y. Li, Z. Li, J. Li, H. Lu, R. Machida, Y. Mashimo, P. Kapli, D.D. McKenna, G. Meng, Y. Nakagaki, J.L. Navarrete-Heredia, M. Ott, Y. Ou, G. Pass, L. Podsiadlowski, H. Pohl, B.M. von Reumont, K. Schutte, K. Sekiya, S. Shimizu, A. Slipinski, A. Stamatakis, W. Song, X. Su, N.U. Szucsich, M. Tan, X. Tan, M. Tang, J. Tang, G. Timelthaler, S. Tomizuka, M. Trautwein, X. Tong, T. Uchifune, M.G. Walzl, B.M. Wiegmann, J. Wilbrandt, B. Wipfler, T.K. Wong, Q. Wu, G. Wu, Y. Xie, S. Yang, Q. Yang, D.K. Yeates, K. Yoshizawa, Q. Zhang, R. Zhang, W. Zhang, Y. Zhang, J. Zhao, C. Zhou, L. Zhou, T. Ziesmann, S. Zou, Y. Li, X. Xu, Y. Zhang, H. Yang, J. Wang, J. Wang, K.M. Kjer and X. Zhou. 2014. Phylogenomics resolves the timing and pattern of insect evolution. Science 346: 763–767.

Miura, T., C. Braendle, A. Shingleton, G. Sisk, S. Kambhampati and D.L. Stern. 2003. A comparison of parthenogenetic and sexual embryogenesis of the pea aphid *Acyrthosiphon pisum* (Hemiptera: Aphidoidea). J. Exp. Zool. B Mol. Dev. Evol. 295: 59–81.

Montllor, C., A. Maxmen and A. Purcell. 2002. Facultative bacterial endosymbionts benefit pea aphids *Acyrthosiphon pisum* under heat stress. Ecol. Entomol. 27: 189–195.

Moran, N.A. and T. Jarvik. 2010. Lateral transfer of genes from fungi underlies carotenoid production in aphids. Science 328: 624–627.

Moran, N.A., J.A. Russell, R. Koga and T. Fukatsu. 2005. Evolutionary relationships of three new species of Enterobacteriaceae living as symbionts of aphids and other insects. Appl. Environ. Microbiol. 71: 3302–3310.

Mutti, N.S., Y. Park, J.C. Reese and G.R. Reeck. 2006. RNAi knockdown of a salivary transcript leading to lethality in the pea aphid, *Acyrthosiphon pisum*. J. Insect Sci. 6: 1–7.

Mutti, N.S., J. Louis, L.K. Pappan, K. Pappan, K. Begum, M.S. Chen, Y. Park, N. Dittmer, J. Marshall, J.C. Reese and G.R. Reeck. 2008. A protein from the salivary glands of the pea aphid, *Acyrthosiphon pisum*, is essential in feeding on a host plant. Proc. Natl. Acad. Sci. USA 105: 9965–9969.

Nicholson, S.J., M.L. Nickerson, M. Dean, Y. Song, P.R. Hoyt, H. Rhee, C. Kim and G.J. Puterka. 2015. The genome of *Diuraphis noxia*, a global aphid pest of small grains. BMC Genomics 16(1): 429.

Nikoh, N., J.P. McCutcheon, T. Kudo, S. Miyagishima, N.A. Moran and A. Nakabachi. 2010. Bacterial genes in the aphid genome: absence of functional gene transfer from *Buchnera* to its host. PLoS Genet. 6: e1000827.

Ogawa, K. and T. Miura. 2014. Aphid polyphenisms: trans-generational developmental regulation through viviparity. Front. Physiol. 5: 1.

Oliver, K.M., J.A. Russell, N.A. Moran and M.S. Hunter. 2003. Facultative bacterial symbionts in aphids confer resistance to parasitic wasps. Proc. Natl. Acad. Sci. USA 100: 1803–1807.

Oliver, K.M., N.A. Moran and M.S. Hunter. 2005. Variation in resistance to parasitism in aphids is due to symbionts not host genotype. Proc. Natl. Acad. Sci. USA 102: 12795–800.

Ollivier, M., T. Gabaldón, J. Poulain, F. Gavory, N. Leterme, J.-P. Gauthier, F. Legeai, D. Tagu, J.-C. Simon and C. Rispe. 2012. Comparison of gene repertoires and patterns of evolutionary rates in eight aphid species that differ by reproductive mode. Genome Biol. Evol. 4: 155–167.

Peccoud, J., A. Ollivier, M. Plantegenest and J.C. Simon. 2009. A continuum of genetic divergence from sympatric host races to species in the pea aphid complex. Proc. Natl. Acad. Sci. USA 106: 7495–7500.

Pitino, M. and S.A. Hogenhout. 2013. Aphid protein effectors promote aphid colonization in a plant species-specific manner. Mol. Plant Microbe Interact. 26: 130–139.

Pitino, M., A.D. Coleman, M.E. Maffei, C.J. Ridout and S.A. Hogenhout. 2011. Silencing of aphid genes by dsRNA feeding from plants. PLoS One 6: e25709.

Powell, G. and J. Hardie. 2000. Host-selection behaviour by genetically identical aphids with different plant preferences. Physiol. Entomol. 25: 54–62.

Prado, E. and W.F. Tjallingii. 2007. Behavioral evidence for local reduction of aphid-induced resistance. J. Insect Sci. 7: 1–8.

Price, D.R.G., R.P. Duncan, S. Shigenobu and A.C.C. Wilson. 2011. Genome expansion and differential expression of amino acid transporters at the aphid/*Buchnera* symbiotic interface. Mol. Biol. Evol. 28: 3113–3126.

Prickett, M.D., M. Page, A.E. Douglas and G.H. Thomas. 2006. BuchneraBASE: a post-genomic resource for *Buchnera* sp. APS. Bioinformatics 22: 641–642.

Pryszcz, L.P., J. Huerta-Cepas and T. Gabaldón. 2011. MetaPhOrs: orthology and paralogy predictions from multiple phylogenetic evidence using a consistency-based confidence score. Nucleic Acids Res. 39: e32–e32.

Puinean, A.M., S.P. Foster, L. Oliphant, I. Denholm, L.M. Field, N.S. Millar, M.S. Williamson and C. Bass. 2010. Amplification of a Cytochrome P450 gene is associated with resistance to neonicotinoid insecticides in the aphid *Myzus persicae*. Plos Genet. 6: e1000999.

Qiao, H.L., E. Tuccori, X.L. He, A. Gazzano, L. Field, J.J. Zhou and P. Pelosi. 2009. Discrimination of alarm pheromone (E)-beta-farnesene by aphid odorant-binding proteins. Insect Biochem. Mol. Biol. 39: 414–419.

Rabatel, A., G. Febvay, K. Gaget, G. Duport, P. Baa-Puyoulet, P. Sapountzis, N. Bendridi, M. Rey, Y. Rahbe, H. Charles, F. Calevro and S. Colella. 2013. Tyrosine pathway regulation is host-mediated in the pea aphid symbiosis during late embryonic and early larval development. BMC Genomics 14: 235.

Ramos, S., A. Moya and D. Martinez-Torres. 2003. Identification of a gene overexpressed in aphids reared under short photoperiod. Insect Biochem. Mol. Biol. 33: 289–298.

Ramsey, J.S., D.S. Rider, T.K. Walsh, M. De Vos, K.H.J. Gordon, L. Ponnala, S.L. Macmil, B.A. Roe and G. Jander. 2010. Comparative analysis of detoxification enzymes in *Acyrthosiphon pisum* and *Myzus persicae*. Insect Mol. Biol. 19: 155–164.

Rao, S.A.K., J.C. Carolan and T.L. Wilkinson. 2013. Proteomic profiling of cereal aphid saliva reveals both ubiquitous and adaptive secreted proteins. PLoS One 8: e57413.

Rider, Jr., S.D., D.G. Srinivasan and R.S. Hilgarth. 2010. Chromatin-remodelling proteins of the pea aphid, *Acyrthosiphon pisum* (Harris). Insect Mol. Biol. 19: 201–214.

Riparbelli, M.G., D. Tagu, J. Bonhomme and G. Callaini. 2005. Aster self-organization at meiosis: a conserved mechanism in insect parthenogenesis? Dev. Biol. 278: 220–230.

Roadmap Epigenomics Consortium, A. Kundaje, W. Meuleman, J. Ernst, M. Bilenky, A. Yen, A. Heravi-Moussavi, P. Kheradpour, Z. Zhang, J. Wang, M.J. Ziller, V. Amin, J.W. Whitaker, M.D. Schultz, L.D. Ward, A. Sarkar, G. Quon, R.S. Sandstrom, M.L. Eaton, Y.C. Wu, A.R. Pfenning, X. Wang, M. Claussnitzer, Y. Liu, C. Coarfa, R.A. Harris, N. Shoresh, C.B. Epstein, E. Gjoneska, D. Leung, W. Xie, R.D. Hawkins, R. Lister, C. Hong, P. Gascard, A.J. Mungall, R. Moore, E. Chuah, A. Tam, T.K. Canfield, R.S. Hansen, R. Kaul, P.J. Sabo, M.S. Bansal, A. Carles, J.R. Dixon, K.H. Farh, S. Feizi, R. Karlic, A.R. Kim, A. Kulkarni, D. Li, R. Lowdon, G. Elliott, T.R. Mercer, S.J. Neph, V. Onuchic, P. Polak, N. Rajagopal, P. Ray, R.C. Sallari, K.T. Siebenthall, N.A. Sinnott-Armstrong, M. Stevens, R.E. Thurman, J. Wu, B. Zhang, X. Zhou, A.E. Beaudet, L.A. Boyer, P.L. De Jager, P.J. Farnham, S.J. Fisher, D. Haussler, S.J. Jones, W. Li, M.A. Marra, M.T. McManus, S. Sunyaev, J.A. Thomson, T.D. Tlsty, L.H. Tsai, W. Wang, R.A. Waterland, M.Q. Zhang, L.H. Chadwick, B.E. Bernstein, J.F. Costello, J.R. Ecker, M. Hirst, A. Meissner, A. Milosavljevic, B. Ren, J.A. Stamatoyannopoulos, T. Wang and M. Kellis. 2015. Integrative analysis of 111 reference human epigenomes. Nature 518: 317–330.

Rossi, M., F.L. Goggin, S.B. Milligan, I. Kaloshian, D.E. Ullman and V.M. Williamson. 1998. The nematode resistance gene Mi of tomato confers resistance against the potato aphid. Proc. Natl. Acad. Sci. USA 95: 9750–9754.

Russell, J.A., A. Latorre, B. Sabater-Muñoz, A. Moya and N.A. Moran. 2003. Side-stepping secondary symbionts: widespread horizontal transfer across and beyond the Aphidoidea. Mol. Ecol. 12: 1061–1075.

Sabater-Munoz, B., F. Legeai, C. Rispe, J. Bonhomme, P. Dearden, C. Dossat, A. Duclert, J.P. Gauthier, D.G. Ducray, W. Hunter, P. Dang, S. Kambhampati, D. Martinez-Torres, T. Cortes, A. Moya, A. Nakabachi, C. Philippe, N. Prunier-Leterme, Y. Rahbe, J.C. Simon, D.L. Stern, P. Wincker and D. Tagu. 2006. Large-scale gene discovery in the pea aphid *Acyrthosiphon pisum* (Hemiptera). Genome Biol. 7: R21.

Sachse, S. and J. Krieger. 2011. Olfaction in insects. The primary processes of odor recognition and coding. e-Neuroforum 2: 49–60.

Sandström, J.P., J.A. Russell, J.P. White and N.A. Moran. 2001. Independent origins and horizontal transfer of bacterial symbionts of aphids. Mol. Ecol. 10: 217–228.

Sattar, S., C. Addo-Quaye, Y. Song, J.A. Anstead, R. Sunkar and G.A. Thompson. 2012. Expression of small RNA in *Aphis gossypii* and its potential role in the resistance interaction with melon. Plos One 7: e48579.

Scarborough, C., J. Ferrari and H.C.J. Godfray. 2005. Aphid protected from pathogen by endosymbiont. Science 310: 1781–1781.

Schmitz, A., C. Anselme, M. Ravallec, C. Rebuf, J.-C. Simon, J.-L. Gatti and M. Poirié. 2012. The cellular immune response of the pea aphid to foreign intrusion and symbiotic challenge. PLoS One 7: e42114.

Schwartzberg, E.G., G. Kunert, S.A. Westerlund, K.H. Hoffmann and W.W. Weisser. 2008. Juvenile hormone titres and winged offspring production do not correlate in the pea aphid, *Acyrthosiphon pisum*. J. Insect Physiol. 54: 1332–1336.

Schwarzkopf, A., D. Rosenberger, M. Niebergall, J. Gershenzon and G. Kunert. 2013. To feed or not to feed: plant factors located in the epidermis, mesophyll, and sieve elements influence pea aphid's ability to feed on legume species. Plos One 8: e75298.

Shigenobu, S. and D.L. Stern. 2013. Aphids evolved novel secreted proteins for symbiosis with bacterial endosymbiont. Proc. R. Soc. Lond. B 280: 20121952. doi: 10.1098/rspb.2012.1952.

Shigenobu, S., H. Watanabe, M. Hattori, Y. Sakaki and H. Ishikawa. 2000. Genome sequence of the endocellular bacterial symbiont of aphids *Buchnera* sp. APS. Nature 407: 81–86.

Shigenobu, S., R.D. Bickel, J.A. Brisson, T. Butts, C.-C. Chang, O. Christiaens, G.K. Davis, E.J. Duncan, D.E.K. Ferrier, M. Iga, R. Janssen, G.-W. Lin, H.-L. Lu, A.P. McGregor, T. Miura, G. Smagghe, J.M. Smith, M. van der Zee, R.A. Velarde, M.J. Wilson, P.K. Dearden and D.L. Stern. 2010. Comprehensive survey of developmental genes in the pea aphid, *Acyrthosiphon pisum*: frequent lineage-specific duplications and losses of developmental genes. Insect Mol. Biol. 19: 47–62.

Simon, J.C., M.E. Pfrender, R. Tollrian, D. Tagu and J.K. Colbourne. 2011. Genomics of environmentally induced phenotypes in 2 extremely plastic arthropods. J. Hered. 102: 512–525.

Simon, J.C., E. d'Alençon, E. Guy, E. Jacquin-Joly, J. Jaquiéry, P. Nouhaud, J. Peccoud, A. Sugio and R. Streiff. 2015. Genomics of adaptation to host-plants in herbivorous insects. Brief. Funct. Genomics. doi.org/10.1093/bfgp/elv015.

Smadja, C., P. Shi, R.K. Butlin and H.M. Robertson. 2009. Large gene family expansions and adaptive evolution for odorant and gustatory receptors in the pea aphid, *Acyrthosiphon pisum*. Mol. Biol. Evol. 26: 2073–2086.

Smadja, C.M., B. Canback, R. Vitalis, M. Gautier, J. Ferrari, J.J. Zhou and R.K. Butlin. 2012. Large-scale candidate scan reveals the role of chemoreceptor genes in host plant specialization and speciation in the pea aphid. Evolution 66: 2723–2738.

Srinivasan, D.G., A. Abdelhady and D.L. Stern. 2014. Gene expression analysis of parthenogenetic embryonic development of the pea aphid, *Acyrthosiphon pisum*, suggests that aphid parthenogenesis evolved from meiotic oogenesis. PLoS One 9: e115099.

Sugio, A., G. Dubreuil, D. Giron and J. Simon. 2015. Plant-insect interactions under bacterial influence: ecological implications and underlying mechanisms. J. Exp. Bot. 66: 467–478.

Sun, Y.F., F. De Biasio, H.L. Qiao, I. Iovinella, S.X. Yang, Y. Ling, L. Riviello, D. Battaglia, P. Falabella, X.L. Yang and P. Pelosi. 2012. Two odorant-binding proteins mediate the behavioural response of aphids to the alarm pheromone (E)-beta-farnesene and structural analogues. Plos One 7: e32759.

Tagu, D., S. Dugravot, Y. Outreman, C. Rispe, J.-C. Simon and S. Colella. 2010. The anatomy of an aphid genome: from sequence to biology. C. R. Biol. 333: 464–473.

Tares, S., L. Arthaud, M. Amichot and A. Robichon. 2013. Environment exploration and colonization behavior of the pea aphid associated with the expression of the foraging gene. PLoS One 8: e65104.

The International Human Genome Sequencing Consortium. 2000. Help in accessing human genome information. Science 289: 1471.

Tjallingii, W.F. 2006. Salivary secretions by aphids interacting with proteins of phloem wound responses. J. Exp. Bot. 57: 739–745.

Tsuchida, T., R. Koga and T. Fukatsu. 2004. Host plant specialization governed by facultative symbiont. Science 303: 1989.

Tsuchida, T., R. Koga, M. Horikawa, T. Tsunoda, T. Maoka, S. Matsumoto, J.-C. Simon and T. Fukatsu. 2010. Symbiotic bacterium modifies aphid body color. Science 330: 1102–1104.

Vandermoten, S., N. Harmel, G. Mazzucchelli, E. De Pauw, E. Haubruge and F. Francis. 2014. Comparative analyses of salivary proteins from three aphid species. Insect Mol. Biol. 23: 67–77.

Vellozo, A.F., A.S. Véron, P. Baa-Puyoulet, J. Huerta-Cepas, L. Cottret, G. Febvay, F. Calevro, Y. Rahbe, A.E. Douglas, T. Gabaldón, M.-F. Sagot, H. Charles and S. Colella. 2011. CycADS: an annotation database system to ease the development and update of BioCyc databases. Database 2011: bar008–bar008.

Walsh, T.K., J.A. Brisson, H.M. Robertson, K. Gordon, S. Jaubert-Possamai, D. Tagu and O.R. Edwards. 2010. A functional DNA methylation system in the pea aphid, *Acyrthosiphon pisum*. Insect Mol. Biol. 19: 215–228.

Wang, X. and L. Kang. 2014. Molecular mechanisms of phase change in locusts. Annu. Rev. Entomol. 59: 225–244.

Wensler, R.J. and B.K. Filshie. 1969. Gustatory sense organs in the food canal of aphids. J. Morphol. 129: 473–492.

Will, T., W.F. Tjallingii, A. Thonnessen and A.J. van Bel. 2007. Molecular sabotage of plant defense by aphid saliva. Proc. Natl. Acad. Sci. USA 104: 10536–10541.

Wilson, A.C.C., P.D. Ashton, F. Calevro, H. Charles, S. Colella, G. Febvay, G. Jander, P.F. Kushlan, S.J. Macdonald, J.F. Schwartz, G.H. Thomas and A.E. Douglas. 2010. Genomic insight into the amino acid relations of the pea aphid, *Acyrthosiphon pisum*, with its symbiotic bacterium *Buchnera aphidicola*. Insect Mol. Biol. 19: 249–258.

Xu, H.J., J. Xue, B. Lu, X.C. Zhang, J.C. Zhuo, S.F. He, X.F. Ma, Y.Q. Jiang, H.W. Fan, J.Y. Xu, Y.X. Ye, P.L. Pan, Q. Li, Y.Y. Bao, H.F. Nijhout and C.X. Zhang. 2015. Two insulin receptors determine alternative wing morphs in planthoppers. Nature 519: 464–467.

Zhang, M., Y.W. Zhou, H. Wang, H.D. Jones, Q. Gao, D.H. Wang, Y.Z. Ma and L.Q. Xia. 2013. Identifying potential RNAi targets in grain aphid (*Sitobion avenae* F.) based on transcriptome profiling of its alimentary canal after feeding on wheat plants. BMC Genomics 14: 560. doi: 10.1186/1471-2164-14-560.

Zhou, J.J., F.G. Vieira, X.L. He, C. Smadja, R. Liu, J. Rozas and L.M. Field. 2010. Genome annotation and comparative analyses of the odorant-binding proteins and chemosensory proteins in the pea aphid *Acyrthosiphon pisum*. Insect Mol. Biol. 19: 113–122.

Zientz, E., T. Dandekar and R. Gross. 2004. Metabolic interdependence of obligate intracellular bacteria and their insect hosts. Microbiol. Mol. Biol. Rev. 68: 745–770.

4

Epigenetic Control of Polyphenism in Aphids

Krishnendu Mukherjee[1,*] and *Arne F. Baudach*[2]

Introduction

The ability to form multiple discrete phenotypes from a single genotype is known as polyphenism. Such phenotypic plasticity is prevalent in social insects, where environmental stimuli trigger distinct phenotypes, e.g., the different castes of honeybees and ants (Miura 2005, Judice et al. 2006, Johnson 2010). Other insects, such as butterflies, respond to seasonal changes in temperature and photoperiod by adapting their pigmentation. The map butterfly *Araschnia levana* provides a striking example of seasonal polyphenism because spring adults are predominantly orange whereas summer adults are black and white (Morehouse et al. 2013). Several pest insects respond to biotic and abiotic stress by modifying physiological parameters including their reproductive strategy, and in this context aphids provide an ideal model system to study seasonal polyphenism (Srinivasan and Brisson 2012). It is becoming increasingly evident that seasonal polyphenism in insects is achieved by the epigenetic regulation of transcriptional reprogramming. The molecular basis of polyphenism in aphids and other insects therefore relies largely on the study of DNA methylation, histone acetylation and the synthesis of microRNAs (miRNAs) that epigenetically regulate gene expression.

Although the analysis of gene sequences has advanced our understanding of insect development, the conversion of genotype to phenotype is governed by complex non-genetic factors commonly grouped under the collective term 'epigenetics'. Waddington

[1] Fraunhofer Institute for Molecular Biology and Applied Ecology, Department of Bioresources, Winchester Str. 2, 35395 Giessen, Germany.
[2] Institute for Insect Biotechnology, Justus-Liebig University, 26-32 Heinrich-Buff-Ring, Giessen 35392, Germany.
* Corresponding author: krishnendu.mukherjee@agrar.uni-giessen.de

(1942) introduced the term to describe the branch of biology which studies the causal interactions between genes and their products which bring the phenotype into being. The term is currently used in different contexts to describe traits that are heritable but not controlled at the level of the DNA sequence, as well as the underlying molecular mechanisms that may not necessarily be heritable. In insects, such mechanisms may include the determination of cell types from pluripotent stem cells, dosage compensation, X chromosome inactivation, imprinting, and position effect variegation (Probst et al. 2009). Here we discuss common epigenetic mechanisms in the context of DNA methylation, histone acetylation and the expression of miRNAs to substantiate their role in establishing heritable gene expression patterns and polyphenism in aphids.

Polyphenism in Aphids

The hemipteran pea aphid (*Acyrthosiphon pisum*) is an attractive model in which to study the causes and consequences of environmentally regulated phenotypic plasticity in insects. First, aphids are economically important because they are major agricultural pests that damage crops by ingesting the sap for nutrition and transmitting viruses (Dixon 1998). Second, their intrinsic responses to biotic and abiotic cues are conveniently amenable to experimental manipulation. For example, they can adapt to different seasonal cues through wing polyphenism (the development of winged and wingless females) and reproductive polyphenism (asexual and sexual individuals) (see Chapter 2 "The Ontogenesis of the Pea Aphid *Acyrthosiphon pisum*" in this volume). Third, aphids also provide a model of caste polyphenism by which soldier aphids are committed to nest defense (Hattori et al. 2013). The phenotypic plasticity of aphids has long been a key area of developmental biology research, but only recently have investigators moved beyond the study of gene functions and interactions.

Polyphenism in insects is regulated by physiological factors such as patterns of hormone secretion or tissue-specific sensitivity to environmental changes and alternating developmental mechanisms (Nijhout 2003, West-Eberhard 2003). Changes in the photoperiod and temperature play a major role in the sex determination of aphids. Depending on the season, they undergo either sexual or asexual reproduction, e.g., short day length and low temperature are necessary for the production of sexual generations. Additionally, sexual reproduction in some aphid species (e.g., *Aphis farinose*) is dependent on physiological changes in the host plant, such as the appearance of stunted shoots (Ogawa and Miura 2014). Asexual reproduction by parthenogenesis has been acquired independently and often secondarily by aphids to increase their population size and to avoid the cost involved in finding males for sexual reproduction (Williams 1975, Maynard Smith 1978, Schön et al. 2009). The typical annual life cycle of aphids consists of recurring parthenogenesis (approximately 10–30 generations) followed by a single sexual stage (Moran 1992, Simon et al. 2002) (see Chapter 2 "The Ontogenesis of the Pea Aphid *Acyrthosiphon pisum*" in this volume).

Wing polyphenism in viviparous and oviparous aphid species results in the development of winged or wingless morphs in response to environmental changes in order to facilitate their migration to new host plants and habitats (Dixon 1998, Braendle et al. 2006, Brisson 2010). Winged morphs are formed in response to environmental factors such as the release of alarm pheromone or the increase in tactile stimulation associated with avoidance behavior from natural enemies (Kunert et al. 2008, Hatano et al. 2010) (see Chapter 9 "Chemical Ecology of Aphids (Hemiptera: Aphididae)" in this volume). Accurate sensing of environmental stimuli is also maternally determined in aphid species such as *Megoura crassicauda* and the pea aphid (Müller et al. 2001, Ishikawa and Miura 2013). The transgenerational induction of winged morphs is independent of the density of the nymph population, i.e., the percentage of winged adults in the same generation does not differ from wingless groups reared under low-density conditions (Müller et al. 2001). Based on such findings, the mechanisms underlying the developmental determination of wing polyphenism are likely to be species-dependent or group-specific. Conversely, the mutualistic association of aphids with ants inhibits the development of winged individuals (Yao 2012). This has been observed in numerous insect species that have lost the ability to fly, resulting in a compensatory increase in fecundity, longevity, and the development of weapons for intraspecific and interspecific competition (Roff and Fairbairn 1991).

Genetic Basis of Polyphenism in Aphids

Phenotypic plasticity in aphids is governed by genes, especially those controlling the synthesis and production of hormones (Nijhout 1999). Genes controlling the production of juvenile hormones play an important role in the sexual and wing polyphenism of pea aphids (see Chapter 3 "Functional and Evolutionary Genomics in Aphids" in this volume). For example, low titers of juvenile hormone III are associated with the high-level expression of juvenile hormone esterase genes in aphids reared under short-day conditions, which produce sexual morph, in contrast to those reared under long-day conditions which produce parthenogenetic morphs. Juvenile hormone esterase catalyzes the degradation of juvenile hormones in insects. In the cricket genus *Gryllus*, an increase in the juvenile hormone/ecdysteroid ratio induces the development of short-winged rather than long-winged morphs (Zera et al. 1989, Zera 2003). Similarly, higher juvenile hormone levels favor the production of queens over workers in honeybees (*Apis mellifera*) fed on royal jelly protein (Hartfelder and Engels 1998). The importance of the juvenile hormones in insect development has been well established in both holometabolous and hemimetabolous insects, and recent evidence suggests that their expression is epigenetically controlled.

Reproductive polyphenism in aphids induced by a change in photoperiod includes the differential expression of candidate genes especially those regulating the insulin pathway. Shorter days favor the expression of genes that promote insulin degradation while simultaneously repressing the insulin gene (Le Trionnaire et al. 2009). The insulin and juvenile hormone signaling cascades appear to be jointly regulated in diapaused insects such as the mosquito *Culex pipiens* under short-day conditions (Tu et al. 2005, Sim and Denlinger 2008). The insulin pathway is thus suppressed in pea aphids reared under short-day conditions, which emphasizes its role in the reproductive polyphenism as an upstream regulator of the juvenile hormone pathway.

The involvement of endocrine signaling in aphid wing development is supported by experiments in which aphids are exposed to precocene II (PII) a chromene derivative produced by *Ageratum* spp. that can interfere with the development of pest insects by inhibiting juvenile hormones. PII induces reversible sterilization and precocious metamorphosis in insects by suppressing the function of the corpora allata gland. The exposure of pea aphids to PII induced the production of winged progeny, possibly by reducing the juvenile hormone titer. Subsequent studies have shown that PII is unable to induce precocious wing development in the pea aphid when juvenile hormone is induced artificially (Hardie et al. 1996). However, juvenile hormone is not known to influence the choice between winged and wingless morphs of aphids, suggesting other regulatory mechanisms may contribute to this form of polyphenism in aphids.

Epigenetic Mechanisms and Polyphenism of Aphids

Chromatin modifications are thought to influence the extraordinary developmental plasticity of aphids in response to environmental stimuli and there is mounting evidence for the epigenetic regulation of transcriptional reprogramming during the formation of different phenotypic morphs in social insects such as honeybees (Guo et al. 2013, Maleszka 2008). Next-generation sequencing and comparative genomics have been combined to search for common epigenetic mechanisms that influence polyphenism in aphids.

DNA Methylation

DNA methylation in eukaryotes typically involves the addition of methyl groups to cytidine residues in the dinucleotide sequence CpG (in animals) and CpNpG (in plants) to create 5-methylcytosine, which behaves normally in terms of DNA base-pairing but changes the way DNA interacts with proteins, thus providing a mechanism for gene regulation (Fig. 1). Because this form of methylation occurs at sites with a two-fold rotational axis of symmetry, the methylation mark can be passed to daughter cells during DNA replication because one strand remains methylated in the daughter duplex, thus explaining how the epigenetic state can be inherited.

The transfer of methyl groups to DNA is mediated by several evolutionarily-conserved enzymes collectively known as DNA methyltransferases (DNMTs). These can be further divided into maintenance methyltransferases, which complete the symmetrical methylation marks on newly-replicated DNA by recognizing the hemimethylated sequences inherited from each parent, and *de novo* methyltransferases which establish new methylation marks on unmethylated DNA (Bestor 2000, Klose and Bird 2006). In mammals, DNMT1 is classified a maintenance methyltransferase, DNMT3 is a *de novo* methyltransferase and DNMT2 was initially misclassified as a DNMT but is now known to methylate transfer RNA (tRNA), which carries a number of constitutive DNA modifications (Goll et al. 2006).

The sequenced genome of the pea aphid is valuable source of genes encoding DNMTs (International Aphid Genomics Consortium 2010). It contains a full complement of DNA methylation genes with orthologs of two maintenance DNA

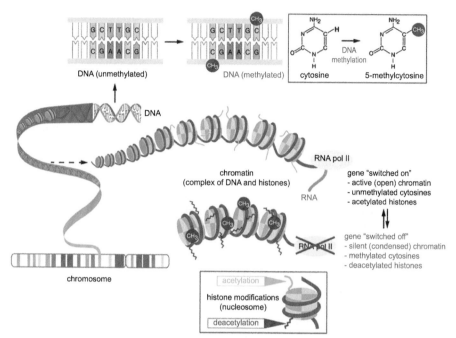

Fig. 1. Study of DNA methylation and histone acetylation mediated chromatin modifications. Gene expression can be regulated prior transcriptional initiation by the chemical modification of DNA or histone proteins comprising the chromatin. Addition of a methyl group (CH3) to the CpG cluster of DNA forms 5-methylcytosine which interferes with DNA interaction to proteins thus providing a mechanism for gene regulation. The chromatin structure can be also controlled by modulating the positive charge density of the core histones, for, e.g., by the removal of acetyl group which leads to compact chromatin inaccessible to RNA polymerase II and transcriptional repression. On the other hand unmethylated cytosines and acetylation of histones impose contrasting effects, i.e., uncoiling of the chromatin and induction of gene expression.

methyltransferases (*Dnmt1a* and *Dnmt1b*), two *de novo* DNA methyltransferases (*Dnmt3a* and *Dnmt3X*), and *Dnmt2*, similar to honeybees. In the red flour beetle (*Tribolium castaneum*) two genes (*Dnmt1* and *Dnmt2*) for DNA methyltransferases were found by whole genome sequencing, whereas the same technique could only identify a *Dnmt2* homolog in the fruit fly *Drosophila melanogaster*. The pea aphid genome also contains genes encoding the methylated-CpG binding proteins MECP2 (one copy) and NP95 (three copies), as well as a *Dnmt1* associated protein that helps to recruit histone deacetylases.

DNA methylation dominates the heterochromatin region of the pea aphid genome (Mandrioli and Borsatti 2007). The frequency of methylated genes in the pea aphid has been predicted *in silico* based on the ratio of observed CpG and expected CpG frequencies (CpG$_{O/E}$). Methylated cytosine nucleotides in CpG dinucleotides are prone to deamination to uracil, which the DNA repair machinery can convert into thymidine. Methylated CpG sites therefore become less abundant over evolutionary timescales, and CpG$_{O/E}$ can be used to predict historically methylated genes. This approach has previously been used to predict methylated genes correctly (Suzuki et al. 2007). Based on this principle, the pea aphid genome was predicted to contain two broad classes of genes differing in methylation status, agreeing with similar observations in honeybees but contrasting to the situation in the fruit fly and red flour beetle. Approximately 0.69% of all cytosine residues are thought to be methylated in the pea aphid genome, predominantly in coding regions. The extent of coding region methylation in insect genes is generally lower than that observed in mammals (Walsh et al. 2010). For example, the honeybee genome contains more than 10 million CpG sites but only 70,000 (0.7%) are methylated (Lyko and Maleszka 2011) compared to 70% occupancy in humans (Strichman-Almashanu et al. 2002). Interestingly, the methylated honeybee genes are predominantly those conserved among arthropods whereas species-restricted genes tend to be non-methylated. Even so, 550 genes showed caste-specific differential methylation between queens and workers (Lyko and Maleszka 2011). Similarly in aphids, $CpG_{O/E}$ values were higher for genes expressed under specific conditions than for constitutive genes, and the methylated genes were often associated with general gene ontology (GO) categories such as metabolic processes. The genes with sparser methylation encompassed a range of functions, including signal transduction, cognition and behavior (Hunt et al. 2010). The sparse methylation of morph-based genes in pea aphids may correlate with the *de novo* methylation of the available CpG sites, allowing them to acquire different methylation patterns and therefore different expression states on a generation-by-generation basis, induced by relevant environmental conditions. This is supported by the experimental suppression of *Dnmt3 de novo* methyltransferase expression in honeybees by RNA interference, which led to changes in reproductive morph specification (Kucharski et al. 2008). There is also a dynamic cycle of methylation and demethylation during the life cycle of model insects such as the honeybee and the red flour beetle with the greatest degree of CpG methylation found in the embryos (Feliciello et al. 2013).

The aphid genome also contains methylated genes encoding enzymes responsible for juvenile hormone synthesis and one encoding a juvenile hormone-binding protein with one methylated site that had a marginally (but still significant) higher level of methylation in winged asexual females compared to wingless peers (Walsh et al. 2010). DNA methylation was also observed in the E4 esterase gene when it was expressed in green peach aphid (*Myzus persicae*) populations that are insecticide resistant, providing an unusual instance of DNA methylation promoting transcription (Field et al. 1996, 1999, Hick et al. 1996).

All previous studies of DNA methylation in aphids have involved the indirect detection of methylation at CpG dinucleotides. The evidence supports the hypothesis that DNA methylation can regulate epigenetic plasticity in aphids as in other social insects. However, the direct analysis of DNA methylation states by techniques such as whole genome bisulfite sequencing is necessary to develop a comprehensive understanding of this phenomenon.

Histone Acetylation

Gene expression in eukaryotes is regulated by specific proteins known as transcription factors which influence the manner in which DNA interacts with the transcriptional apparatus. In addition to these specific interactions, the behavior of DNA can be influenced more generally by conformational changes in the proteins that make up chromatin, i.e., the complex of DNA and protein which is structural basis of chromosomes in the eukaryotic cell nucleus (Fig. 1). The basic repeat element of chromatin is the nucleosome, which comprises DNA wrapped around an octamer of core histones (two each of H2A, H2B, H3 and H4). The nucleosomes are linked together like beads on a string by segments of linker DNA paired with linker histones (H1, H5). The structure of chromatin, and hence the accessibility of the DNA, can be controlled by modulating the positive charge density of the core histones, principally by the addition or removal of acetyl groups (Fig. 1).

Acetylated histones form a loose and accessible type of chromatin that promotes gene expression, whereas deacetylated histones bind DNA more tightly and render it inaccessible and transcriptionally silent. This property of histones is controlled by the opposing activities of two families of enzymes: histone acetyltransferases (HATs) and histone deacetylases (HDACs). The pea aphid genome contains multiple genes representing HATs and HDACs (Rider et al. 2010) and these abundant histone modifying enzymes may facilitate the phenotypic plasticity which is necessary to maintain their complex life cycle. HATs representing the GNAT and MYST superfamilies are conserved between the pea aphid and fruit fly, but whereas the fruit fly genome contains only one PCAF/GCN5 gene, there are two paralogs in the pea aphid. Similar expansion is evident among the pea aphid HDAC gene families, many of which are not present in the fruit fly. Multiple copies of RPD3-like HDACs and sirtuins are also present in the pea aphid. Interestingly, there is also evidence that some epigenetic regulators have been lost from the aphid genome, e.g., class IV HDACs have not been identified despite their presence in dipteran and coleopteran insects. Histones and histone variants are conserved in the aphid and fruit fly genomes, although histone variants such as Cenp-A and protamines have been lost (Rider et al. 2010). The pea aphid genome sequence also suggests that antagonistic chromatin modifying and remodeling pathways have expanded, including the expansion of gene families involved in histone acetylation and histone deacetylation (Rider et al. 2010). A similar situation is observed for genes involved in histone methylation and histone demethylation (Rider et al. 2010, Krauss et al. 2006). It is unclear whether particular histone modification states are heritable per se, but there is significant crosstalk among histone modification, DNA methylation and even post-transcriptional regulation which helps to reinforce and perpetuate the consequences of histone-based epigenetic mechanisms. This aphid-specific loss and gain of the histone acetylation pathway may have a significant impact on the regulation of chromatin structure and gene expression in the context of adaptive phenotypic plasticity.

Because the effect of chromatin remodeling and gene expression is context-dependent, these multiple antagonistic activities suggest that chromatin structure in aphids is subject to complex regulation. Chromatin immunoprecipitation (ChIP), expression analysis and evolutionary analysis of these genes should help distinguish among the competing hypotheses, and next-generation sequencing could be used to analyze morph-specific chromatin modifications (Zhou et al. 2011).

MicroRNAs

MicroRNAs are non-coding RNAs 18–22 nucleotides in length which regulate gene expression at the post-transcriptional level by binding to complementary mRNAs.

They are involved in the regulation of gene expression during many physiological processes (e.g., development, immunity, the cell cycle and apoptosis) and are also associated with a number of diseases (Ambros 2004, Bartel 2004, Lu and Liston 2009). The genes encoding miRNAs are found individually and as polycistronic clusters (Lagos-Quintana et al. 2001). In the nucleus, the transcription of miRNA genes by RNA polymerase II/III produces double-stranded transcripts known as primary miRNAs (pri-miRNAs). These are trimmed by the RNase III enzyme Drosha to form double-stranded precursor miRNAs (pre-miRNAs) that are transported to the cytoplasm by Exportin-5. This results in the formation of an RNA-induced silencing complex (RISC), comprising the pre-miRNA, the RNase III enzyme Dicer, the double-stranded RNA binding domain proteins TRBP (Tar RNA binding protein), PACT (protein activator of PKR) and the core component Argonaute-2 (Ago2). The pre-miRNA is processed by Dicer to form a mature single-stranded miRNA which guides the RISC to complementary mRNAs. If the sequence complementarity is perfect the mRNA target is cleaved and degraded, whereas imperfectly-matched mRNAs are deadenylated or arrested in the complex resulting in translational repression (Bartel 2004, Macfarlane and Murphy 2010).

Because the function of miRNAs is sequence-dependent, individual miRNAs can target multiple mRNAs as long as they contain the matching complementary sequence, and individual mRNAs can be regulated by multiple miRNAs if the mRNA contains more than one target sequence. The detailed comparison of genome sequences has revealed a surprising degree of conservation in the miRNA repertoires of insects and mammals. However, many further miRNAs are limited to particular species and sophisticated bioinformatics algorithms are needed to identify them *de novo*, which can be expensive and time consuming. If a complete genome sequence is not available, miRNAs can be identified using microarray technology and validated by quantitative RT-PCR. This has been highly effective as a strategy for the identification of greater wax moth miRNAs that are active at different developmental stages or induced by pathogens, but it is important to validate the sequences and control for non-specific hybridization (Mukherjee and Vilcinskas 2014).

The pea aphid sequencing project identified 163 putative miRNAs, including 52 conserved genes and 111 orphans. There was also evidence for the remarkable expansion of protein-encoding gene families associated with miRNA-based gene regulation, including Pasha (the RNA-binding co-factor of Drosha), Dicer-1 and Argonaute-1. Such expansion has also been observed in other aphid species but not thus far in any other metazoan taxa. The identification of aphid miRNAs was followed by the elucidation of the corresponding active small RNA pathways (Legeai et al. 2010). This comprehensive series of experiments included the use of homology searches, deep sequencing and *in silico* prediction to analyze 149 pea aphid miRNAs, 55 of which were conserved among insects whereas 94 were aphid-specific. The experiments revealed that 17 miRNAs were differentially expressed among asexual females producing asexual progeny, asexual females producing sexual progeny, and sexual females, concurring with the detection of morph-specific miRNA expression in polyphenic locusts (Wei et al. 2009) and honeybees (Weaver et al. 2007). The functional roles of these morph-specific aphid miRNAs can now be investigated in detail.

Conclusions

There is increasing evidence that complex processes such as development, reproduction and immunity are epigenetically regulated in response to different environmental stimuli. Insects have emerged as an attractive model system for the analysis of such phenomena and the conserved epigenetic mechanisms that allow the same genotype to be expressed as multiple environment-dependent phenotypes, including DNA methylation, histone modification and the expression of miRNAs (Mukherjee et al. 2015). The expansion of gene families representing epigenetic control mechanisms in the pea aphid make this species an ideal model host for the investigation of seasonal phenotypic plasticity in insects. DNA methylation, histone acetylation and miRNAs may all play a role in the regulation of reproductive and wing polyphenism in aphids, but much of the evidence is based on *in silico* predictions and empirical verification is now required. More detailed investigation of the direct roles of specific DNA methyltransferases, chromatin remodeling proteins and miRNAs will improve our understanding of the epigenetic regulation of polyphenism in aphids, and eventually may facilitate the development of novel pest control strategies based on epigenetic triggers of non-productive developmental pathways.

Keywords: Aphid polyphenism, phenotypic plasticity, endocrine system, epigenetics, DNA methylation, histone acetylation, miRNAs

References

Ambros, V. 2004. The functions of animal microRNAs. Nature 431: 350–355.

Bartel, D.P. 2004. MicroRNAs: genomics, biogenesis, mechanism, and function. Cell 116: 281–297.

Bestor, T.H. 2000. The DNA methyltransferases of mammals. Hum. Mol. Genet. 9: 2395–2402.

Braendle, C., G.K. Davis, J.A. Brisson and D.L. Stern. 2006. Wing dimorphism in aphids. Heredity 97: 192–199.

Brisson, J.A. 2010. Aphid wing dimorphisms: linking environmental and genetic control of trait variation. Philos. Trans. R. Soc. Lond. B, Biol. Sci. 365: 605–616.

Dixon, A.F.G. 1998. Aphid Ecology. Chapman & Hall, London, U.K.

Feliciello, I., J. Parazajder, I. Akrap and D. Ugarković. 2013. First evidence of DNA methylation in insect *Tribolium castaneum*: environmental regulation of DNA methylation within heterochromatin. Epigenetics 8: 534–541.

Field, L.M., S.E. Crick and A.L. Devonshire. 1996. Polymerase chain reaction-based identification of insecticide resistance genes and DNA methylation in the aphid *Myzus persicae* (Sulzer). Insect Mol. Biol. 5: 197–202.

Field, L.M., R.L. Blackman, C. Tyler-Smith and A.L. Devonshire. 1999. Relationship between amount of esterase and gene copy number in insecticide-resistant *Myzus persicae* (Sulzer). Biochemical J. 339: 737–742.

Goll, M.G., F. Kirpekar, K.A. Maggert, J.A. Yoder, C.L. Hsieh, X. Zhang, K.G. Golic, S.E. Jacobsen and T.H. Bestor. 2006. Methylation of tRNAAsp by the DNA methyltransferase homolog Dnmt2. Science 311: 395–398.

Guo, X., S. Su, G. Skogerboe, S. Dai, W. Li, Z. Li, F. Liu, R. Ni, Y. Guo, S. Chen, S. Zhang and R. Chen. 2013. Recipe for a busy bee: microRNAs in Honey Bee caste determination. PLoS One 8(12): e81661.

Hardie, J., N. Gao, T. Timár, P. Sebók and K. Honda. 1996. Precocene derivatives and aphid morphogenesis. Arch. Ins. Biochem. Physiol. 32: 493–501.

Hartfelder, K. and W. Engels. 1998. Social insect polymorphism: hormonal regulation of plasticity in development and reproduction in the honeybee. Curr. Top. Dev. Biol. 40: 45–77.

Hatano, E., G. Kunert and W.W. Weisser. 2010. Aphid wing induction and ecological costs of alarm pheromone emission under field conditions. PLoS One 5(6): e11188.

Hattori, M., O. Kishida and T. Itino. 2013. Soldiers with large weapons in predator-abundant midsummer: reproductive plasticity in a eusocial aphid. Evol. Ecol. 27: 847–862.

Hick, C.A., L.M. Field and A.L. Devonshire. 1996. Changes in the methylation of amplified esterase DNA during loss and reselection of insecticide resistance in peach-potato aphids, *Myzus persicae*. Insect Biochem. Mol. 26: 41–47.

Hunt, B.G., J.A. Brisson, S.V. Yi and M.A. Goodisman. 2010. Functional conservation of DNA methylation in the pea aphid and the honeybee. Genome Biol. Evol. 2: 719–728.

International Aphid Genomics Consortium. 2010. Genome sequence of the pea aphid *Acyrthosiphon pisum*. PLoS Biol. 8(2): e1000313.

Ishikawa, A. and T. Miura. 2013. Transduction of high-density signals across generations in aphid wing polyphenism. Physiol. Entomol. 38: 150–156.

Johnson, B.R. 2010. Division of labor in honeybees: form, function, and proximate mechanisms. Behav. Ecol. Sociobiol. 64: 305–316.

Judice, C.C., M.F. Carazzole, F. Festa, M.C. Sogayar, K. Hartfelder and G.A. Pereira. 2006. Gene expression profiles underlying alternative caste phenotypes in a highly eusocial bee, *Melipona quadrifasciata*. Insect Mol. Biol. 15: 33–44.

Klose, R.J. and A.P. Bird. 2006. Genomic DNA methylation: the mark and its mediators. Trends Biochem. Sci. 31: 89–97.

Krauss, V., A. Fassl, P. Fiebig, I. Patties and H. Sass. 2006. The evolution of the histone methyltransferase gene Su(var)3-9 in metazoans includes a fusion with and a re-fission from a functionally unrelated gene. BMC Evol. Biol. 6: 18.

Kucharski, R., J. Maleszka, S. Foret and R. Maleszka. 2008. Nutritional control of reproductive status in honeybees via DNA methylation. Science 319: 1827–1830.

Kunert, G., K. Schmoock-Ortlepp, U. Reissmann, S. Creutzburg and W.W. Weisser. 2008. The influence of natural enemies on wing induction in *Aphis fabae* and *Megoura viciae* (Hemiptera: Aphididae). Bull. Entomol. Res. 98: 59–62.

Lagos-Quintana, M., R. Rauhut, W. Lendeckel and T. Tuschl. 2001. Identification of novel genes coding for small expressed RNAs. Science 294: 853–858.

Le Trionnaire, G., F. Francis, S. Jaubert-Possamai, J. Bonhomme, E. De Pauw, J.P. Gauthier, E. Haubruge, F. Legeai, N. Prunier-Leterme, J.C. Simon, S. Tanguy and D. Tagu. 2009. Transcriptomic and proteomic analyses of seasonal photoperiodism in the pea aphid. BMC Genomics 10: 456. doi: 10.1186/1471-2164-10-456.

Legeai, F., G. Rizk, T. Walsh, O. Edwards, K. Gordon, D. Lavenier, N. Leterme, A. Méreau, J. Nicolas, D. Tagu and S. Jaubert-Possamai. 2010. Bioinformatic prediction, deep sequencing of microRNAs and expression analysis during phenotypic plasticity in the pea aphid, *Acyrthosiphon pisum*. BMC Genomics 11: 281. doi: 10.1186/1471-2164-11-281.

Lu, L.F. and A. Liston. 2009. MicroRNA in the immune system, microRNA as an immune system. Immunology 127: 291–298.

Lyko, F. and R. Maleszka. 2011. Insects as innovative models for functional studies of DNA methylation. Trends Genet. 27: 127–131.

Macfarlane, L.A. and P.R. Murphy. 2010. MicroRNA: biogenesis, function and role in cancer. Curr. Genomics 11: 537–561.

Maleszka, R. 2008. Epigenetic integration of environmental and genomic signals in honey bees: the critical interplay of nutritional, brain and reproductive networks. Epigenetics 3: 188–192.

Mandrioli, M. and F. Borsatti. 2007. Analysis of heterochromatic epigenetic markers in the holocentric chromosomes of the aphid *Acyrthosiphon pisum*. Chromosome Res. 15: 1015–1022.

Maynard Smith, J. 1978. The Evolution of Sex. Cambridge University Press, Cambridge.

Miura, T. 2005. Developmental regulation of caste-specific characters in social-insect polyphenism. Evol. Dev. 7(2): 122–129.

Moran, N.A. 1992. The evolution of aphid life cycles. Annu. Rev. Entomol. 37: 321–348.

Morehouse, N.I., N. Mandon, J.P. Christides, M. Body, G. Bimbard and J. Casas. 2013. Seasonal selection and resource dynamics in a seasonally polyphenic butterfly. J. Evol. Biol. 26: 175–185.

Mukherjee, K. and A. Vilcinskas. 2014. Development and immunity-related microRNAs of the lepidopteran model host *Galleria mellonella*. BMC Genomics 15: 705. doi: 10.1186/1471-2164-15-705.

Mukherjee, K., R.M. Twyman and A. Vilcinskas. 2015. Insects as models to study the epigenetic basis of disease. Prog. Biophys. Mol. Biol. doi: 10.1016/j.pbiomolbio.2015.02.009.

Müller, C.B., I.S. Williams and J. Hardie. 2001. The role of nutrition, crowding and interspecific interactions in the development of winged aphids. Ecol. Entomol. 26: 330–340.

Nijhout, H.F. 1999. Control mechanisms of phenotypic development in insects. Bioscience 49: 181–192.

Nijhout, H.F. 2003. Development and evolution of adaptive polyphenisms. Evol. Dev. 5: 9–18.

Ogawa, K. and T. Miura. 2014. Aphid polyphenisms: trans-generational developmental regulation through viviparity. Front. Physiol. 5: 1. doi: 10.3389/fphys.2014.00001.

Probst, A.V., E. Dunleavy and G. Almouzni. 2009. Epigenetic inheritance during the cell cycle. Nat. Rev. Mol. Cell Biol. 10: 192–206.

Rider, S.D., Jr., D.G. Srinivasan and R.S. Hilgarth. 2010. Chromatin-remodelling proteins of the pea aphid, *Acyrthosiphon pisum* (Harris). Insect. Mol. Biol. 2: 201–214.

Roff, D.A. and J.D. Fairbairn. 1991. Wing dimorphisms and the evolution of migratory polymorphisms among the Insecta. Am. Zool. 31: 243–251.

Schön, I., K. Martens and P. van Dijk. 2009. Lost Sex: The Evolutionary Biology of Parthenogenesis. Dordrecht: Springer. doi: 10.1007/978-90-481-2770-2.

Sim, C. and D.L. Denlinger. 2008. Insulin signaling and FOXO regulate the overwintering diapause of the mosquito *Culex pipiens*. Proc. Natl. Acad. Sci. USA 105: 6777–6781.

Simon, J.C., C. Rispe and P. Sunnucks. 2002. Ecology and evolution of sex in aphids. Trends Ecol. Evol. 17: 34–39.

Srinivasan, D.G. and J.A. Brisson. 2012. Aphids: a model for polyphenism and epigenetics. Genet. Res. Int. 2012: 431531. doi: 10.1155/2012/431531.

Strichman-Almashanu, L.Z., R.S. Lee, P.O. Onyango, E. Perlman, F. Flam, M.B. Frieman and A.P. Feinberg. 2002. A genome-wide screen for normally methylated human CpG islands that can identify novel imprinted genes. Genome Res. 12: 543–554.

Suzuki, M.M., A.R. Kerr, D. De Sousa and A. Bird. 2007. CpG methylation is targeted to transcription units in an invertebrate genome. Genome Res. 17: 625–631.

Tu, M.P., C.M. Yin and M. Tatar. 2005. Mutations in insulin signaling pathway alter juvenile hormone synthesis in *Drosophila melanogaster*. Gen. Comp. Endocrinol. 142: 347–356.

Waddington, C.H. 1942. The epigenotype. Endeavour 1: 18–20.

Walsh, T.K., J.A. Brisson, H.M. Robertson, K. Gordon, S. Jaubert-Possamai, D. Tagu and O.R. Edwards. 2010. A functional DNA methylation system in the pea aphid, *Acyrthosiphon pisum*. Insect Mol. Biol. 19: 215–228.

Weaver, D.B., J.M. Anzola, J.D. Evans, J.G. Reid, J.T. Reese, K.L. Childs, E.M. Zdobnov, M.P. Samanta, J. Miller and C.G. Elsik. 2007. Computational and transcriptional evidence for microRNAs in the honey bee genome. Genome Biology 8: R97.

Wei, Y., S. Chen, P. Yang, Z. Ma and L. Kang. 2009. Characterization and comparative profiling of the small RNA transcriptomes in two phases of locust. Genome Biology 10: R6.

West-Eberhard, M.J. 2003. Developmental Plasticity and Evolution. Oxford University Press, Oxford.

Williams, G.C. 1975. Sex and Evolution. Princeton University Press, Princeton.

Yao, I. 2012. Ant attendance reduces flight muscle and wing size in the aphid *Tuberculatus quercicola*. Biol. Lett. 8: 624–627.

Zera, A.J. 2003. The endocrine regulation of wing polymorphism in insects: state of the art, recent surprises, and future directions. Integr. Comp. Biol. 43: 607–616.

Zera, A.J., C. Strambi, K.C. Tiebel, A. Strambi and M.A. Rankin. 1989. Juvenile hormone and ecdysteroid titers during critical periods of wing morph determination in *Gryllus rubens*. J. Insect Physiol. 35: 501–511.

Zhou, V.W., A. Goren and B.E. Bernstein. 2011. Charting histone modifications and the functional organization of mammalian genomes. Nature Rev. Genet. 12: 7–18.

5

Bacterial Symbionts of Aphids (Hemiptera: Aphididae)

Marisa Skaljac

Introduction

Symbiosis is an intimate association between unrelated organisms, which has been essential in the evolutionary diversification of eukaryotes (Moran and Yun 2015). Insects frequently share long-term and stable relationships with heritable bacteria which has enabled them to become the most diverse and successful animals on earth (Dale and Moran 2006, Brownlie and Johnson 2009, Gibson and Hunter 2009). Most bacterial symbionts in insects cannot be cultivated independently and they may be present in the low titer in the host (Oliver et al. 2010). The symbionts may be difficult to cultivate because of their slow growth, microaerophilic lifestyle, requirement for specific host metabolites and/or the loss and inactivation of genes, which reduces their independence (Pontes and Dale 2006).

Microbial diversity and ecology can be studied without cultivation using the 16S ribosomal DNA (rDNA) sequencing approach (Klindworth et al. 2013, Vetrovsky and Baldrian 2013). The 16S rDNA is particularly suitable because it is ubiquitous in eukaryotes and only weakly affected by horizontal gene transfer, so conserved regions can be used to design taxon-specific primers and hybridization probes (Daubin et al. 2003). In addition, variable regions within the 16S rDNA can be used to classify microbial diversity (Vetrovsky and Baldrian 2013). The rapid development of next-generation sequencing technologies has therefore provided a new approach for the analysis of biodiversity, revealing the "rare biosphere" within different host organisms (Klindworth et al. 2013).

Fraunhofer Institute for Molecular Biology and Applied Ecology (IME), Bioresources Project Group, Winchesterstr. 2, 35394 Gießen, Germany.
E-mail: marisa.skaljac@ime.fraunhofer.de

Insects as Holobionts

Until recently, symbiosis research focused on individual host–symbiont relationships, but next-generation sequencing has revealed that animals and plants live in an association with hundreds of thousands of different microbial species (Rosenberg and Zilber-Rosenberg 2011). The genetic information contained within these numerous microorganisms often exceeds that of their hosts, and this diverse microbiota is now known to play a substantial role in the lives of many plants and animals (Zilber-Rosenberg and Rosenberg 2008, Rosenberg and Zilber-Rosenberg 2011).

The hologenome theory of evolution considers the holobiont (the host organism plus its entire consortium of microbes) as a unit of evolutionary selection (Rosenberg et al. 2007, Zilber-Rosenberg and Rosenberg 2008, Sharon et al. 2010). The hologenome is therefore the sum of genetic information in the host genome and all the associated microbial genomes. The best-characterized holobionts are insect–microbe systems (Guerrero et al. 2013). The hologenome theory is based on several principles: (1) all higher organisms are in a symbiotic relationship with microorganisms, (2) these microorganisms can be transmitted to the subsequent generation, (3) the association between the host and the symbiotic microbes influences the fitness of the holobiont within its environment, and (4) variation in the hologenome can be cause by changes in the host and microbial genomes. In addition to recombination and mutation within each species, genetic variation in holobionts can also involve microbial amplification, the acquisition of novel symbionts from the environment, and horizontal gene transfer between microbial species (Zilber-Rosenberg and Rosenberg 2008, Rosenberg and Zilber-Rosenberg 2011).

Environmental factors such as temperature, nutrient availability and disease can affect the relative abundance of different symbionts, causing certain species to proliferate and others to decline. The proliferation of microbes within the context of a hologenome is functionally equivalent to gene amplification and has a strong impact on adaptation. Higher organisms live in an environment inhabited with billions of microorganisms and occasionally some microbes can find a niche and become established in the host. Novel symbionts can therefore affect the phenotype of the holobiont and can also introduce new genes in this system. Furthermore, horizontal gene transfer can easily occur between bacteria, especially when they are abundant within the holobiont. This process is often mediated by transposons, plasmids, bacteriophages or genomic islands.

The holobiont is a therefore complex system that evolves by adaptive processes in order to increase fitness in its environment and increase the chances of passing acquired genetic traits to subsequent generations (Fig. 1). In addition, the inheritance of acquired traits by holobionts may help them to survive, reproduce and gain the time necessary for the host genome to evolve (Zilber-Rosenberg and Rosenberg 2008).

Bacterial Symbionts Associated with Insect Hosts

Although some bacterial symbionts are parasitic and reduce the fitness of their host, others provide their insect hosts with fitness benefits (Oliver et al. 2003,

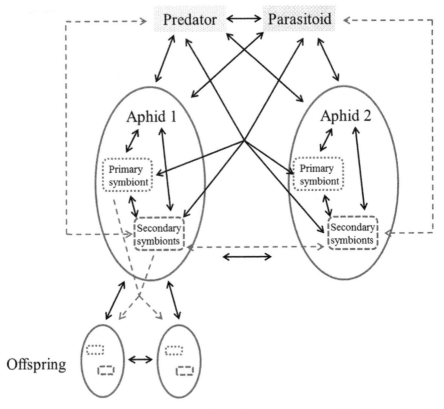

Fig. 1. The multi-trophic interactions within aphid–symbiont–enemy systems. The effects of bacterial symbionts extend beyond their aphid host to the natural enemies of aphids. Double arrows represent multiple interactions (e.g., aphid–symbiont, symbiont–symbiont, or predator–symbiont). Dashed arrows represent the vertical transmission of symbionts between mother (Aphid 1) and offspring, the horizontal transmission of secondary symbionts between two aphid individuals, or the horizontal transmission of secondary symbionts between aphids and their enemies.

Moran et al. 2008). In general, these symbionts can affect insect development, nutrition, reproduction, thermal tolerance, defense and immunity (Dale and Moran 2006, Dunbar et al. 2007, Tsuchida et al. 2011).

Primary or obligatory bacterial symbionts are essential for host survival. Primary symbionts are found in 10–15% of insect species, where they often supplement the host diet with amino acids or vitamins that not found in sufficient quantities in the food source (Baumann 2005, Ferrari and Vavre 2011). This enables the host to survive on nutritionally deficient diets (e.g., phloem sap) and thus in ecological niches which may otherwise be inaccessible. Primary symbionts are typically localized in organs known as bacteriosomes which consist of bacteriocytes. Depending on the host group, the bacteriosomes may consist of fat body cells, gut wall cells, or highly specialized cells already present in the embryo (Buchner 1965, Braendle et al. 2003).

Secondary or facultative bacterial symbionts are not essential, but they provide fitness-enhancing functions and are present in the majority of terrestrial arthropods

(Moran et al. 2008). Secondary symbionts may act as mutualists or reproductive manipulators (Stouthamer et al. 1999, Feldhaar 2011). They do not reside in specialized organs and instead inhabit a variety of cells and tissues, and may also colonize body cavities (e.g., gut) and hemolymph, with variable persistence (Feldhaar 2011). In insects with bacteriosomes, the secondary symbionts may invade the bacteriocytes and co-reside with or even displace the primary symbionts (Buchner 1965, Gottlieb et al. 2008). Secondary bacterial symbionts are often distributed in an irregular manner among host tissues and species (Skaljac et al. 2010, 2013).

The categorization of primary and secondary symbionts is not always certain because the same microbial species may range from obligate to facultative and from parasitic to mutualistic depending on the environment (Feldhaar 2011). Most bacterial symbionts, but all primary symbionts, are vertically transmitted from mother to offspring and the transmission rates are often near 100% (Ferrari and Vavre 2011). Secondary symbionts can also be transferred horizontally between individuals and between species (Russell et al. 2003). Primary symbionts may evolve from secondary symbionts by losing their potential for horizontal transmission (Moran et al. 2008). Bacterial symbionts that are strictly vertically transmitted must either increase the fitness of their host or manipulate host reproduction in ways that benefit their own transmission (Haine 2008). In many cases, secondary symbionts experimentally introduced (e.g., by microinjection) into previously uninfected hosts establish stable, maternally-inherited infections indicating that persistence is largely based on bacterial capabilities rather than host adaptations to maintain the infection (Moran et al. 2008).

The best-studied secondary bacterial symbiont is *Wolbachia pipientis*, which is widely distributed in arthropods and is usually associated with reproductive alterations. This makes it a promising target for disease/pest management strategies (Werren 1997, Stouthamer et al. 1999). Recent studies describe fascinating mutualistic relationships between *Wolbachia* spp. and their hosts, as a result of continuous genetic exchange between the host and multiple symbiont genomes present in this system (Saridaki and Bourtzis 2010, Nikoh et al. 2014).

Aphids—Models for the Investigation of Symbiosis

Aphids (Hemiptera, Sternorrhyncha, Aphidoidea) are notorious agricultural pests that cause damage to numerous plants by feeding on the phloem sap and vectoring phytoviruses (Will and Vilcinskas 2015). They are also biological models for the investigation of symbiosis, insect–plant interactions and virus transmission (The International Aphid Genomics Consortium 2010). Aphids are hosts for multiple bacterial symbionts, including *Buchnera aphidicola* which supplements their unbalanced diet with essential amino acids (Buchner 1965). Other bacterial symbionts may protect them from biological threats present in the environment (Oliver et al. 2010, 2014).

Aphids have a complex life cycle which can include several different phenotypes (e.g., asexual or sexual, wingless or winged), each perfectly adapted to specific ecological niches (The International Aphid Genomics Consortium 2010). During the spring and summer months, the life cycle of many aphids includes only asexual females which give birth to live clonal offspring, whereas shorter autumn days trigger the production of sexual wingless females and males. These produce overwintering eggs that yield a new generation of asexual females in the spring (The International Aphid Genomics Consortium 2010, Novakova et al. 2013). Other aphid species have lineages that reproduce permanently by parthenogenesis (Simon et al. 2011). Among almost 5000 aphid species, the pea aphid *Acyrthosiphon pisum* (Harris) was chosen for the first aphid genome sequencing project, because it is widely used in laboratory studies, attacks important crops and it is closely related to other deleterious pests such as the green peach aphid (*Myzus persicae*) (The International Aphid Genomics Consortium 2010).

In nature, aphids face diverse challenges such as parasitoid wasps, bacterial, viral and fungal pathogens, and predators (Gerardo et al. 2010, Costopoulos et al. 2014). Surprisingly, the pea aphid genome sequence revealed the absence of essential components of the immune system that are normally present in other insects. For example, it lacks genes for encoding antimicrobial peptides (AMPs) that target bacteria, and also components of the immunity-related immune deficient (IMD) pathway (Gerardo et al. 2010). Antibacterial activity was not induced in naïve pea aphids injected with bacteria, but several genes encoding thaumathins may confer protection against pathogenic fungi (Altincicek et al. 2008).

It is unclear whether aphids lost AMPs in order to protect their mutualistic bacterial symbionts, because AMPs are present in other insect–symbiont systems where their role is to control bacterial abundance (Login et al. 2011, Vilcinskas 2013). However, aphids may rely on cellular mechanisms to control their symbionts, given that macrophage migration inhibitory factors absent in other insects are present and differentially expressed in aphids (Laughton et al. 2011, Schmitz et al. 2012, Dubreuil et al. 2014).

Primary (obligatory) Bacterial Symbionts of Aphids

The pea aphid and its nutrient-providing symbiont *B. aphidicola* offer an excellent model of obligatory symbiosis. This relationship has been maintained by maternal transmission for 100–200 million years and neither of the associates can reproduce independently (Moran et al. 2008, Moran and Yun 2015). *Buchnera*-free aphids grow slowly, are sterile and often die (Tsuchida et al. 2005). The bacteria are localized in the aphid bacteriosome, which consists of 60–80 bacteriocytes that cannot survive in the hemocoel or in other cells. The abundance of *B. aphidicola* and the number of bacteriocytes decline when the aphid reaches adulthood (Nishikori et al. 2009). The AT-rich *Buchnera* genome is only 641 kb in length and encodes genes for the biosynthesis of several essential amino acids while lacking those for non-essential amino acids (Shigenobu et al. 2000). Several *Buchnera* genes of were found in the pea aphid genome, and some are strongly expressed in bacteriocytes suggesting they may regulate symbiosis with *Buchnera* (The International Aphid Genomics Consortium 2010). This indicates complementarity and syntrophy between the host and the primary symbiont. The *Buchnera* genome also lacks genes for the biosynthesis of cell-surface components, regulator genes and genes involved in cellular defense (Shigenobu et al. 2000).

Buchnera is vertically transmitted to the next generation by colonizing developing embryos in the female abdomen (Braendle et al. 2003, Koga et al. 2012, Moran and Yun 2015). Bacteriocytes are spatially close to ovarioles containing developing embryos, and transmission occurs when *Buchnera* cells are exocytosed from the bacteriocytes near the embryo (Koga et al. 2012) and then endocytosed by the posterior syncytial cytoplasm of embryo, before repackaging into new bacteriocytes (Moran and Yun 2015).

A single nucleotide deletion in the promoter of a *Buchnera* small heat shock protein gene causes a significant decline in bacterial numbers on exposure to heat, e.g., 4 h at 35°C (Dunbar et al. 2007). Moran and Yun (2015) managed to disrupt the long-lasting and stable relationship between *Buchnera* and the pea aphid by replacing the native, heat-sensitive primary symbiont with a heat-tolerant mutant. After successful replacement by heat shock and microinjection, the pea aphids also became much more heat tolerant, confirming that symbiont genotype can affect host ecology. The high mutation rate in the heat-tolerance allele allowed *Buchnera* to evolve to the heat-sensitive genotype in the most pea aphid lines under cool conditions (Dunbar et al. 2007).

Several studies provide insight into the evolutionary transition from secondary to primary symbiosis. The *Buchnera* BCc genome from the cedar aphid (*Cinara cedri*, subfamily Lachninae) is 416 kb in length, the smallest known genome from this bacterial species (Lamelas et al. 2011). Compared to the equivalent *Buchnera* strain in the pea aphid, *Buchnera* BCc has also lost its ability to synthesize tryptophan and riboflavin (vitamin B12) which abolishes its primary symbiotic role. *C. cedri* additionally carries the co-obligatory bacterial symbiont *Serratia symbiotica* SCc, whose genome sequence revealed its ability to carry out both tryptophan and riboflavin biosynthesis.

The comparison between genomes of *S. symbiotica* SCc and facultative *S. symbiotica* from the pea aphid, it was revealed that the former suffered reduction and that it is closer to a size of an obligate symbiont (e.g., *Buchnera* from the pea aphid) (Lamelas et al. 2011, Manzano-Marin and Latorre 2014).

Recently, the *S. symbiotica* SCt-VLC genome from the aphid *C. tujafilina* revealed a missing link in the transition between facultative *S. symbiotica* from the pea aphid and the co-obligatory *S. symbiotica* SCc from *C. cedri* (Manzano-Marin and Latorre 2014). *S. symbiotica* SCt-VLC can synthesize vitamins and cofactors, especially riboflavin, but not amino acids. In contrast, the *Buchnera* species that infect Lachninae such as *C. cedri* and *C. tujafilina* have lost the ability to synthesize riboflavin, thus the presence of *S. symbiotica* SCc and SCt-VLC is obligatory. However, genes for the biosynthesis of riboflavin are still present in *Buchnera* genome in the pea aphid (Manzano-Marin and Latorre 2014).

Secondary (facultative) Bacterial Symbionts of Aphids

Functions of Secondary Bacterial Symbionts in Aphids

There is growing evidence that aphids are superinfected with multiple facultative symbionts that have a significant impact on fitness (Table 1) (Oliver et al. 2010,

Table 1. Bacterial symbionts of aphids and their associated effect on the host.

	Bacterial symbionts	Aphid (genera and/or species) positive for symbionts	Effect on the host	References
Primary	*Buchnera aphidicola*	all	• Supplies nutrients	Shigenobu et al. 2000, Douglas 2003, Oliver et al. 2010
	Serratia symbiotica strains SCc, SCt-VLC (co-obligatory with *B. aphidicola*)	*Cinara cedri, Cinara tujafilina*		Lamelas et al. 2011, Augustinos et al. 2011, Manzano-Marin and Latorre 2014
Secondary	*Hamiltonella defensa*	*Acyrthosiphon pisum, Aphis craccivora, Aphis fabae, Periphyllus bulgaricus, Essigella californica, Eulachnus brevipilosus, Hormaphis cornu, Hyperomyzus lactucae, Hysteroneura setariae, Geopemphigus* sp., *Melaphis rhois, Schlechtendalia chinensis, Melanocallis caryaefoliae, Monellia caryella, Uroleucon, Macrosiphum, Pemphigus, Nippolachnus, Sitobion avenae, Sitobion fragariae*	• Protection against parasitoids and predators • Tolerance to heat stress • Host plant utilization • Production of alarm pheromones • Supplies nutrients	Russell et al. 2003, Russell and Moran 2006, Burke et al. 2009, Degnan et al. 2009, Oliver et al. 2010, 2012, 2014, Costopoulos et al. 2014
	Serratia symbiotica	*Acyrthosiphon pisum, Acyrthosiphon lactucae, Aphis craccivora, Aphis fabae, Periphyllus bulgaricus, Chaitophorus populifolii, Essigella californica, Hormaphis cornu, Macrosiphoniella helichrysi, Hysteroneura setariae, Geopemphigus* sp., *Melaphis rhois, Schlechtendalia chinensis, Smynthurodes betae, Melanocallis caryaefoliae, Monellia caryella, Panaphis juglandis, Tuberolachnus salignus, Trama troglodytes, Macrosiphum, Lachnus, Stomaphis, Uroleucon, Pemphigus, Periphyllus, Pterochloroides, Cinara*	• Protection against parasitoids and predators • Tolerance to heat stress • Host plant utilization • Supplies nutrients	Russell et al. 2003, Burke et al. 2009, Oliver et al. 2010, 2014, Brady et al. 2014, Costopoulos et al. 2014, Foray et al. 2014

	Bacterial symbionts	Aphid (genera and/or species) positive for symbionts	Effect on the host	References
Secondary	*Regiella insecticola*	*Acyrthosiphon pisum, Acyrthosiphon lactucae, Aphis craccivora, Aphis citricola, Aphis nerii, Colopha kansugei, Myzus persicae, Chaitophorus populeti, Essigella californica, Drepanosiphum oregonensis, Hormaphis cornu, Brachycaudus cardui, Hyste roneura setariae, Melaphis rhois, Schlechtendalia chinensis, Melanocallis caryaefoliae, Monellia caryella, Macrosiphum, Uroleucon, Pemphigus*	• Protection against fungal pathogens and parasitoids • Host plant utilization	Russell et al. 2003, Tsuchida et al. 2005, Vorburger et al. 2010, Lukasik et al. 2013
	Arsenophonus sp.	*Aphis craccivora, Aphis gossypii, Aphis glycines, Aphis idaei, Aphis ruborum, Aphis spiraecola, Melanaphis donacis, Stomaphis*	• Mostly unknown • Protection against parasitoids (?) • Host plant utilization	Moran et al. 2005, Burke et al. 2009, Jousselin et al. 2013, Wulff et al. 2013, Brady et al. 2014
	Sodalis sp.	*Eulachnus pallidnus, Eulachnus rileyi, Nippolachnus piri*	• Unknown	Burke et al. 2009
	Rickettsia sp.	*Acyrthosiphon pisum, Amphorophora rubi, Aphis gossypii, Sitobion miscanthi*	• Protection against fungal pathogens	Haynes et al. 2003, Oliver et al. 2010, Lukasik et al. 2013
	Wolbachia sp.	*Acyrthosiphon pisum, Cinara cedri, Aphis gossypii, Aphis nerii, Aphis fabae, Pentalonia caladii, Sitobion miscanthi, Tuberolachnus salignus, Maculolachnus submacula, Cavariella sp., Metopolophium dirhodum, Toxoptera citricida, Sipha maydis, Baizongia pistaciae, Neophyllaphis podocarpi*	• Mostly unknown • Supplements nutrients (?) • Reproductive manipulation (?)	Oliver et al. 2010, Augustinos et al. 2011, Brady et al. 2014
	Spiroplasma sp.	*Acyrthosiphon pisum*	• Protection against fungal pathogens • Reproductive manipulation	Oliver et al. 2010, Simon et al. 2011, Lukasik et al. 2013
	Candidatus Rickettsiella viridis	*Acyrthosiphon pisum*	• Protection against fungal pathogens and parasitoids • Modification of body colour	Tsuchida et al. 2010, 2014, Lukasik et al. 2013, Martinez et al. 2014
	Pea aphid X-type	*Acyrthosiphon pisum*	• Protection against parasitoids under fluctuating temperatures and host plant utilization when co-infected with *H. defensa* • Lower protection from predation	Guay et al. 2009, Oliver et al. 2014, Polin et al. 2014

2014). The pea aphid is infected with one or two facultative symbiont genera per individual among at least eight bacterial genera detected in this host (Oliver et al. 2010, Oliver et al. 2014). The aphid species in which the largest numbers of specimens have been examined for secondary bacterial infections are *Sitobion miscanthi*, *Aphis fabae*, *A. gossypii* and *A. craccivora* (Vorburger et al. 2009, Carletto et al. 2008, Najar-Rodríguez et al. 2009, Li et al. 2011, Jones et al. 2011, Brady et al. 2014, Wang et al. 2009). Some aphid species are comparable to the pea aphid in terms of symbiont heterogeneity and abundance (e.g., *A. craccivora*, *Microlophum carnosum*, *Sitobion avenae*), whereas others have less diverse microbial community (e.g., *Megoura crassicauda*) (Haynes et al. 2003, Brady et al. 2014, Tsuchida et al. 2006). The recorded symbiont diversity and abundance depends on the sampling method, but several studies suggest that many aphid species are infected with the same or similar bacterial symbionts (Burke et al. 2009, Brady et al. 2014). In hemipetran insects (e.g., aphids and whiteflies), the presence of similar symbionts often reflects horizontal transfer, possibly via common enemies or host plants (Caspi-Fluger et al. 2012, Gehrer and Vorburger 2012, Skaljac et al. 2013).

Moran et al. (2005) characterized the three members of the *Enterobacteriaceae* that infect a wide range of aphids, including the pea aphid, and names these species *S. symbiotica*, *Hamiltonella defensa* and *Regiella insecticola*. *H. defensa* is known to defend the pea aphid against the dominant parasitoid wasp *Aphidius ervi* by blocking the development of wasp eggs deposited in the aphid hemocoel, independent of aphid genotype (Oliver et al. 2003, 2005). *S. symbiotica* provides a lower level of protection against the same parasitoid wasp, but aphids carrying both symbionts are more protected than those carrying either individual species (Oliver et al. 2006). The closely-related species *R. insecticola* does not protect against parasitism. Furthermore, parasitized aphids carrying *H. defensa* alone produce more offspring, which also indicates the direct mutualistic benefits of this symbiont when the host population is under parasitoid attack (Oliver et al. 2006).

In the pea aphid, part of the protection conferred by *H. defensa* is due to a presence of the bacteriophage APSE (*A. pisum* secondary endosymbiont) which produces multiple toxins (Moran et al. 2005, Degnan and Moran 2008, Degnan et al. 2009, Oliver et al. 2009). There are three variants of this bacteriophage and the degree of protection is lowest with APSE 1 and highest with APSE 3 (Oliver et al. 2010). Pea aphids simultaneously carrying *H. defensa* and APSE 3 can spontaneously lose the bacteriophage, resulting is the loss of parasitoid resistance (Oliver et al. 2009).

Martinez et al. (2014) showed how the parasitoid wasp *A. ervi* challenges the ecology of nutritional and protective symbionts in the pea aphid in a system where resistance is dependent on the host genotype, the *H. defensa* strain and the associated APSE. When a susceptible host aphid is infected with a protective symbiont, parasitism tends to increase the replication of APSE but reduce the abundance of *H. defensa* as a result of bacteriophage-mediated lysis, whereas the number of *Buchnera* increased. This suggests that *H. defensa* or host factors inhibit the ability of parasitoids to affect *Buchnera*, which in turn creates a host environment suitable for larval development due to the accumulation of nutritional components (Martinez et al. 2014).

H. defensa infects approximately 14% of aphid species but its impact on parasitism has only been studied in a few of them (Oliver et al. 2010). One study showed that

H. defensa protects black bean aphids (*A. fabae*) from the parasitoid wasp *Lysiphlebus fabarum*, and another showed that *H. defensa* experimentally transferred from *A. craccivora* to *A. pisum* conferred protection against the parasitoid wasp *A. ervi* in the new host (Oliver et al. 2010, Schmid et al. 2012). The ability of *H. defensa* to invade new hosts reflects the presence of assorted virulence loci related to type III secretion systems that are also found in pathogenic bacteria such as *Salmonella* and *Yersinia* spp. (Degnan et al. 2009). The *H. defensa* genome suggests it can synthesize only two essential amino acids, seven non-essential amino acids, and most vitamins, the exceptions being thiamine (B1) and pantothenate (B5). Therefore, the successful persistence of *H. defensa* in the pea aphid requires the essential amino acids produced by *Buchnera* (Degnan et al. 2009).

As well as its protective role against parasitoids, *H. defensa* co-infected with *S. symbiotica* protects the pea aphid against a major predator, the ladybird beetle *Hippodamia convergens* (Costopoulos et al. 2014). The presence of *H. defensa* and *S. symbiotica* did not deter the predator feeding, but instead significantly reduced the survival of the predatory larvae, survival from egg hatching to pupation and survival to adult emergence. Furthermore, ladybird beetles consuming infected aphids were significantly heavier that those consuming uninfected aphids indicating that the two symbionts protect the entire aphid community by reducing the beetle population and therefore the risk of predation for clone mates (Costopoulos et al. 2014).

The natural enemies of aphids have evolved different counterstrategies as a response to symbiont-mediated protection (Oliver et al. 2014). The parasitoid wasp *A. ervi* can discriminate between naïve aphids and those infected with *H. defensa* in order to avoid hosts that are protected against the wasp larvae (Oliver et al. 2012). This ability may involve the lower level of alarm pheromones produced by aphids infected with *H. defensa* (Oliver et al. 2014). In addition, aphids infected with *H. defensa* are less aggressive and show weaker escape responses than uninfected aphids (Dion et al. 2011). In natural aphid populations, the frequency of *H. defensa* infection is moderate and usually does not become fixed (Vorburger et al. 2009, Russell et al. 2013, Ferrari et al. 2012). A moderate frequency of infection might be less costly for the aphid population when the pressure of parasitism is absent, whereas the number of infected individuals tends to increase when the aphid population is under the threat of parasitoid attack (Oliver et al. 2008).

As stated above, higher temperatures have a negative impact on the primary symbiont *Buchnera* and reduces the number of bacteriocytes, thus directly affecting host fitness (Dunbar et al. 2007). Some aphid species lose fitness at temperatures as low as 25–28°C, but the most are negatively affected at temperatures of 38–40°C (Oliver et al. 2010). Chen and Purcell (1997) found that pea aphids in the hot Central Valley of California were heavily infected (up to 80%) with the symbiont now known as *S. symbiotica*, providing early evidence that symbionts can confer heat tolerance on their host. Experimental pea aphid lines infected with *S. symbiotica* after exposure to 39°C for 4 h showed a 50% reduction in fecundity compared to aphids raised at normal temperatures, whereas uninfected aphids produced almost no offspring under the same heat-stress conditions (Chen et al. 2000, Montllor et al. 2002). Heat-treated young aphids infected with *S. symbiotica* also retained 70% of their bacteriocytes whereas uninfected heat-treated aphids retained only 7% of their bacteriocytes. These data suggest that *S. symbiotica* also protects the primary symbiont *Buchnera* from heat damage (Montllor et al. 2002).

Russell and Moran (2006) reported the beneficial impact of *S. symbiotica* on pea aphid survival, development and fecundity during heat stress. They also observed strain-dependent levels of protection. A higher proportion of infected aphids was observed in warmer seasons, but *S. symbiotica* strains from colder regions of the US such as New York conferred less protection from heat stress than strains from warmer regions such as Arizona (Russell and Moran 2006). The same study confirmed that *H. defensa* also protects against heat stress, but it only improves survival, not fecundity or development. In contrast to *S. symbiotica*, *R. insecticola* did not increase aphid fecundity or the number of bacteriocytes under heat stress (Oliver et al. 2010). Heat tolerance generally increased the density of aphid colonies, making them more susceptible to predation, and higher temperatures also reduced the protection of pea aphids against parasitoids conferred by *H. defensa* (Bensadia et al. 2006, Oliver et al. 2010). This highlights the complexity of environmental effects on aphid populations and their symbionts (Fig. 1).

Several studies have revealed a correlation between aphids, their host plants and secondary symbionts (Oliver et al. 2010, 2014). The bacterial symbionts may affect host plant choice by supplementing the aphid diet with nutrients or influencing their performance on the certain host plants (Oliver et al. 2010). However, there is a complex association between symbiont type/strain, aphid genotype and the host plant. Some genetically different pea aphid strains are adapted to host plants such as alfalfa (*Medicago sativa*) and red clover (*Trifolium pratense*) (Via 1999, Via et al. 2000). There is a correlation between the infection of pea aphids with *R. insecticola* and the choice of clover as a host plant, but it is important to emphasize that this depends mostly on the aphid genotype (Oliver et al. 2010, 2014, Russell et al. 2013). For example, Leonardo and Muiru (2003) and Tsuchida et al. (2004) have shown that *R. insecticola* confers fitness benefits on pea aphids feeding on white clover (*Trifolium repens*), whereas no such association was shown in another study carried out in California (Leonardo 2004). Several factors could influence the association between clover plants and infection with *R. insecticola* such as the protection conferred by this symbiont against the aphid-specific fungal pathogen *Pandora neophidis* which is often found on clover plants (Oliver et al. 2010). Furthermore, titers of *S. symbiotica* in pea aphids increase with declining levels of dietary nitrogen, which could expand the range of suitable host plants (Wilkinson et al. 2007). Another study reported the poor performance of *A. fabae* feeding on suboptimal and low-nitrogen host plants, combined with higher titers of *S. symbiotica* and *R. insecticola*, suggesting that bacterial symbionts can also restrict the host plant range (Chandler et al. 2008). Brady et al. (2014) found that *H. defensa* associated with the aphid *A. craccivora* was exclusively collected from *M. satvia*, whereas the same aphid species infected with bacterial symbiont *Arsenophonus* sp. and was collected mostly from locust *Robnia* sp. In addition, phylogenetic analysis also revealed different haplotypes of *A. craccivora* infecting these two host plants, suggesting that specialized aphid lineages become adapted to certain host plants (Brady et al. 2014).

Aphid populations are susceptible to fungal infections and, as mentioned above, one of the major fungal pathogens of aphids is *P. neophidis*. Ferrari et al. (2001) reported that pea aphid clones differed in terms of their resistance to *P. neophidis*, and later this was shown to correlate with the presence of *R. insecticola* (Ferrari et al. 2004). Lukasik et al. (2013) reported that *P. neophidis* resistance is conferred by *R. insecticola* and three other bacterial symbionts (representing the genera *Rickettsia*, *Rickettsiella* and *Spiroplasma*). Interestingly, these symbionts increased aphid survival and reduced the sporulation of the fungal pathogen, thus protecting the entire aphid community. The strongest protection was conferred by *Rickettsia* and *Rickettsiella*, whose symbiotic function was not previously understood (Tsuchida et al. 2010, Simon et al. 2007). *R. insecticola* and one of the three tested strains of *Spiroplasma* conferred only partial resistance to *P. neophidis*, whereas *H. defensa* had no effect on aphid survival or fungal sporulation (Lukasik et al. 2013). Parker et al. (2013) revealed that *R. insecticola* also confers protection against *Zoophthora occidentalis*, another fungal pathogen of aphids, but not against the general entomopathogenic fungus *Beauveria bassiana*. The strains that confer only partial protection nevertheless reduce the transmission of fungi by inhibiting and delaying sporulation, thus offering community-wide protection.

    R. insecticola was shown to confer nearly complete protection in *M. persicae* against parasitoids and the same effect was achieved when it was injected into *A. fabae* (Vorburger et al. 2010). However, the strain of *R. insecticola* used in these experiments was found to be genetically distinct from those that can protect pea aphid against *P. neophidis* following the comparative sequencing of the protective 5.15 and non-protective LSR1 strains (Hansen et al. 2012). The protective *R. insecticola* 5.15 genome encodes several categories of pathogenicity factors (e.g., the O-antigen biosynthesis pathway, a type 1 secretion system and RTX toxins, a type 3 secretion system and effectors, hemin transport, and the two component system PhoPQ). These factors were missing or inactivated in the *R. insecticola* LSR1 genome and influence the virulence of the symbiont against parasitoids (Hansen et al. 2012). Furthermore, *R. insecticola* 5.15 also conferred a general protective phenotype against *Aphidius* parasitoids in different aphid host species. Hansen et al. (2012) did not detect the presence of bacteriophage in *R. insecticola* 5.15, which is a part of protective phenotype in *H. defensa*. It is not clear why eukaryotic pathogenicity factors, present in protective symbiont genomes, selectively target and kill parasitoid wasps but not the aphid host. The bacterial pathogens may achieve virulence by disabling, resisting or interfering with the host innate immune system, and *R. insecticola* 5.15 may interact differentially with the wasp and aphid immune system components party reflecting the differential regulation of pathogenicity-related factors in different host cells (Hansen et al. 2012).

    Several genera of symbiotic bacteria found in aphids have not yet been assigned a function, including *Wolbachia*, *Arsenophonus* and *Sodalis* (Table 1). Despite the frequency of infections in arthropods generally, *Wolbachia* spp. are rare in aphids except *C. cedri* (Oliver et al. 2010). Augustinos et al. (2011) suggested that the presence of *Wolbachia* spp. has been overlooked due to the diverse species but overall low titer in aphids. They tested for *Wolbachia* infection among 153 aphid species collected across Mediterranean and Middle-Eastern countries varied significantly, and found 18 aphid species carrying *Wolbachia* that were not previously known to be infected with this symbiont (Table 1). The frequency of *Wolbachia* infection was highest in *C. cedri*, which has a predominantly parthenogenetic lifestyle, suggesting *Wolbachia* may increase the prevalence of asexual lineages (Augustinos et al. 2011). *Wolbachia* is best known for its ability to induce reproductive alterations such as parthenogenesis, cytoplasmic incompatibility, male killing and feminization (Saridaki and Bourtzis 2010). *Wolbachia* may also fulfil a nutritional function in *C. cedri* as it does in the bedbug *Cimex lectularius* (Hosokawa et al. 2010, Nikoh et al. 2014). Furthermore, *Wolbachia* may promote resistance against RNA viruses in several *Drosophila* species and it can reduce infection by *West Nile virus*, *Dengue virus* and *Chikungunya virus* in mosquitoes (Oliver et al. 2014).

The genus *Arsenophonus* is considered one of the most diverse symbiotic lineages and is widespread among different insect taxa (Wilkes et al. 2010, Duron et al. 2010, Skaljac et al. 2010, 2013), but it is rare in aphids (Oliver et al. 2010). Jousselin et al. (2013) conducted extensive screening for the presence and diversity of *Arsenophonus* in 86 aphid species, revealing that it was present in 7% of the tested species, particularly the genus *Aphis*. Based on phylogenetic analysis, the diversity of *Arsenophonus* was shown to be influenced by vertical and horizontal transmission. Genetically identical strains were present not only in different aphid species, but also in more distantly-related insects, such as the whitefly *Bemisia tabaci* (Jousselin et al. 2013, Mouton et al. 2012). Furthermore, one clade of *Arsenophonus* (C) in aphids was showed to be phylogenetically close to *Arsenophonus* found in planthoppers and the plants on which they feed (Zreik et al. 1998, Bressan et al. 2008). A similar case was *B. tabaci* and its associated symbiont *Rickettsia,* which was also found in the phloem of its host plants (Caspi-Fluger et al. 2012). These studies suggest that plants serve as a reservoir for symbiont horizontal transmission thus explaining the presence of genetically similar *Arsenophonus* strains in unrelated phloem-feeding insect pests (Jousselin et al. 2013). The functions of *Arsenophonus* species in aphids are mostly unknown, but in other insects this bacterium can manipulate the reproduction of parasitoid wasps, provide nutritional functions in haematophagous insects, participate in virus transmission in whiteflies, etc. (Dale et al. 2006, Hansen et al. 2007, Duron et al. 2010, Rana et al. 2012, Bressan et al. 2008). Because *Wolbachia* and *Arsenophonus* show both mutualistic and parasitic phenotypes (e.g., reproductive manipulation) in many insect hosts, these symbionts may use both tactics to achieve a high frequency of infection in host populations (Oliver et al. 2014).

Most studies have focused on the biological roles of bacterial symbionts in aphids during the parthenogenetic life cycle, whereas the impact on sexual reproduction has not been investigated in detail (Simon et al. 2011). These missing data would provide further insight into the roles of aphid symbionts given that some of the abovementioned bacteria (e.g., *Wolbachia, Arsenophonus, Rickettsia* and *Spiroplasma*) are known to cause reproductive modifications in other arthropods (Engelstädter and Hurst 2009). Simon et al. (2011) investigated whether secondary symbionts affected the production of sexual forms in different pea aphid genetic backgrounds, including natural populations. *Spiroplasma* conferred a strong male-killing phenotype at the early nymphal stage, whereas *Hamiltonella, Regiella, Rickettsia, Serratia* and *Ricketsiella* did not affect the reproductive mode of any of the aphid strains. Previous studies have reported phylogenetically similar male-killing *Spiroplasma* from ladybird beetles, which are major aphid predators (Fukatsu et al. 2001). This may indicate the horizontal transmission of *Spiroplasma* between the predator and aphid prey (Chiel et al. 2009). *Spiroplasma* infection frequencies in natural aphid populations range from 0% to 40%

and tend to be higher on perennial than annual host plants (Hurst and Jiggins 2000, Frantz et al. 2009). The breeding of the aphid clone mates significantly reduces their fitness, so male-killing *Spiroplasma* in the pea aphid may help to reduce the frequency of such behavior (Simon et al. 2011).

Guay et al. (2009) reported the presence of a pea aphid X-type symbiont (PAXS) that was often concurrent with *H. defensa* infections in natural populations. This symbiotic combination proved to be highly beneficial for the aphid host. The association between PAXS and *H. defensa* conferred strong and early resistance to the parasitoid wasp *A. ervi*, but PAXS also conferred heat tolerance (unlike *H. defensa* acting alone) allowing the pea aphid to remain highly resistant against parasitoids even under heat stress (Guay et al. 2009). PAXS may protect *H. defensa* from the heat stress in the same way that *S. symbiotica* protects bacteriocytes containing *Buchnera* (Montllor et al. 2002). Guay et al. (2009) also found that UV-B stress had a minor impact on aphid fecundity, and no effect on resistance to parasitoids, regardless of the aphid or symbiont strain.

The secondary symbiont *Rickettsiella* "*Candidatus* Rickettsiella viridis" infects European and North American pea aphids, but rarely those from Japan (Tsuchida et al. 2010, 2014). Interestingly, *Rickettsiella* changes the aphid body color from red to green, which may affect selection by predators and therefore the evolutionary ecology of the aphids (Tsuchida et al. 2010). Ladybird beetles prefer to consume red aphids, whereas the parasitoid wasp *A. ervi* preferentially attacks green aphids (Bilodeau et al. 2013, Losey et al. 1997, Libbrecht et al. 2007). The red color arises from the accumulation of carotenoid pigments, whereas green is attributed to polycyclic quinone pigments that tend to accumulate in aphids infected with *Rickettsiella* (Tsuchida et al. 2010). Aphids infected with *Rickettsiella* retain normal fitness whereas aphids co-infected with *Rickettsiella* and *H. defensa* suffer a fitness penalty (Tsuchida et al. 2014). Such co-infections are common in nature and they intensify the green color of the aphid body (Tsuchida et al. 2010, 2014, Russell et al. 2013). The presence of both symbionts may be more beneficial than individual infections because *H. defensa* confers protection against parasitoids while *Rickettsiella* simultaneously protects against fungi (Lukasik et al. 2013, Tsuchida et al. 2014). Such combination-dependent increases in fitness have previously been reported for superinfections involving *S. symbiotica* and *H. defensa,* or with PAXS and *H. defensa* (Oliver et al. 2014).

Infections with Multiple Strains of Secondary Symbionts

Most secondary bacterial symbionts exist as multiple strains within the same aphid population, but there has been little effort to investigate this strain diversity (Russell et al. 2013). Interestingly, the same symbiont may exist as strains which confer a range of phenotypes on the host, including protective and non-protective strains that have different effects on host fitness (Oliver et al. 2014). Protective symbionts can also vary in the strength of protection against particular genotypes of parasitoids. Such bacterial diversity may allow the host to adapt to selective pressure from natural enemies and also contributes to the diversity of the aphid cytoplasmic genotypes (Russell et al. 2013).

The biggest strain diversity was observed for *H. defensa* and the associated APSE in the pea aphid, but also other aphids such as *Sitobion avenae* and *S. fragariae* (Russell et al. 2013). In the pea aphid and *A. fabae*, several strains of *H. defensa* conferred different levels of protection against the parasitoid wasp *A. ervi*, but were also associated with different fitness costs to the host (Martinez et al. 2014, Cayetano et al. 2015). Brady et al. (2014) detected multiple strains for four of the six secondary symbionts infecting *A. craccivora*: two strains each for *Arsenophonus*, *H. defensa* and *S. symbiotica*, and three strains of *Rickettsia*. Two clades of *Arsenophonus* (A and C) were found in several aphid species and each clade contained different strains of the symbiont (Jousselin et al. 2013). The large-scale screening of *Wolbachia* in different aphid species revealed the presence of bacterial strains belonging to the known supergroups A and B as well as two new supergroups (Augustinos et al. 2011). Lukasik et al. (2013) showed that one strain of *Spiroplasma* reduced the sporulation and virulence of a fungal pathogen even though *Spiroplasma* was previously thought solely to affect aphid reproduction (Simon et al. 2011).

The diversity of symbionts in pea aphids may be maintained by balancing selection among different symbiont species, strains and superinfections conferring advantages under a limited range of environmental conditions (Russell et al. 2013). Furthermore, symbiont-driven co-evolution in this system may reflect the presence of the multiple symbiont strains with defensive roles, the existence of genetic variation for parasitoid virulence, the emergence of parasitoids that are resistant against protection conferred by *H. defensa*, and the likelihood of strain-enemy specificity (Oliver et al. 2014).

Transmission of Bacterial Symbionts in Aphids

As previously mentioned, the primary symbionts of aphids are strictly maternally transmitted whereas secondary symbionts may be transmitted vertically and/or horizontally between individuals and among species (Russell et al. 2003). Rare horizontal transmission events are necessary for the spread of symbionts to novel host lineages (Chiel et al. 2009). This could explain the extremely high diversity of secondary symbionts in aphids. The horizontal transmission mechanisms in aphids are not completely understood, although transmission through shared host plants, natural enemies, mating and cannibalism have been reported (Moran and Dunbar 2006, Chiel et al. 2009, Caspi-Fluger et al. 2012). Phylogenetic analysis indicates that the same secondary symbionts can be found in distantly related hosts (Russell et al. 2003). Sometimes, in superinfected aphids, the failure of symbiont transmission stabilizes single or double infections (Moran and Dunbar 2006, Oliver et al. 2006).

R. insecticola and *H. defensa* are normally paternally transmitted in aphids (Moran and Dunbar 2006). However, successful horizontal transmission by the microinjection of hemolymph has been achieved for *R. insecticola*, *H. defensa*, *S. symbiotica*, *Rickettsia*, *Arsenophonus* and *Spiroplasma* in several aphid species (Oliver et al. 2010, Simon et al. 2011). *R. insecticola* and *H. defensa* are also horizontally transmitted via parasitoids in *A. fabae*, where the parasitoid acts as a 'dirty needle' for the transfer of bacteria (Gehrer and Vorburger 2012). Persistent infection was also established by feeding pea aphids on an artificial diet containing *H. defensa* (Darby and Douglas 2003). Interspecies symbiont transfer by microinjection has met with mixed results. Sometimes, the transmission of secondary symbionts between distantly related aphid species has been successful (e.g., between *A. craccivora* or *Myzocallis* and *A. pisum*), whereas other cases even transfer between closely related species such as *Acyrthosiphon kondoi* and *A. pisum* was not possible (Oliver et al. 2010).

Several studies have shown that symbionts can be transmitted via plants, but this is rarely the case for aphids (Moran and Dunbar 2006, Oliver et al. 2010, Caspi-Fluger et al. 2012). Genetically identical strains of *Arsenophonus* were recently found in conspecific aphids from geographically distant regions, which suggests that successful vertical and horizontal transmission caused the widespread distribution of this symbiont across different aphid species (Oliver et al. 2010, Jousselin et al. 2013). Guldemond et al. (1994) reported that some male aphids attempted to mate with sexual females from related species, which could lead to the transmission of symbionts without reproduction. The horizontal acquisition of novel symbionts can instantly transform host ecological traits (Oliver et al. 2014, Moran and Yun 2015). For example, Simon et al. (2011) reported the strong and direct negative impact of transfected symbionts (e.g., *H. defensa* and *Spiroplasma*) on the fitness and reproduction of previously uninfected pea aphid populations during the sexual reproduction phase.

Costs of Living in a Symbiotic Relationship

An individual aphid represents a habitat with limited resources, which may lead to competition between symbionts and therefore significant fitness costs for the host (Oliver et al. 2006). So called 'direct costs' arise from a trade-off between allocating resources to protective symbiosis rather than traits such as survival and reproduction (Polin et al. 2014). Further ecological costs may arise if symbiosis negatively affects interactions between the host and other organisms in the environment.

Host–symbiont interactions require on a delicate balance that ensures the fitness of both partners (Laughton et al. 2014). Furthermore, the host must maintain control of the symbiont populations and little is known about the differential control of primary and secondary symbionts. In some insect–symbiont systems, bacteriocytes and sheath cells contain high concentrations of AMPs and lysozymes, which could help to control the population of symbionts (Nakabachi et al. 2005, Login et al. 2011). Aphids may control their symbionts using the phagocytic activity of their immune cells (hemocytes), although the regulation of this process is poorly understood (Schmitz et al. 2012, Laughton et al. 2014). A host may need to trade-off between the control of symbiont populations and investment in development, survival and reproduction when faced with limited resources (Laughton et al. 2014).

H. defensa is a mutualist in aphids but even this kind of defensive symbiosis has its costs for the host, such as lower competitive ability, reduced lifespan and more limited reproduction (Oliver et al. 2008, Simon et al. 2011, Vorburger and Gouskov 2011, Cayetano and Vorburger 2013). The cost and benefits of defensive symbiosis were recently compared in *A. fabae* infected with *H. defensa* isolates conferring variable levels of resistance to the parasitoid *L. fabarum* (Cayetano et al. 2015). Remarkably, this study revealed that strongly protective isolates of *H. defensa* have a lower cost than weaker isolates in terms of host lifespan and lifetime reproduction, although the underlying mechanism is unclear. It would be interesting to determine whether these more costly but less protective symbionts affect additional ecological features (Cayetano et al. 2015).

In nature, pea aphids are rarely co-infected with *H. defensa* and *S. symbiotica* even though this combination of symbionts confers greater resistance to parasitoids than individual infections (Oliver et al. 2006). The greater resistance may reflect the high density of *S. symbiotica* in this system, whereas the density of *H. defensa* is similar in single or double infections. Infection with these symbionts causes severe fitness costs, such as low fecundity, longer generation times and lower weight at adulthood (Oliver et al. 2006). Laughton et al. (2014) measured the fitness costs of *H. defensa*, *S. symbiotica* and *R. insecticola* in the pea aphid over time. The densities of secondary symbiont populations increased with the host age, perhaps reflecting the rupture of cells housing these symbionts. Bacteriocytes and sheath cells become more fragile as the host ages, and the latter are rarely seen in old aphids (Douglas and Dixon 1987, Nishikori et al. 2009). Laughton et al. (2014) showed that *S. symbiotica* conferred the highest fitness costs on pea aphid survival, fecundity and development. *H. defensa* infections were also costly but variable whereas *R. insecticola* showed lower levels of virulence and the fitness costs were also lower.

As stated earlier, aphids infected with *H. defensa* are less aggressive and show less intense escape responses (Dion et al. 2011). Polin et al. (2014) reported the same behavior in pea aphids infected with *H. defensa* alone or in a combination with PAXS, which resulted in higher predation by ladybird beetles. This suggests that protective symbiosis results in associated ecological costs, but the underlying mechanism is not known.

Spatial Localization of Secondary Bacterial Symbionts in Aphids

The localization of bacterial symbionts in host tissues helps to determine their functions and mechanisms of transmission as well as providing insight into their interactions with the host and with other symbionts in the same host (Koga et al. 2009, Kliot et al. 2014). Fluorescence *in situ* hybridization (FISH) targeting rDNA sequences is a powerful procedure for the detection and analysis of symbionts without cultivation (Fig. 2B, C) (Kliot et al. 2014, Skaljac et al. 2010, 2013).

Buchnera is consistently found in the primary bacteriocytes of the pea aphid, whereas secondary bacterial symbionts usually reside in the secondary bacteriocytes, sheath cells and hemolymph (Fig. 2; Table 2) (Oliver et al. 2010). Secondary bacteriocytes comprise a small number of large cells that are intercalated between the primary bacteriocytes, whereas sheath cells are small and flat cells located at periphery of the primary bacteriocytes (Moran et al. 2005). *H. defensa*, *S. symbiotica*, *R. insecticola* and *Rickettsiella* are localized in the cytoplasm of secondary bacteriocytes and sheath cells, but they are also found in the hemolymph of the pea aphid (Fukatsu et al. 2000, Moran et al. 2005, Tsuchida et al. 2005). *Rickettsiella* is the only symbiont localized in other tissues, including oenocytes, ovariole pedicels and the posterior ovary (Tsuchida et al. 2014). *H. defensa* and *Rickettsiella* often co-exist in the secondary bacteriocytes, sheath cells and hemolymph.

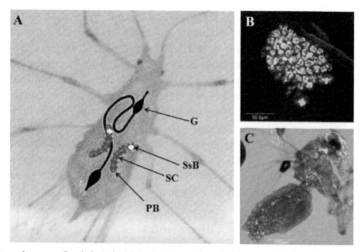

Fig. 2. Internal organs often infected with bacterial symbionts in insects. (A) Schematic illustration of the aphid intestinal system and a pair of bacteriosomes with uninucleate primary bacteriocytes (PB, grey), syncytial secondary bacteriocytes (SsB, yellow) and tiny sheath cells (SC, orange). (B) Fluorescent *in situ* hybridization (FISH) showing a bacteriosome co-infected with primary (red, cy3) and secondary (green, cy5) bacterial symbionts (Skaljac et al. 2010). (C) FISH of the primary symbiont (red, cy3) located in the bacteriosome, and a secondary symbiont (green, cy5) infecting various tissues in the whitefly (Skaljac et al. 2013).

Table 2. Classification of bacterial symbionts, with alternative names and tissue localization in aphid hosts.

	Bacterial symbionts of aphids	Bacterial classification (Class/Family)	Alternative names	Tissue localization of symbionts in aphids
Primary	*Buchnera aphidicola*	Gammaproteobacteria/ Enterobacteriaceae	/	Primary bacteriocytes
Secondary	*Hamiltonella defensa*		PABS, T-type	Secondary bacteriocytes, sheath cells, hemolymph
	Serratia symbiotica		S-symbiont, PASS, R-type	
	Regiella insecticola		PAUS, U-type	
	Arsenophonus sp.		/	Unknown
	Sodalis sp.		/	Unknown
	Pea aphid X-type		X-type, PAXS	Unknown
	"*Candidatus Rickettsiella viridis*"	Gammaproteobacteria/ Legionellaceae	/	Secondary bacteriocytes, sheath cells, hemolymph, oenocytes, ovariole pedicels, posterior ovary
	Rickettsia sp.	Alphaproteobacteria/ Rickettsiaceae	PAR	Secondary bacteriocytes, sheath cells, hemolymph
	Wolbachia sp.	Alphaproteobacteria/ Anaplasmataceae	/	Secondary bacteriocytes,
	Spiroplasma sp.	Mollicutes/ Spiroplasmataceae	/	Hemolymph

Abbreviations: PABS, pea aphid *Bemisia*-like symbiont; PAR, pea aphid rickettsia; PASS, pea aphid secondary symbiont; PAUS, pea aphid U-type symbiont; PAXS, pea aphid X-type symbiont.

In the pea aphid, *Rickettsia* is found with the same distribution as *H. defensa, S. symbiotica* and *R. insecticola* suggesting common mechanisms for infection and maintenance in the host (Sakurai et al. 2005). Because the bacteriosome cells are found close together, biological interactions such as competition for resources, energy and space may occur between primary and secondary symbionts. Several studies have shown that *Buchnera* is inhibited in the presence of secondary symbionts such as *Rickettsia* and *S. symbiotica* (Sakurai et al. 2005, Oliver et al. 2010). Furthermore, Gómez-Valero et al. (2004) reported that *Wolbachia* is co-localized with *S. symbiotica* in the secondary bacteriocytes of the cedar aphid, whereas *Spiroplasma* was only found in the hemolymph (Fukatsu et al. 2001). The localization of secondary bacterial symbionts in aphids can also vary over time, and additional tissues can be infected as symbionts proliferate during host development (Oliver et al. 2010).

Conclusions

Microorganisms play a crucial role in various organisms, including insects, zebrafish, mice and human and colonize every surface that is exposed to the external environment (Ley et al. 2008, Sekirov et al. 2010). The importance of microbes is highlighted by international projects such as the NIH Human Microbiome Project and the Earth Microbiome Project which seek a better understanding of the roles played by microbiota (Gilbert et al. 2014, The NIH HMP Working Group et al. 2009). These project also aim to develop strategies for the manipulation of microbiota to improve human health and benefit ecosystems (Turnbaugh et al. 2007).

Insects have less diverse microbiota than mammals and this makes them ideal models for studying the function of bacterial symbionts and host–symbiont interactions (Douglas 2011). Furthermore, symbiosis can be used as a powerful tool for the management of pest species, by disrupting and manipulating symbionts with a significant impact on insect traits (Douglas 2007).

The pea aphid is the best-characterized model of multiple heritable symbionts and most of the general principles of insect–symbiont relationships have been inferred from this particular system (Oliver et al. 2010, 2014). More studies are necessary to determine the phenotypic effects of secondary bacterial symbionts on other aphid species, as recently shown for *A. fabae, M. persicae* and several other species (Cayetano et al. 2015, Vorburger et al. 2010, Oliver et al. 2010). Sometimes, the same bacterial symbionts are present in unrelated insect species and the effects of defensive symbionts may extend beyond their insect hosts and enemies to natural communities.

Recent studies show that aphid–symbiont relationships are strongly affected by the host and symbiont genotypes, host-symbiont and symbiont-symbiont interactions. Another important challenge is to understand how a diverse community of multiple bacterial species can evolve and persist in the same host, despite competition for host resources (Lukasik et al. 2013).

Sequencing technologies are becoming more sensitive and less expensive, offering the potential to discover new and unknown bacterial symbionts in aphids, as well as other potentially widespread symbionts such as yeasts and fungi (Feldhaar 2011). The sequenced genomes of secondary symbionts together with ecological, evolutionary and physiological studies will increase our understanding of the interactions between multiple microbial symbionts and their host insects.

Acknowledgements

I would like to thank Andreas Vilcinskas for the invitation and support in writing this book chapter, to *Richard M. Twyman* for *editing the chapter,* and to LOEWE Center for Insect Biotechnology & Bioresources (Giessen, Germany) for valuable help.

Keywords: aphids, primary symbionts, secondary symbionts, holobiont, multitrophic interactions

References

Altincicek, B., J. Gross and A. Vilcinskas. 2008. Wounding-mediated gene expression and accelerated viviparous reproduction of the pea aphid *Acyrthosiphon pisum*. Insect Mol. Biol. 17(6): 711–716.

Augustinos, A.A., D. Santos-Garcia, E. Dionyssopoulou, M. Moreira, A. Papapanagiotou, M. Scarvelakis, V. Doudoumis, S. Ramos, A.F. Aguiar, P.A.V. Borges, M. Khadem, A. Latorre, G. Tsiamis and K. Bourtzis. 2011. Detection and characterization of *Wolbachia* infections in natural populations of aphids: is the hidden diversity fully unraveled? PLoS One 6(12): e28695.

Baumann, P. 2005. Biology bacteriocyte-associated endosymbionts of plant sap-sucking insects. Annu. Rev. Microbiol. 59: 155–189.

Bensadia, F., S. Boudreault, J.F. Guay, D. Michaud and C. Cloutier. 2006. Aphid clonal resistance to a parasitoid fails under heat stress. J. Insect Physiol. 52(2): 146–157.

Bilodeau, E., J.F. Guay, J. Turgeon and C. Cloutier. 2013. Survival to parasitoids in an insect hosting defensive symbionts: a multivariate approach to polymorphic traits affecting host use by its natural enemy. PLoS One 8(4): e60708.

Brady, C.M., M.K. Asplen, N. Desneux, G.H. Heimpel, K.R. Hopper and C.R. Linnen. 2014. Worldwide populations of the aphid *Aphis craccivora* are infected with diverse facultative bacterial symbionts. Microb. Ecol. 67(1): 195–204.

Braendle, C., T. Miura, R. Bickel, A.W. Shingleton, S. Kambhampati and D.L. Stern. 2003. Developmental origin and evolution of bacteriocytes in the aphid-*Buchnera* symbiosis. PLoS Biol. 1(1): E21.

Bressan, A., O. Semetey, B. Nusillard, D. Clair and E. Boudon-Padieu. 2008. Insect vectors (Hemiptera: Cixiidae) and pathogens associated with the disease syndrome 'basses richesses' of sugar beet in France. Plant Disease 92: 113–119.

Brownlie, J.C. and K.N. Johnson. 2009. Symbiont-mediated protection in insect hosts. Trends Microbiol. 17(8): 348–354.

Buchner, P. 1965. Endosymbiosis of Animals with Plant Microorganisms. John Wiley, New York.

Burke, G.R., B.B. Normark, C. Favret and N.A. Moran. 2009. Evolution and diversity of facultative symbionts from the aphid subfamily Lachninae. Appl. Environ. Microbiol. 75(16): 5328–5335.

Carletto, J., G. Gueguen, F. Fleury and F. Vanlerberghe-Masutti. 2008. Screening the bacterial endosymbiotic community of sap-feeding insects by terminal-restriction fragment length polymorphism analysis. Entomologia Experimentalis et Applicata 129(2): 228–234.

Caspi-Fluger, A., M. Inbar, N. Mozes-Daube, N. Katzir, V. Portnoy and E. Belausov. 2012. Horizontal transmission of the insect symbiont *Rickettsia* is plant-mediated. Proc. Biol. Sci. 279(1734): 1791–1796.

Cayetano, L. and C. Vorburger. 2013. Genotype-by-genotype specificity remains robust to average temperature variation in an aphid/endosymbiont/parasitoid system. J. Evol. Biol. 26(7): 1603–1610.

Cayetano, L., L. Rothacher, J.-C. Simon and C. Vorburger. 2015. Cheaper is not always worse: strongly protective isolates of a defensive symbiont are less costly to the aphid host. Proc. Biol. Sci. 282(1799): 20142333.

Chandler, S.M., T.L. Wilkinson and A.E. Douglas. 2008. Impact of plant nutrients on the relationship between a herbivorous insect and its symbiotic bacteria. Proc. Biol. Sci. 275(1634): 565–570.

Chen, D.-Q. and A.H. Purcell. 1997. Occurrence and transmission of facultative endosymbionts in aphids. Curr. Microbiol. 34: 220–225.

Chen, D.-Q., C.B. Montllor and A.H. Purcell. 2000. Fitness effects of two facultative endosymbiotic bacteria on the pea aphid, *Acyrthosiphon pisum*, and the blue alfalfa aphid, *A. kondoi*. Entomol. Exp. Appl. 95(3): 315–323.

Chiel, E., E. Zchori-Fein, M. Inbar, Y. Gottlieb, T. Adachi-Hagimori, S.E. Kelly, M.K. Asplen and M.S. Hunter. 2009. Almost there: transmission routes of bacterial symbionts between trophic levels. PLoS One 4(3): e4767.

Costopoulos, K., J.L. Kovacs, A. Kamins and N.M. Gerardo. 2014. Aphid facultative symbionts reduce survival of the predatory lady beetle *Hippodamia convergens*. BMC Ecol. 14: 5.

Dale, C. and N.A. Moran. 2006. Molecular interactions between bacterial symbionts and their hosts. Cell 126(3): 453–465.

Dale, C., M. Beeton, C. Harbison, T. Jones and M. Pontes. 2006. Isolation, pure culture, and characterization of "*Candidatus Arsenophonus arthropodicus*", an intracellular secondary endosymbiont from the hippoboscid louse fly *Pseudolynchia canariensis*. Appl. Environ. Microbiol. 72(4): 2997–3004.

Darby, A.C. and A.E. Douglas. 2003. Elucidation of the transmission patterns of an insect-borne bacterium. Appl. Environ. Microbiol. 69(8): 4403–4407.

Daubin, V., N.A. Moran and H. Ochman. 2003. Phylogenetics and the cohesion of bacterial genomes. Science 301(5634): 829–832.

Degnan, P.H. and N.A. Moran. 2008. Diverse phage-encoded toxins in a protective insect endosymbiont. Appl. Environ. Microbiol. 74(21): 6782–6791.

Degnan, P.H., Y. Yu, N. Sisneros, R.A. Wing and N.A. Moran. 2009. *Hamiltonella defensa*, genome evolution of protective bacterial endosymbiont from pathogenic ancestors. Proc. Natl. Acad. Sci. USA 106(22): 9063–9068.

Dion, E., S.E. Polin, J.C. Simon and Y. Outreman. 2011. Symbiont infection affects aphid defensive behaviours. Biol. Lett. 7(5): 743–746.

Douglas, A.E. 2007. Symbiotic microorganisms: untapped resources for insect pest control. Trends Biotechnol. 25(8): 338–342.

Douglas, A.E. 2003. Buchnera bacteria and other symbionts of aphids. pp. 23–38. *In*: K. Bourtzis and T.A. Miller (eds.). Insect Symbiosis. CRC Press, Boca Raton.

Douglas, A.E. 2011. Lessons from studying insect symbioses. Cell Host Microbe 10(4): 359–367.

Douglas, A.E. and A.F.G. Dixon. 1987. The mycetocyte symbiosis of aphids: variation with age and morph in virginoparae of *Megoura viciae* and *Acyrthosiphon pisum*. J. Insect Physiol. 33(2): 109–113.

Dubreuil, G., E. Deleury, D. Crochard, J.C. Simon and C. Coustau. 2014. Diversification of MIF immune regulators in aphids: link with agonistic and antagonistic interactions. BMC Genomics 15: 762.

Dunbar, H.E., A.C. Wilson, N.R. Ferguson and N.A. Moran. 2007. Aphid thermal tolerance is governed by a point mutation in bacterial symbionts. PLoS Biol. 5(5): e96.

Duron, O., T.E. Wilkes and G.D. Hurst. 2010. Interspecific transmission of a male-killing bacterium on an ecological timescale. Ecol. Lett. 13(9): 1139–1148.

Engelstädter, J. and G.D.D. Hurst. 2009. The ecology and evolution of microbes that manipulate host reproduction. Annu. Rev. Ecol. Evol. Syst. 40(1): 127–149.

Feldhaar, H. 2011. Bacterial symbionts as mediators of ecologically important traits of insect hosts. Ecol. Entomol. 36: 533–543.

Ferrari, J. and F. Vavre. 2011. Bacterial symbionts in insects or the story of communities affecting communities. Philos. Trans. R. Soc. Lond. B Biol. Sci. 366(1569): 1389–400.

Ferrari, J., C.B. Muller, A.R. Kraaijeveld and H.C.J. Godfray. 2001. Clonal variation and covariation in aphid resistance to parasitoids and a pathogen. Evolution 55: 1805–1814.

Ferrari, J., A.C. Darby, T.J. Daniell, H.C.J. Godfray and A.E. Douglas. 2004. Linking the bacterial community in pea aphids with host-plant use and natural enemy resistance. Ecol. Entomol. 29(1): 60–65.

Ferrari, J., J.A. West, S. Via and H.C.J. Godfray. 2012. Population genetic structure and secondary symbionts in host-associated populations of the pea aphid complex. Evolution 66(2): 375–390.

Foray, V., A.S. Grigorescu, A. Sabri, E. Haubruge, G. Lognay, F. Francis and P. Thonart. 2014. Whole-Genome Sequence of *Serratia symbiotica* Strain CWBI-2.3T, a Free-Living Symbiont of the Black Bean Aphid *Aphis fabae*. Genome Announcements 2(4): e00767–14. http://doi.org/10.1128/genomeA.00767-14.

Frantz, A., V. Calcagno, L. Mieuzet, M. Plantegenest and J.-C. Simon. 2009. Complex trait differentiation between host-populations of the pea aphid *Acyrthosiphon pisum* (Harris): implications for the evolution of ecological specialisation. Biol. J. Linnean Soc. 97(4): 718–727.

Fukatsu, T., N. Nikoh, R. Kawai and R. Koga. 2000. The secondary endosymbiotic bacterium of the pea aphid *Acyrthosiphon pisum* (Insecta: Homoptera). Appl. Environ. Microbiol. 66(7): 2748–2758.

Fukatsu, T., T. Tsuchida, N. Nikoh and R. Koga. 2001. *Spiroplasma* symbiont of the pea aphid, *Acyrthosiphon pisum* (Insecta: Homoptera). Appl. Environ. Microbiol. 67(3): 1284–1291.

Gehrer, L. and C. Vorburger. 2012. Parasitoids as vectors of facultative bacterial endosymbionts in aphids. Biol. Lett. 8(4): 613–615.

Gerardo, N.M., B. Altincicek, C. Anselme, H. Atamian, S.M. Barribeau, M. de Vos, E.J. Duncan, J.D. Evans, T. Gabaldón, M. Ghanim, A. Heddi, I. Kaloshian, A. Latorre, A. Moya, A. Nakabachi, B.J. Parker, V. Pérez-Brocal, M. Pignatelli, Y. Rahbé, J.S. Ramsey, C.J. Spragg, J. Tamames, D. Tamarit, C. Tamborindeguy, C. Vincent-Monegat and A. Vilcinskas. 2010. Immunity and other defenses in pea aphids, *Acyrthosiphon pisum*. Genome Biol. 11(2): R21.

Gibson, C.M. and M.S. Hunter. 2009. Negative fitness consequences and transmission dynamics of a heritable fungal symbiont of a parasitic wasp. Appl. Environ. Microbiol. 75(10): 3115–3119.

Gilbert, J.A., J.K. Jansson and R. Knight. 2014. The earth microbiome project: successes and aspirations. BMC Biol. 12: 69.

Gómez-Valero, L., M. Soriano-Navarro, V. Pérez-Brocal, A. Heddi, A. Moya, J.M. García-Verdugo and A. Latorre. 2004. Coexistence of *Wolbachia* with *Buchnera aphidicola* and a secondary symbiont in the aphid *Cinara cedri*. J. Bacteriol. 186(19): 6626–6633.

Gottlieb, Y., M. Ghanim, G. Gueguen, S. Kontsedalov, F. Vavre, F. Fleury and E. Zchori-Fein. 2008. Inherited intracellular ecosystem: symbiotic bacteria share bacteriocytes in whiteflies. Faseb J. 22(7): 2591–2599.

Guay, J.F., S. Boudreault, D. Michaud and C. Cloutier. 2009. Impact of environmental stress on aphid clonal resistance to parasitoids: role of *Hamiltonella defensa* bacterial symbiosis in association with a new facultative symbiont of the pea aphid. J. Insect Physiol. 55(10): 919–926.

Guerrero, R., L. Margulis and M. Berlanga. 2013. Symbiogenesis: the holobiont as a unit of evolution. Int. Microbiol. 16(3): 133–143.

Guldemond, J.A., A.F.G. Dixon and W.T. Tigges. 1994. Mate recognition in *Cryptomyzus* aphids: copulation and insemination. Entomol. Exp. Appl. 73(1): 67–75.

Haine, E.R. 2008. Symbiont-mediated protection. Proc. Biol. Sci. 275(1633): 353–361.

Hansen, A.K., G. Jeong, T.D. Paine and R. Stouthamer. 2007. Frequency of secondary symbiont infection in an invasive psyllid relates to parasitism pressure on a geographic scale in California. Appl. Environ. Microbiol. 73(23): 7531–7535.

Hansen, A.K., C. Vorburger and N.A. Moran. 2012. Genomic basis of endosymbiont-conferred protection against an insect parasitoid. Genome Res. 22(1): 106–114.

Haynes, S., A.C. Darby, T.J. Daniell, G. Webster, F.J.F. van Veen, H.C.J. Godfray, J.I. Prosser and A.E. Douglas. 2003. Diversity of bacteria associated with natural aphid populations. Appl. Environ. Microbiol. 69(12): 7216–7223.

Hosokawa, T., R. Koga, Y. Kikuchi, X.Y. Meng and T. Fukatsu. 2010. *Wolbachia* as a bacteriocyte-associated nutritional mutualist. Proc. Natl. Acad. Sci. USA 107(2): 769–774.

Hurst, G.D. and F.M. Jiggins. 2000. Male-killing bacteria in insects: mechanisms, incidence, and implications. Emerg. Infect. Dis. 6(4): 329–336.

Jones, R.T., A. Bressan, A.M. Greenwell and N. Fierer. 2011. Bacterial communities of two parthenogenetic aphid species cocolonizing two host plants across the Hawaiian Islands. Appl. Environ. Microbiol. 77(23): 8345–8349.

Jousselin, E., A. Coeur d'Acier, F. Vanlerberghe-Masutti and O. Duron. 2013. Evolution and diversity of *Arsenophonus* endosymbionts in aphids. Mol. Ecol. 22(1): 260–270.

Klindworth, A., E. Pruesse, T. Schweer, J. Peplies, C. Quast, M. Horn and F.O. Glöckner. 2013. Evaluation of general 16S ribosomal RNA gene PCR primers for classical and next-generation sequencing-based diversity studies. Nucleic Acids Res. 41(1): e1.

Kliot, A., S. Kontsedalov, G. Lebedev, M. Brumin, P.B. Cathrin, J.M. Marubayashi, M. Skaljac, E. Belausov, H. Czosnek and M. Ghanim. 2014. Fluorescence *in situ* hybridizations (FISH) for the localization of viruses and endosymbiotic bacteria in plant and insect tissues. J. Vis. Exp. (84): e51030. doi: 10.3791/51030.

Koga, R., T. Tsuchida and T. Fukatsu. 2009. Quenching autofluorescence of insect tissues for *in situ* detection of endosymbionts. Appl. Entomol. Zool. 44(2): 281–291.

Koga, R., X.Y. Meng, T. Tsuchida and T. Fukatsu. 2012. Cellular mechanism for selective vertical transmission of an obligate insect symbiont at the bacteriocyte-embryo interface. Proc. Natl. Acad. Sci. USA 109(20): E1230–E1237.

Lamelas, A., M.J. Gosalbes, A. Manzano-Marin, J. Pereto, A. Moya and A. Latorre. 2011. *Serratia symbiotica* from the aphid *Cinara cedri*: a missing link from facultative to obligate insect endosymbiont. PLoS Genet. 7(11): e1002357.

Laughton, A.M., J.R. Garcia, B. Altincicek, M.R. Strand and N.M. Gerardo. 2011. Characterisation of immune responses in the pea aphid, *Acyrthosiphon pisum*. J. Insect Physiol. 57(6): 830–839.

Laughton, A.M., M.H. Fan and N.M. Gerardo. 2014. The combined effects of bacterial symbionts and aging on life history traits in the pea aphid, *Acyrthosiphon pisum*. Appl. Environ. Microbiol. 80(2): 470–477.

Leonardo, T.E. 2004. Removal of a specialization-associated symbiont does not affect aphid fitness. Ecol. Lett. 7(6): 461–468.

Leonardo, T.E. and G.T. Muiru. 2003. Facultative symbionts are associated with host plant specialization in pea aphid populations. Proc. Biol. Sci. 270 (Suppl. 2): S209–212.

Ley, R.E., C.A. Lozupone, M. Hamady, R. Knight and J.I. Gordon. 2008. Worlds within worlds: evolution of the vertebrate gut microbiota. Nat. Rev. Micro. 6(10): 776–788.

Li, T., J.H. Xiao, Z.H. Xu, R.W. Murphy and D.W. Huang. 2011. A possibly new *Rickettsia*-like genus symbiont is found in Chinese wheat pest aphid, *Sitobion miscanthi* (Hemiptera: Aphididae). J. Invertebr. Pathol. 106(3): 418–421.

Libbrecht, R., D.M. Gwynn and M.D.E. Fellowes. 2007. *Aphidius ervi* preferentially attacks the green morph of the pea aphid, *Acyrthosiphon pisum*. J. Insect Behav. 20(1): 25–32.

Login, F.H., S. Balmand, A. Vallier, C. Vincent-Monégat, A. Vigneron, M. Weiss-Gayet, D. Rochat and A. Heddi. 2011. Antimicrobial peptides keep insect endosymbionts under control. Science 334(6054): 362–365.

Losey, J.E., J. Harmon, F. Ballantyne and C. Brown. 1997. A polymorphism maintained by opposite patterns of parasitism and predation. Nature 388(6639): 269–272.

Lukasik, P., M. van Asch, H. Guo, J. Ferrari and H.C. Godfray. 2013. Unrelated facultative endosymbionts protect aphids against a fungal pathogen. Ecol. Lett. 16(2): 214–218.

Manzano-Marin, A. and A. Latorre. 2014. Settling down: the genome of *Serratia symbiotica* from the aphid *Cinara tujafilina* zooms in on the process of accommodation to a cooperative intracellular life. Genome Biol. Evol. 6(7): 1683–1698.

Martinez, A.J., S.R. Weldon and K.M. Oliver. 2014. Effects of parasitism on aphid nutritional and protective symbioses. Mol. Ecol. 23(6): 1594–1607.

Montllor, C.B., A. Maxmen and A.H. Purcell. 2002. Facultative bacterial endosymbionts benefit pea aphids *Acyrthosiphon pisum* under heat stress. Ecol. Entomol. 27(2): 189–195.

Moran, N.A. and H.E. Dunbar. 2006. Sexual acquisition of beneficial symbionts in aphids. Proc. Natl. Acad. Sci. USA 103(34): 12803–12806.

Moran, N.A. and Y. Yun. 2015. Experimental replacement of an obligate insect symbiont. Proc. Natl. Acad. Sci. USA 112(7): 2093–2096.

Moran, N.A., P.H. Degnan, S.R. Santos, H.E. Dunbar and H. Ochman. 2005. The players in a mutualistic symbiosis: insects, bacteria, viruses, and virulence genes. Proc. Natl. Acad. Sci. USA 102(47): 16919–16926.

Moran, N.A., J.P. McCutcheon and A. Nakabachi. 2008. Genomics and evolution of heritable bacterial symbionts. Annu. Rev. Genet. 42: 165–190.

Mouton, L., M. Thierry, H. Henri, R. Baudin, O. Gnankine, B. Reynaud, E. Zchori-Fein, N. Becker, F. Fleury and H. Delatte. 2012. Evidence of diversity and recombination in *Arsenophonus* symbionts of the *Bemisia tabaci* species complex. BMC Microbiol. 12 (Suppl. 1): S10.

Najar-Rodríguez, A.J., E.A. McGraw, R.K. Mensah, G.W. Pittman and G.H. Walter. 2009. The microbial flora of *Aphis gossypii*: patterns across host plants and geographical space. J. Invertebr. Pathol. 100(2): 123–126.

Nakabachi, A., S. Shigenobu, N. Sakazume, O. Shiraki, Y. Hayashizaki, P. Carninci, H. Ishikawa, T. Kudo and T. Fukatsu. 2005. Transcriptome analysis of the aphid bacteriocyte, the symbiotic host cell that harbors an endocellular mutualistic bacterium, *Buchnera*. Proc. Natl. Acad. Sci. USA 102(15): 5477–5482.

Nikoh, N., T. Hosokawa, M. Moriyama, K. Oshima, M. Hattori and T. Fukatsu. 2014. Evolutionary origin of insect-*Wolbachia* nutritional mutualism. Proc. Natl. Acad. Sci. USA 111(28): 10257–10262.

Nishikori, K., K. Morioka, T. Kubo and M. Morioka. 2009. Age- and morph-dependent activation of the lysosomal system and *Buchnera* degradation in aphid endosymbiosis. J. Insect Physiol. 55(4): 351–357.

Novakova, E., V. Hypsa, J. Klein, R.G. Foottit, C.D. von Dohlen and N.A. Moran. 2013. Reconstructing the phylogeny of aphids (Hemiptera: Aphididae) using DNA of the obligate symbiont *Buchnera aphidicola*. Mol. Phylogenet. Evol. 68(1): 42–54.

Oliver, K.M., J.A. Russell, N.A. Moran and M.S. Hunter. 2003. Facultative bacterial symbionts in aphids confer resistance to parasitic wasps. Proc. Natl. Acad. Sci. USA 100(4): 1803–1807.

Oliver, K.M., N.A. Moran and M.S. Hunter. 2005. Variation in resistance to parasitism in aphids is due to symbionts not host genotype. Proc. Natl. Acad. Sci. USA 102(36): 12795–12800.

Oliver, K.M., N.A. Moran and M.S. Hunter. 2006. Costs and benefits of a superinfection of facultative symbionts in aphids. Proc. Biol. Sci. 273(1591): 1273–1280.

Oliver, K.M., J. Campos, N.A. Moran and M.S. Hunter. 2008. Population dynamics of defensive symbionts in aphids. Proc. Biol. Sci. 275(1632): 293–299.

Oliver, K.M., P.H. Degnan, M.S. Hunter and N.A. Moran. 2009. Bacteriophages encode factors required for protection in a symbiotic mutualism. Science 325(5943): 992–994.

Oliver, K.M., P.H. Degnan, G.R. Burke and N.A. Moran. 2010. Facultative symbionts in aphids and the horizontal transfer of ecologically important traits. Annu. Rev. Entomol. 55: 247–266.

Oliver, K.M., K. Noge, E.M. Huang, J.M. Campos, J.X. Becerra and M.S. Hunter. 2012. Parasitic wasp responses to symbiont-based defense in aphids. BMC Biol. 10: 11.

Oliver, K.M., A.H. Smith and J.A. Russell. 2014. Defensive symbiosis in the real world – advancing ecological studies of heritable, protective bacteria in aphids and beyond. Functional Ecology 28: 341–355. doi: 10.1111/1365-2435.12133.

Parker, B.J., C.J. Spragg, B. Altincicek and N.M. Gerardo. 2013. Symbiont-mediated protection against fungal pathogens in pea aphids: a role for pathogen specificity? Appl. Environ. Microbiol. 79(7): 2455–2458.

Polin, S., J.C. Simon and Y. Outreman. 2014. An ecological cost associated with protective symbionts of aphids. Ecol. Evol. 4(6): 826–830.

Pontes, M.H. and C. Dale. 2006. Culture and manipulation of insect facultative symbionts. Trends Microbiol. 14(9): 406–412.

Rana, V.S., S.T. Singh, N.G. Priya, J. Kumar and R. Rajagopal. 2012. *Arsenophonus* GroEL interacts with CLCuV and is localized in midgut and salivary gland of whitefly *B. tabaci*. PLoS One 7(8): e42168.

Rosenberg, E. and I. Zilber-Rosenberg. 2011. Symbiosis and development: the hologenome concept. Birth Defects Res. C. Embryo Today 93(1): 56–66.

Rosenberg, E., O. Koren, L. Reshef, R. Efrony and I. Zilber-Rosenberg. 2007. The role of microorganisms in coral health, disease and evolution. Nat. Rev. Microbiol. 5(5): 3553–3562.

Russell, J.A. and N.A. Moran. 2006. Costs and benefits of symbiont infection in aphids: variation among symbionts and across temperatures. Proc. Biol. Sci. 273(1586): 603–610.

Russell, J.A., A. Latorre, B. Sabater-Munoz, A. Moya and N.A. Moran. 2003. Side-stepping secondary symbionts: widespread horizontal transfer across and beyond the *Aphidoidea*. Mol. Ecol. 12(4): 1061–1075.

Russell, J.A., S. Weldon, A.H. Smith, K.L. Kim, Y. Hu, P. Lukasik et al. 2013. Uncovering symbiont-driven genetic diversity across North American pea aphids. Mol. Ecol. 22(7): 2045–2059.

Sakurai, M., R. Koga, T. Tsuchida, X.-Y. Meng and T. Fukatsu. 2005. *Rickettsia* symbiont in the pea aphid *Acyrthosiphon pisum*: novel cellular tropism, effect on host fitness, and interaction with the essential symbiont *Buchnera*. Appl. Environ. Microbiol. 71(7): 4069–4075.

Saridaki, A. and K. Bourtzis. 2010. *Wolbachia*: more than just a bug in insects genitals. Curr. Opin. Microbiol. 13(1): 67–72.

Schmid, M., R. Sieber, Y.-S. Zimmermann and C. Vorburger. 2012. Development, specificity and sublethal effects of symbiont-conferred resistance to parasitoids in aphids. Funct. Ecol. 26(1): 207–215.

Schmitz, A., C. Anselme, M. Ravallec, C. Rebuf, J.-C. Simon, J.-L. Gatti and M. Poirié. 2012. The cellular immune response of the pea aphid to foreign intrusion and symbiotic challenge. PLoS One 7(7): e42114.

Sekirov, I., S.L. Russell, L.C. Antunes and B.B. Finlay. 2010. Gut microbiota in health and disease. Physiol. Rev. 90(3): 859–904.

Sharon, G., D. Segal, J.M. Ringo, A. Hefetz, I. Zilber-Rosenberg and E. Rosenberg. 2010. Commensal bacteria play a role in mating preference of *Drosophila melanogaster*. Proc. Natl. Acad. Sci. USA 107(46): 20051–20056.

Shigenobu, S., H. Watanabe, M. Hattori, Y. Sakaki and H. Ishikawa. 2000. Genome sequence of the endocellular bacterial symbiont of aphids *Buchnera* sp. APS. Nature 407(6800): 81–86.

Simon, J.C., M. Sakurai, J. Bonhomme, T. Tsuchida, R. Koga and T. Fukatsu. 2007. Elimination of a specialised facultative symbiont does not affect the reproductive mode of its aphid host. Ecol. Entomol. 32(3): 296–301.

Simon, J.C., S. Boutin, T. Tsuchida, R. Koga, J.-F. Le Gallic, A. Frantz, Y. Outreman and T. Fukatsu. 2011. Facultative symbiont infections affect aphid reproduction. PLoS One 6(7): e21831.

Skaljac, M., K. Zanic, S.G. Ban, S. Kontsedalov and M. Ghanim. 2010. Co-infection and localization of secondary symbionts in two whitefly species. BMC Microbiol. 10: 142.

Skaljac, M., K. Zanic, S. Hrncic, S. Radonjic, T. Perovic and M. Ghanim. 2013. Diversity and localization of bacterial symbionts in three whitefly species (Hemiptera: Aleyrodidae) from the east coast of the Adriatic Sea. Bull. Entomol. Res. 103(1): 48–59.

Stouthamer, R., J.A. Breeuwer and G.D. Hurst. 1999. *Wolbachia pipientis*: microbial manipulator of arthropod reproduction. Annu. Rev. Microbiol. 53: 71–102.

The International Aphid Genomics Consortium. 2010. Genome sequence of the pea aphid *Acyrthosiphon pisum*. PLoS Biol. 8 (2): e1000313.

The NIH HMP Working Group, Jane Peterson, Susan Garges, M. Giovanni, P. McInnes, L. Wang, J.A. Schloss, V. Bonazzi, J.E. McEwen, K.A. Wetterstrand, C. Deal, C.C. Baker, V. Di Francesco, T.K. Howcroft, R.W. Karp, R.D. Lunsford, C.R. Wellington, T. Belachew, M. Wright, C. Giblin, H. David, M. Mills, R. Salomon, C. Mullins, B. Akolkar, L. Begg, C. Davis, L. Grandison, M. Humble, J. Khalsa, A.R. Little, H. Peavy, C. Pontzer, M. Portnoy, M.H. Sayre, P. Starke-Reed, S. Zakhari, J. Read, B. Watson and M. Guyer. 2009. The NIH Human Microbiome Project. Genome Res. 19(12): 2317–2323.

Tsuchida, T., R. Koga and T. Fukatsu. 2004. Host plant specialization governed by facultative symbiont. Science 303(5666): 1989.

Tsuchida, T., R. Koga, X.Y. Meng, T. Matsumoto and T. Fukatsu. 2005. Characterization of a facultative endosymbiotic bacterium of the pea aphid *Acyrthosiphon pisum*. Microb. Ecol. 49(1): 126–133.

Tsuchida, T., R. Koga, M. Sakurai and T. Fukatsu. 2006. Facultative bacterial endosymbionts of three aphid species, *Aphis craccivora*, *Megoura crassicauda* and *Acyrthosiphon pisum*, sympatrically found on the same host plants. Appl. Entomol. Zool. 41: 129–137.

Tsuchida, T., R. Koga, M. Horikawa, T. Tsunoda, T. Maoka and S. Matsumoto. 2010. Symbiotic bacterium modifies aphid body color. Science 330(6007): 1102–1104.

Tsuchida, T., R. Koga, S. Matsumoto and T. Fukatsu. 2011. Interspecific symbiont transfection confers a novel ecological trait to the recipient insect. Biol. Lett. 7(2): 245–248.

Tsuchida, T., R. Koga, A. Fujiwara and T. Fukatsu. 2014. Phenotypic effect of "*Candidatus Rickettsiella viridis*", a facultative symbiont of the pea aphid (*Acyrthosiphon pisum*), and its interaction with a coexisting symbiont. Appl. Environ. Microbiol. 80(2): 525–533.

Turnbaugh, P.J., R.E. Ley, M. Hamady, C.M. Fraser-Liggett, R. Knight and J.I. Gordon. 2007. The Human Microbiome Project. Nature 449(7164): 804–810.

Vetrovsky, T. and P. Baldrian. 2013. The variability of the 16S rRNA gene in bacterial genomes and its consequences for bacterial community analyses. PLoS One 8(2): e57923.

Via, S. 1999. Reproductive isolation between sympatric races of pea aphids. I. Gene flow restriction and habitat choice. Evolution 53(5): 1446–1457.

Via, S., A.C. Bouck and S. Skillman. 2000. Reproductive isolation between divergent races of pea aphids on two hosts. II. Selection against migrants and hybrids in the parental environments. Evolution 54(5): 1626–1637.

Vilcinskas, A. 2013. Evolutionary plasticity of insect immunity. J. Insect Physiol. 59(2): 123–129.

Vorburger, C. and A. Gouskov. 2011. Only helpful when required: a longevity cost of harbouring defensive symbionts. J. Evol. Biol. 24(7): 1611–1617.

Vorburger, C., C. Sandrock, A. Gouskov, L.E. Castaneda and J. Ferrari. 2009. Genotypic variation and the role of defensive endosymbionts in an all-parthenogenetic host-parasitoid interaction. Evolution 63(6): 1439–1450.

Vorburger, C., L. Gehrer and P. Rodriguez. 2010. A strain of the bacterial symbiont *Regiella insecticola* protects aphids against parasitoids. Biol. Lett. 6(1): 109–111.

Wang, Z., Z.R. Shen, Y. Song, H.Y. Liu and Z.X. Li. 2009. Distribution and diversity of *Wolbachia* in different populations of the wheat aphid *Sitobion miscanthi* (Hemiptera: Aphididae) in China. Europ. J. Entomol. 106: 49–55.

Werren, J.H. 1997. Biology of *Wolbachia*. Annu. Rev. Entomol. 42: 587–609.

Wilkes, T.E., A.C. Darby, J.H. Choi, J.K. Colbourne, J.H. Werren and G.D. Hurst. 2010. The draft genome sequence of *Arsenophonus nasoniae*, son-killer bacterium of *Nasonia vitripennis*, reveals genes associated with virulence and symbiosis. Insect Mol. Biol. 19 Suppl. 1: 59–73.

Wilkinson, T.L., R. Koga and T. Fukatsu. 2007. Role of host nutrition in symbiont regulation: impact of dietary nitrogen on proliferation of obligate and facultative bacterial endosymbionts of the pea aphid *Acyrthosiphon pisum*. Appl. Environ. Microbiol. 73(4): 1362–1366.

Will, T. and A. Vilcinskas. 2015. The structural sheath protein of aphids is required for phloem feeding. Insect Biochem. Mol. Biol. 57: 34–40.

Wulff, J.A., K.A. Buckman, K. Wu, G.E. Heimpel and J.A. White. 2013. The Endosymbiont Arsenophonus Is Widespread in Soybean Aphid, Aphis glycines, but Does Not Provide Protection from Parasitoids or a Fungal Pathogen. PLoS ONE 8(4): e62145. doi:10.1371/journal.pone.0062145.

Zilber-Rosenberg, I. and E. Rosenberg. 2008. Role of microorganisms in the evolution of animals and plants: the hologenome theory of evolution. FEMS Microbiol. Rev. 32(5): 723–735.

Zreik, L., J.M. Bove and M. Garnier. 1998. Phylogenetic characterization of the bacterium-like organism associated with marginal chlorosis of strawberry and proposition of a *Candidatus* taxon for the organism, '*Candidatus Phlomobacter fragariae*'. Int. J. Syst. Bacteriol. 1: 257–261.

6

Aphid Immunity

Andreas Vilcinskas

Introduction

Immunity describes the ability of organisms to resist infections and diseases. The immune system encompasses all structures and molecules contributing to the recognition of pathogens and defense against them. Insects are the most successful group of organisms in terms of biodiversity, and among many other factors their powerful immune system has contributed substantially to this ecological dominance. Insects are protected by staggered defense mechanisms that provide immunity against microbial pathogens. Their sclerotized chitinous integument forms an efficient primary barrier against most microorganisms, and only parasitic fungi can directly infect insects by breaching the exoskeleton because they secrete enzymes that degrade proteins and chitin to facilitate penetration (Gillespie et al. 2000). Insects lack an antibody-based adaptive immune system as found in vertebrates, so antimicrobial defense relies on an innate immune system encompassing cellular and humoral components. The cellular defense mechanisms include phagocytosis and the multicellular encapsulation of microbes and parasites that have gained access to the hemocoel. Immunity-related cells circulating in the hemolymph can distinguish between host and non-host surfaces, adhering specifically to pathogen-associated molecular patterns that are not found on insect cells. For example, lipopolysaccharides or peptidoglycans in bacterial cell walls, or β-1,3 glucans representing the major components of fungal cells, are recognized by insect receptors that in turn trigger further immune responses.

Institute for Insect Biotechnology, Justus-Liebig University of Giessen, Heinrich-Buff-Ring 26-32, 35392 Giessen, Germany; and Fraunhofer Institute for Molecular Biology and Applied Ecology, Department Bioresources, Winchesterstrasse 2, 35394 Giessen, Germany.
E-mail: Andreas.Vilcinskas@agrar.uni-giessen.de; Vilcinskas@ime.fraunhofer.de

Microbial cells are ingested and destroyed by phagocytic cells such as plasmatocytes (Jiang et al. 2010). Phagocytosis achieves the rapid and efficient removal of most microbes, but pathogens or parasites that are too large or too numerous for ingestion can instead be entrapped within multi-layered sheets of immune-competent hemocytes. This encapsulation process is complex and is accomplished by the melanization of cell aggregates resulting in the formation of black nodules. The melanization process is controlled by a tightly-regulated proteolytic cascade that converts prophenoloxidase into phenoloxidase thus catalyzing the formation of chemically inert melanin. Melanin synthesis intermediates such as chinons help to kill pathogens and parasites trapped within the nodules. Multicellular encapsulation is also the major defense mechanism against parasitoids, i.e., parasitic hymenoptera or diptera which lay their eggs in or on other insects and use them as hosts for the developing larvae.

Genetic analysis in the powerful fruit fly model *Drosophila melanogaster* has revealed four major immune signaling pathways that are evolutionarily conserved among insects: the Toll pathway, the immunodeficiency (IMD) pathway, the c-Jun N-terminal kinase (JNK) pathway, and the Janus kinase/signal transducers and activators of transcription (JAK/STAT) pathway. Each pathway is activated in response to particular microbes or their molecules (Agaisse and Perrimon 2004, Delaney et al. 2006, Tauszig et al. 2000) and ultimately controls the expression of immunity-related effector molecules (Vilcinskas 2013).

Peptidic immunity-related effectors are known as antimicrobial peptides (AMPs). Immune-competent hemocytes and other tissues such as the fat body (a functional analog of the vertebrate liver) are known to produce at least a subset of these AMPs. Insect AMPs can be classified according to their activities or grouped depending on structural characteristics, including the prominence of particular amino acid residues (Vilcinskas 2011, Yi et al. 2014). Different insect species show remarkable differences in the number of AMPs they produce. The invasive ladybird *Harmonia axyridis* produces more than 50 AMPs, the largest repertoire reported thus far, whereas aphids produce few AMPs with a limited range of targets (Vilcinskas et al. 2013). Lysozymes are evolutionarily-conserved enzymes that are often classed as immunity-related defense molecules because some are active against bacteria (Wiesner and Vilcinskas 2010). They include c-type lysozymes, which are structurally similar to the well-characterized lysozyme found in the albumin of hens' eggs and possess muramidase activity that allows them to break down bacterial cell walls (Callewaert and Michiels 2010), and i-type lysozymes, which typically comprise separate domains with muramidase and isopeptidase activity, although the latter is infrequently confirmed (Beckert et al. 2015).

Integument, Cellular Immune Responses and Melanization in Aphids

Aphids are soft-bodied insects with a relatively thin cuticle that is often translucent. This does not provide strong protection against parasitoids or fungi. The latter often generate small black spots at penetration sites because the proteases secreted by fungi not only breach the exoskeleton but also activate the prophenoloxidase system resulting in melanization (Gillespie et al. 2000). We observed only limited local melanization when we pierced the pea aphid (*Acyrthosiphon pisum*) with a bacteria-contaminated needle. Even the coagulation of the hemolymph upon piercing was weaker than in other insects such as the larvae of the red flour beetle *Tribolium castaneum*. This limited capacity to seal wounds by coagulation and melanization suggests that other defense mechanisms play a more active role (Altincicek et al. 2008a).

Cellular immunity in insects is established by immune-competent hemocytes circulating in the hemocoel. Three morphologically distinct cell types were identified in the pea aphid by Laughton et al. (2011), namely prohemocytes which are the precursors of other hemocytes, granulocytes which are responsible for the phagocytosis of bacteria, and oenocytoids which display melanotic activity (Fig. 1). A more detailed morphological and functional characterization of aphid hemocytes by Schmitz et al. (2012) resulted in the further identification of plasmatocytes, which along with granulocytes are responsible for the phagocytosis of bacteria (including primary and secondary symbionts present in the hemolymph), adherence to foreign objects and the expression of phenoloxidase. Two pea aphid genes encode putative prophenoloxidases supporting the hypothesis that aphids possess a functioning melanization system (Gerardo et al. 2010) although our data suggest that neither the aphid integument nor its phenoloxidase system provide a strong physical barrier or chemical protection against parasitic fungi. Schmitz et al. (2012) also identified spherulocytes which may promote coagulation in a similar manner to coagulocytes in other species.

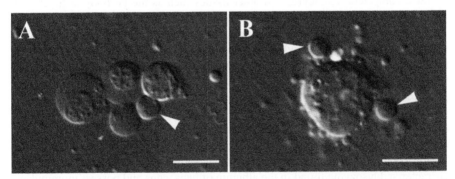

Fig. 1. Hemocyte-types of the pea aphid characterized according to Schmitz et al. 2012. (A) Granulocytes and prohemocyte (arrowhead). (B) Oenocytoid with two attached prohemocytes (arrowheads). Scale bars 25 µm.

One of the most intriguing findings reported by Schmitz et al. (2012) was the presence of symbionts in the circulating hemocytes, which may explain how the vertical transfer of symbionts is achieved. We have previously shown in the greater wax moth *Galleria mellonella* that bacteria can translocate from the gut into the hemocoel, where they are ingested or encapsulated, and from there to the eggs. This transfer of bacteria or bacterial fragments may facilitate maternal transgenerational immune priming (Freitak et al. 2014). A similar mechanism may allow the transgenerational delivery of primary and secondary symbionts in female aphids during sexual reproduction. The transmission of bacterial symbionts to parthenogenetic embryos has been well documented (Wilkinson et al. 2003).

Loss of Antibacterial Immunity-related Genes in Aphids

Aphids have emerged as models of innate immunity in organisms that depend on obligate and facultative bacterial symbionts. Up to 20% of the dry weight of an aphid comprises diverse bacterial symbionts such as *Buchnera aphidicola* (Baumann et al. 1995), which are maintained within specialized cells known as bacteriocytes (Chen et al. 1996, Koga et al. 2003, Tsuchida et al. 2005). These Gram-negative bacteria provide aphids with essential metabolites, including amino acids (The International Aphid Genomics Consortium 2010) that are not found in the sugar-rich phloem sap (Akman Gündüz and Douglas 2009). Nevertheless, aphids are threatened by a broad spectrum of microbial pathogens and parasites, which means they have to protect themselves from microbes while simultaneously avoiding damage to their essential endosymbiont populations.

The pea aphid was selected among more than 4700 known aphid species for the first aphid genome sequencing project. The sequence of this basal hemimetabolous insect provides a useful outgroup that can be used for the phylogenetic analysis of the multiple genome sequences available for holometabolous insects (The International Aphid Genomics Consortium 2010). Before the pea aphid genome sequence was available, we screened experimentally for immunity-related genes in this species using suppression subtractive hybridization (SSH). The injection of bacteria or soluble elicitors of immune responses such as bacterial lipopolysaccharides into insects induces the expression of defense-related genes. SSH is a PCR-based method that amplifies differentially expressed cDNAs while simultaneously suppressing the amplification of common cDNAs. It has been used to identify immune-inducible genes in insects for which no genomic resources are available, and has revealed new genes and known genes with an unanticipated role in immunity (Altincicek and Vilcinskas 2007a,b, Vogel et al. 2011). For example, SSH and bioinformatics analysis identified a gene encoding the antifungal peptide thaumatin in the red flour beetle, even though the same gene was overlooked during the analysis of the genome sequence (Altincicek et al. 2008b).

In the pea aphid, SSH revealed many differentially expressed genes but none of them were homologous to the antibacterial AMPs found in other insects. Furthermore, there was no activity against living bacteria in hemolymph samples from untreated or aphids or those pierced with a bacteria-contaminated needle. However, lytic zone assays using freeze-dried bacteria (*Micrococcus luteus*) revealed the presence of an inducible lysozyme-like (muramidase) activity in the pea aphid hemolymph (Altincicek et al. 2008a). Subsequent bioinformatic analysis of the pea aphid genome confirmed the absence of antibacterial AMP sequences homologous to those from other insects, and the presence of three inducible genes encoding lysozymes (Gerardo et al. 2010). One of these genes encoded a c-type lysozyme whereas the others encoded i-type lysozymes, one of which was strongly expressed in bacteriocytes, suggesting a role in the regulation of symbionts (Gerardo et al. 2010). We concluded that antibacterial AMPs have been lost during the evolution of aphids because, for example, defensins are found even in ancient aptergote insects such as the firebrat *Thermobia domestica* (Altincicek and Vilcinskas 2007a). Furthermore, other hemipteran species also possess immune-inducible AMPs including defensins, including *Pyrrhocoris apterus* (Cociancich et al. 1994) and *Rhodnius prolixus* (Ursic-Bedoya and Lowenberger 2007).

Interestingly, we identified six pea aphid genes encoding antifungal thaumatins homologous to the single thaumatin gene identified by SSH in the red flour beetle. This confirmed that bioinformatics applied to the analysis of genomic and transcriptomic data is most suitable for the identification of homologs of known genes whereas experimental approaches such as SSH allow the identification of novel or unexpected genes involved in innate immunity (Vilcinskas 2013).

Additional bioinformatics analysis of the pea aphid genome revealed that the evolutionary losses extended beyond the known antibacterial AMPs to include other genes that were considered essential for recognition of microbes and immunity-related signaling. Combined gene annotation and functional assays confirmed the presence of some insect immunity-related signaling pathways, such as the Toll and JAK/STAT pathways, whereas the IMD signaling pathway and peptidoglycan recognition proteins were missing (Gerardo et al. 2010). More recently, the JNK pathway was shown to defend pea aphids against bacteria in an IMD-independent manner (Renoz et al. 2015).

Whereas much of the antibacterial arsenal has been lost in aphids, other immunity-related genes which are not common among insects are not only present in aphids, but have expanded, diversified and appear to be differentially regulated during immune responses. For example, the pea aphid genome encodes five members of the macrophage migration inhibitory factor (MIF) family, which are key regulators of innate immune responses in vertebrates (Dubreuil et al. 2014).

Peptidoglycan receptor proteins bind to peptidoglycans on the cell walls of Gram-positive and Gram-negative bacteria and thus allow the immune system to sense invading bacteria. They have been found in all insect genomes sequenced thus far except aphids (Gerardo et al. 2010). In contrast, the pea aphid genome contains two genes encoding Gram-negative binding proteins (GNBPs), which despite their name also mediate the sensing of Gram-positive bacteria. The c-type lectins also act as microbial pattern recognition receptors and five such genes are found in the pea aphid genome (Gerardo et al. 2010).

The specific loss of antibacterial peptides has been interpreted as an adaptation reflecting the dependence of aphids on their bacterial symbionts, which enable them to feed exclusively on phloem sap despite the lack of essential amino acids and other nutrients (Gerardo et al. 2010, Vilcinskas 2013, The International Aphid Genomics Consortium 2010). Because aphids feed exclusively on phloem, any harm to their symbiotic bacteria would inevitably impact on the aphids themselves (Altincicek et al. 2008a). Phloem sap is normally sterile, which limits the uptake of microbes compared to insects that feed on leaves that are typically covered in bacteria and fungi (Freitak et al. 2007), and this limited exposure may have promoted the loss of antibacterial AMPs whose synthesis is costly in terms of fitness associated resources (Vilcinskas 2013). Therefore, it appears that antibacterial defenses can be lost from the genome if their loss costs less in terms of overall fitness than their maintenance. Such trade-offs inevitably result in negative selection against genes that do not increase fitness in a given environment (Vilcinskas 2013). The absence of antibacterial defenses may therefore reflect the coevolution of aphids with their primary and secondary bacterial symbionts, which in turn are known to provide protection against fungi and parasitic wasps, respectively (Oliver et al. 2005, Scarborough et al. 2005). This conclusion is supported by the observation that lysozyme genes and lysozyme-like activity has been retained in the pea aphid because lysozymes predominantly target Gram-positive bacteria, whereas aphid symbionts are Gram-negative (Gerardo et al. 2010).

Several studies have shown that the elimination of bacterial endosymbionts by the oral administration of antibiotics results in delayed aphid development, smaller body size and a lower rate of reproduction (Wilkinson 1998). To test our hypothesis that aphids have lost antibacterial AMPs to prevent their primary and secondary symbionts coming to harm, we added 11 insect AMPs individually to an aphid artificial diet and tested their activity against the pea aphid (Will et al. 2015). We found that seven of the AMPs affected aphid fitness. Cecropin A, Et-AMP15 and apidaecin affected both reproduction and survival, cecropin A1 affected neither but did have the effect of reducing the aphid body size, and finally stomoxyn, eristalin and apidaecin Ia affected survival alone without reducing the reproductive rate. The selected AMPs represented three classes and five families but none precisely replicated the effects of the antibiotic chlortetracycline, which reduces the body size and reproductive rate but has no effect on survival. The most common effect of the AMPs was to reduce survival (6 AMPs had this property). The characteristics of the four inactive AMPs suggest that α-helical and proline-rich AMPs are the least active against aphid microbiota (Will et al. 2015).

Antifungal Immunity in Aphids

The pea aphid is not completely defenseless against pathogens because the genome does contain several thaumatin genes, encoding peptides with ~200 residues stabilized by disulfide bridges. Thaumatins were originally found in plants and a synthetic version of the thaumatin found in the red flour beetle was accordingly shown to be against fungi infecting either plants or insects (Altincicek et al. 2008b). Phylogenetic analysis revealed that the pea aphid thaumatins form a monophyletic group closely related to beetle thaumatins (Gerardo et al. 2010) suggesting that aphid thaumatins provide a defense against parasitic fungi. An immune response requires the allocation of fitness-related resources resulting in trade-offs between investment into different fitness components. Such trade-offs between immunity and other complex parameters have recently been observed in aphids, but interestingly only in response to fungal elicitors (Barribeau et al. 2014).

Conclusion

Aphids appear to possess a weaker immune system than other insects. They have a thin exoskeleton which does not provide a strong physical barrier against fungi, and a limited capacity to seal wounds by coagulation and melanization. The first fully sequenced aphid genome and relevant functional tests have demonstrated that aphids have lost several components of the insect innate immune system that were once considered essential for defense against microbes. These include peptidoglycan recognition proteins, components of the IMD signaling pathway and antibacterial AMPs including defensins. The loss of antibacterial AMPs has been attributed to the allocation of fitness costs related to immune responses, particularly the dependence of aphids on their primary and secondary symbionts. Aphids rely primarily on cellular defenses based on immune-competent hemocytes circulating in the hemolymph which can attach to foreign surfaces and phagocytose bacteria, including primary and secondary symbionts. Furthermore, microbial symbionts in aphids can provide protection against fungi and parasitoids. The dependence of aphids on beneficial microbes is can be exploited to develop sustainable approaches for aphid control in agricultural and horticultural settings. Transgenic plants expressing antibacterial AMPs, which have been lost during aphid evolution, offer a promising strategy to target aphids via their symbionts.

Acknowledgements

The author acknowledges generous financial support from the Hessen State Ministry of Higher Education, Research and the Arts (HMWK) via the LOEWE research center "Insect Biotechnology and Bioresources" and thanks Dr. Richard M. Twyman for editing the manuscript.

Keywords: Immunity, antimicrobial peptides, symbiosis, melanization, pathogens, parasitoids

References

Agaisse, H. and N. Perrimon. 2004. The roles of JAK/STAT signaling in *Drosophila* immune responses. Immunol. Rev. 198: 72–82.

Akman Gündüz, E. and A.E. Douglas. 2009. Symbiotic bacteria enable insect to use a nutritionally inadequate diet. Proc. Biol. Sci. 276: 987–991.

Altincicek, B. and A. Vilcinskas. 2007a. Identification of immune-related genes from an apterygote insect, the firebrat *Thermobia domestica*. Insect Biochem. Mol. Biol. 37: 726–731.

Altincicek, B. and A. Vilcinskas. 2007b. Analysis of the immune-inducible transcriptome from microbial stress resistant, rat-tailed maggots of the drone fly *Eristalis tenax*. BMC Genomics 8: 326.

Altincicek, B., J. Gross and A. Vilcinskas. 2008a. Wounding-mediated gene expression and accelerated viviparous reproduction of the pea aphid *Acyrthosiphon pisum*. Insect Mol. Biol. 17: 711–716.

Altincicek, B., E. Knorr and A. Vilcinskas. 2008b. Beetle immunity: identification of immune-inducible genes from the model insect *Tribolium castaneum*. Dev. Comp. Immunol. 32: 585–595.

Barribeau, S.M., B.J. Parker and N.M. Gerardo. 2014. Exposure to natural pathogens reveals costly aphid response to fungi but not bacteria. Ecol. Evol. 4: 488–493.

Baumann, P., L. Baumann, C.Y. Lai, D. Rouhbakhsh, N.A. Moran and M.A. Clark. 1995. Clark. Genetics, physiology, and evolutionary relationships of the genus *Buchnera*: intracellular symbionts of aphids. Annu. Rev. Microbiol. 49: 55–94.

Beckert, A., J. Wiesner, A. Baumann, A.-K. Pöppel, H. Vogel and A. Vilcinskas. 2015. Two c-type lysozymes boost the innate immune system of the invasive ladybird *Harmonia axyridis*. Dev. Comp. Immunol. 49: 303–312.

Callewaert, L. and C.W. Michiels. 2010. Lysozymes in the animal kingdom. J. Biosci. 35: 127–60.

Chen, D.Q., B.C. Campbell and A.H. Purcell. 1996. A new rickettsia from a herbivorous insect, the pea aphid *Acyrthosiphon pisum* (Harris). Curr. Microbiol. 33: 123–128.

Cociancich, S., A. Dupont, G. Hegy, R. Lanot, F. Holder, C. Hetru, J.A. Hoffmann and P. Bulet. 1994. Novel inducible antibacterial peptides from a hemipteran insect, the sap-sucking bug *Pyrrhocoris apterus*. Biochem. J. 300: 567–575.

Delaney, J.R., S. Stoven, H. Uvell, K.V. Anderson, Y. Engstrom and M. Mlodzik. 2006. Cooperative control of *Drosophila* immune responses by the JNK and NF-kappa B signaling pathways. EMBO J. 25: 3068–3077.

Douglas, A.E. 2015. Multiorganismal insects: Diversity and function of resident microorganisms. Annu. Rev. Entomol. 60: 17–34.

Dubreuil, G., E. Deleury, D. Crochard, J.C. Simon and C. Coustau. 2014. Diversification of MIF immune regulators in aphids: link with agonistic and antagonistic interactions. BMC Genomics 15: 762.

Freitak, D., C.W. Wheat, D.G. Heckel and H. Vogel. 2007. Immune system responses and fitness costs associated with consumption of bacteria in larvae of *Trichoplusia ni*. BMC Biol. 5: 56.

Freitak, D., H. Schmidtberg, F. Dickel, G. Lochnit, H. Vogel and A. Vilcinskas. 2014. The maternal transfer of bacteria can mediate trans-generational immune priming in insects. Virulence 5: 547–554.

Gerardo, N.M., B. Altincicek, C. Anselme, H. Atamian, S.M. Barribeau, M. de Vos, E.J. Duncan, J.D. Evans, T. Gabaldón, M. Ghanim, A. Heddi, I. Kaloshian, A. Latorre, A. Moya, A. Nakabachi, B.J. Parker, V. Pérez-Brocal, M. Pignatelli, Y. Rahbé, J.S. Ramsey, C.J. Spragg, J. Tamames, D. Tamarit, C. Tamborindeguy, C. Vincent-Monegat and A. Vilcinskas. 2010. Immunity and other defenses in pea aphids, *Acyrthosiphon pisum*. Genome Biol. 11: R21.

Gillespie, J., A. Bailey, B. Cobb and A. Vilcinskas. 2000. Fungal elicitors of insect immune responses. Arch. Insect Biochem. Physiol. 44: 49–68.

Jiang, H., A. Vilcinskas and M. Kanost. 2010. Immunity in lepidopteran insects. Adv. Exp. Med. Biol. 708: 181–204.

Koga, R., T. Tsuchida and T. Fukatsu. 2003. Changing partners in an obligate symbiosis: a facultative endosymbiont can compensate for loss of the essential endosymbiont *Buchnera* in an aphid. Proc. Biol. Sci. 270: 2543–2550.

Laughton, A.M., J.R. Garcia, B. Altincicek, M.R. Strand and N.M. Gerardo. 2011. Characterisation of immune responses in the pea aphid, *Acyrthosiphon pisum*. J. Insect Physiol. 57: 830–839.

Lukasik, P., M. van Asch, H. Guo, J. Ferrari and H.C. Godfray. 2013. Unrelated facultative endosymbionts protect aphids against a fungal pathogen. Ecol. Lett. 16: 214–218.

Oliver, K.M., N.A. Moran and M.S. Hunter. 2005. Variation in resistance to parasitism in aphids is due to symbionts not host genotype. Proc. Natl. Acad. Sci. USA 102: 12795–12800.

Renoz, F., C. Noël, A. Errachid, V. Foray and T. Hance. 2015. Infection dynamic of symbiotic bacteria in the pea aphid *Acyrthosiphon pisum* gut and host immune response at the early steps in the infection process. PLoS ONE 10: e0122099.

Scarborough, C.L., J. Ferrari and H.C. Godfray. 2005. Aphid protected from pathogen by endosymbiont. Science 310: 1781.

Schmitz, A., C. Anselme, M. Ravallec, C. Rebuf, J.C. Simon, J.L. Gatti and M. Poirié. 2012. The cellular immune response of the pea aphid to foreign intrusion and symbiotic challenge. PLoS One 7: e42114.

Tauszig, S., E. Jouanguy, J.A. Hoffmann and J.L. Imler. 2000. Toll-related receptors and the control of antimicrobial peptide expression in *Drosophila*. Proc. Natl. Acad. Sci. USA 97: 10520–10525.

The International Aphid Genomics Consortium. 2010. Genome sequence of the pea aphid *Acyrthosiphon pisum*. PLoS Biol. 8: e1000313.

Tsuchida, T., R. Koga, X.Y. Meng, T. Matsumoto and T. Fukatsu. 2005. Characterization of a facultative endosymbiotic bacterium of the pea aphid *Acyrthosiphon pisum*. Microb. Ecol. 49: 126–133.

Ursic-Bedoya, R. and C.A. Lowenberger. 2007. *Rhodnius prolixus*: identification of immune-related genes up-regulated in response to pathogens and parasites using suppressive subtractive hybridization. Dev. Comp. Immunol. 31: 109–120.

Vilcinskas, A. 2011. Anti-infective therapeutics from the Lepidopteran model host *Galleria mellonella*. Curr. Pharm. Des. 17: 1240–1245.

Vilcinskas, A. 2013. Evolutionary plasticity of insect immunity. J. Insect Physiol. 59: 123–129.

Vilcinskas, A., K. Mukherjee and H. Vogel. 2013. Expansion of the antimicrobial peptide repertoire in the invasive ladybird *Harmonia axyridis*. Proc. R. Soc. Lond. Ser. B Biol. Sci. 280: 20122113. doi: 10.1098/rsp.2012.2113.

Vogel, H., B. Altincicek, G. Glöckner and A. Vilcinskas. 2011. A comprehensive transcriptome and immune-gene repertoire of the lepidopteran model host *Galleria mellonella*. BMC Genomics 12: 308.

Wiesner, J. and A. Vilcinskas. 2010. Antimicrobial peptides: the ancient arm of the human immune system. Virulence 1: 440–464.

Wilkinson, T.L. 1998. The elimination of intracellular microorganisms from insects: an analysis of antibiotic-treatment in the pea aphid (*Acyrthosiphon pisum*). Comp. Biochem. Physiol. 119: 871–881.

Wilkinson, T.L., T. Fukatsu and H. Ishikawa. 2003. Transmission of symbiotic bacteria *Buchnera* to parthenogenetic embryos in the aphid *Acyrthosiphon pisum* (Hemiptera: Aphidoidea). Arthropod Struct. Dev. 32: 241–245.

Will, T., C. Flohr and A. Vilcinskas. 2015. Insect-derived antimicrobial peptides affect performance and fecundity in the pea aphid *Acyrthosiphon pisum*. J. Insect Physiol. In press.

Yi, H.Y., M. Chowdhury, Y.D. Huang and X.Q Yu. 2014. Insect antimicrobial peptides and their applications. Appl. Microbiol. Biotechnol. 98: 5807–5822.

7

Aphid Molecular Stress Biology

Laramy S. Enders* and Nicholas J. Miller

Introduction

Stress is prevalent in nature. Organisms are constantly challenged by a multitude of environmental factors that adversely affect basic physiological functions, development and reproductive potential, which ultimately threaten long-term population persistence. As a result, it is well recognized that environmental stress has broad reaching effects that influence ecological interactions and shape the evolutionary trajectory of many taxa (Hoffmann and Hercus 2000, Hoffmann and Parsons 1991, Schulte 2014). Although stress has been defined in many ways, the underlying theme across disciplines is that stress has negative consequences for an organism relative to benign conditions. In the context of this chapter stress is functionally defined as any environmental factor that reduces population growth or components of Darwinian fitness when compared to non-stressful conditions (Hoffmann and Parsons 1991, Sibly and Calow 1989).

Types of Stress and Consequences for Aphid Populations

Stress biology of aphid-plant systems is complex because both partners in these systems suffer from direct effects of external environmental stressors, but each also serves as a source of biotic stress for the other. Aphids therefore experience stress directly but also indirectly via changes to the host plant. Many aphids are vulnerable to temperature extremes, showing decreases in survival, reproduction, dispersal and flight capability (McCornack et al. 2004, Walters and Dixon 1984, Wang and

Department of Entomology, University of Nebraska-Lincoln, 103 Entomology Hall, Lincoln, NE 68583-8687, USA.
 E-mail: lenders2@unl.edu
* Corresponding author: laramy.enders@gmail.com

Tsai 2001). Although thermal limits vary between species, most adults cannot survive above 40°C and a minimum of 4°C is generally required for development (Broadbent and Hollings 1951, Harrington et al. 1995, Hullé et al. 2010). Aphids also suffer attack from a range of natural enemies, including parasitoid wasps and fungal pathogens (Hagen and VanDenBosch 1968, Van Veen et al. 2008). However, the predominant source of biotic stress for aphids arises from the intimate and often antagonistic relationship with the host plant. Plants possess morphological and chemical defenses that impose considerable stress on aphids, directly leading to decreased survival and inhibiting growth and reproduction (Chen 2008, Howe and Jander 2008, Smith and Chuang 2014) (see Chapter 8 "The Effect of Plant Within-Species Variation on Aphid Ecology" in this volume).

An additional plant-mediated dimension of stress for aphids results from altered physiology, biochemistry, and resource allocation within stressed hosts (Bale et al. 2007). Drought conditions, poor soil quality, and herbivory can profoundly affect whole plant chemistry and development, which indirectly influence aphid health and population growth (Bale et al. 2007, Nguyen et al. 2007, Tariq et al. 2012). For example, water stress induces changes in the nutritional profile of phloem and defense metabolism that can negatively affect aphid feeding behavior and performance (Hale et al. 2003, Nguyen et al. 2007, Ponder et al. 2001). However, in some cases aphids may actually benefit from feeding on drought stressed plants (Tariq et al. 2012). While most indirect host plant-mediated stress results from changes in phloem quality or induction of chemical defenses, antagonistic interactions also exist between plant pathogens and aphids. Aphid population growth can be reduced on infected plants (Donaldson and Gratton 2007) and in some cases plant viruses may impose a cost when vectored by aphids (de Oliveira et al. 2014).

Whether stress is imposed directly or indirectly through changes in the host plant, prolonged exposure ultimately constrains aphid population growth. The environmental factors that govern aphid population dynamics have long been a central focus of research, particularly within agro-ecosystems where aphids pose a threat to plant health and agricultural production. One well-studied phenomenon involves the mid-summer collapse of many aphid populations on agricultural crops and illustrates how multiple interacting stressors drive seasonal patterns in aphid ecological dynamics (Karley et al. 2004). Declining host plant nutritional quality, severe weather events, and increased pressure from predators and parasites each can reduce aphid populations. In combination these factors contribute to rapid local extinction (Karley et al. 2003, 2004). In addition to naturally stressful environmental processes, human-mediated forms of stress are important determinants of aphid population persistence. Many common agricultural practices, such as the use of insecticides and release of biological control agents, are designed to control aphid pest species and as a consequence impose stress on aphid populations (see review by Van Emden and Harrington 2007). Human activities that contribute to global climate change also have the potential to profoundly influence the distribution and survival of aphid populations through changes in atmospheric CO_2, precipitation and temperature (Hullé et al. 2010, Pritchard et al. 2007).

How do Aphids Cope with Stress?

Aphids exhibit an impressive flexibility in behavioral and physiological responses to stress. Included in the aphid's diverse repertoire for coping with stress are changes in feeding behavior (Ponder et al. 2001), dispersal (Hodgson 1991, Roitberg et al. 1979), selective resorption of embryos (Ward and Dixon 1982), and associations with beneficial symbiotic microbes (Oliver et al. 2010). One of the most well recognized stress responses in aphids is the switch to production of winged offspring in reaction to crowding, predators, and lowered host plant quality (Müller et al. 2001, Purandare et al. 2014, Sutherland 1969). It is hypothesized this polyphenism provides adaptive benefits by enabling populations to quickly respond to suboptimal conditions and disperse to more suitable environments (Müller et al. 2001). In addition, phenotypic plasticity and a diversity of life cycles further contribute to the remarkable ability of aphids to respond to a variety of adverse conditions (Bale et al. 2007). The adaptive flexibility of aphids has culminated in their ability to successfully colonize at least 5,000 plant species and thrive in a range of environments from the tropics to arctic regions (Blackman and Eastop 2000, 2008, 2013).

In general, aphids demonstrate a spectrum of tolerance to stress, ranging from extremely sensitive (fitness decreases significantly) to highly resistant (little to no effect on fitness). For example, quantitative variation in virulence to stressful host plant defenses has given rise to the many described "biotypes" in agro-ecosystems, where several aphid species, such as the greenbug (*Schizaphis graminum*), are pests of barley, sorghum, soybean, and wheat (Burd and Porter 2006, Burd et al. 2006, Hill et al. 2012, Shufran et al. 2000). Aphids also show variation in tolerance to insecticides, with some species rapidly adapting to chemical compounds through increased expression of detoxification mechanisms and gene duplication (Bass et al. 2014). The considerable plasticity in stress response exhibited by aphids makes them intriguing systems in which to study the effects of stress on multiple levels.

Genomic Resources and the Aphid Molecular Stress Reponses

Since the start of the 21st century the field of molecular biology has experienced unprecedented growth in high-throughput techniques for measuring global changes in gene expression, protein abundance, and metabolite profiling (Boerjan et al. 2012). Many techniques are no longer cost prohibitive or time consuming, opening avenues of research into the distinctive biological processes that enable aphids to persist in the face of a multitude of natural and human induced forms of stress. Briefly described below are three core areas of research used to measure aphid molecular stress responses.

Transcriptomics: Measurement of changes in gene expression from whole organisms or specific tissues can be performed using (1) quantitative reverse transcription PCR (RT-qPCR) on targeted genes or (2) high-throughput technology (e.g., microarrays, next generation sequencing (NGS)) for whole genome expression. See review by Wang et al. (2009) of commonly used RNA sequencing (RNAseq) techniques.

Proteomics: Several gel-based and gel-free methods are available for identification and quantification of changes in protein abundance (see reviews by Bantscheff et al. 2012, Baggerman et al. 2005). Use of two-dimensional gel electrophoresis for separation of proteins and mass spectrometry based techniques for protein identification is common. Recent high throughput approaches couple liquid chromatography and mass spectrometry.

Metabolomics: The comprehensive analysis of all metabolites synthesized by an organism or specific tissue, which measures products of cellular processes downstream of transcription and translation. See review by Macel et al. (2010) of commonly used platforms and applications in various fields.

Early progress in aphid genomics included development of expressed sequence tag libraries (EST) for the pea aphid (*Acrythosiphon pisum*), green peach aphid (*Myzus persicae*), brown citrus aphid (*Toxoptera citricida*), bird cherry-oat aphid (*Rhopalosiphum padi*) and the cotton-melon aphid (*Aphis gossypii*) (Hunter et al. 2003, Lee et al. 2005, Ramsey et al. 2007, Sabater-Muñoz et al. 2006, Tagu et al. 2004). Micoarrays were also developed for *A. pisum* and *M. persicae*, including a dual array of *A. pisum* and its obligate bacterial symbiont (*Buchnera aphidicola*) (Ramsey et al. 2007, Wilson et al. 2006). The boom of the 'omics era' has rapidly expanded the molecular tools available for aphids. Complete genome sequences are now available for the pea aphid and *B. aphidicola* strains associated with several species (International Aphid Genomics Consortium 2010, van Ham et al. 2003). Next generation sequencing has also produced extensive transcriptomes for several important aphid pest species, including *A. gossypii* (Li et al. 2013) and the soybean aphid, *Aphis glycines* (Liu et al. 2012). Continuous growth of aphid genomics resources lead to the creation of AphidBase, a centralized bioinformatics database set up initially for the *A. pisum* genome (Legeai et al. 2010) (see also Chapter 3 "Functional and Evolutionary Genomics in Aphids" in this volume).

As 'omics' techniques continue to progress, a picture of the molecular interactions between aphids, their host plants, and the external environment is emerging. The remainder of this chapter focuses on aphid molecular responses to common forms of naturally occurring and human-mediated stress. We describe changes in gene expression, protein abundance and biochemical processes that occur when aphids are exposed to stress. Table 1 provides a summary of studies examining various levels of the aphid molecular stress response. We conclude with an overview of how ongoing 'omics' advancements have contributed to our understanding of the core components of the aphid molecular stress response.

Temperature Stress

Temperature is the primary abiotic factor affecting aphid physiology, population growth, and species distribution (Harrington et al. 1995, Hullé et al. 2010). Exposure to suboptimal temperatures can be lethal, but also produces a range of sub-lethal effects that alter metabolism, development and reproductive potential (Bale et al. 2002, 2007). There is growing interest in the mechanisms underlying tolerance to extreme temperatures associated with climate change (Bale et al. 2002, Hullé et al.

Table 1. Summary of studies examining metabolomic, proteomic, and transcriptomic changes caused by exposure to various forms of environmental and human mediated forms of stress.

STRESS TYPE	APHID SPECIES	STRESS TREATMENT	METHODS†	REFERENCE
Temperature				
	Acyrthosiphon pisum	heat shock (36°C)	microarray	Wilson et al. 2006
	Acyrthosiphon pisum	heat shock (39°C)	metabolomic profiling	Burke et al. 2010
	Aphis glycines	heat (34°C)	NGS	Enders et al. 2014
	Macrosiphum euphorbiae	heat (35°C)	2-DE/MS proteomic profiling	Nguyen et al. 2009
Insecticides				
	Aphis gossypii	neonicitinoid (thiamethoxam)	NGS	Pan et al. 2015
	Aphis gossypii	tetronic acid derivative (spirotetramat)	2-DE/MS proteomic profiling	Xi et al. 2015
	Myzus persicae	carbamate (pirimicarb)	microarray	Silva et al. 2012
	Myzus persicae nicotianae	carbamate (pirimicarb)	microarray/RT-qPCR	Cabrera-Brandt et al. 2014
Parasites and Pathogens				
	Acyrthosiphon pisum	bacterial infection/wounding	SSH Method/EST	Altincicek et al. 2008
	Acyrthosiphon pisum	viral infection	microarray	Brault et al. 2010
	Acyrthosiphon pisum	bacterial infection	EST/SSH/RT-qPCR	Gerardo et al. 2010
	Acyrthosiphon pisum	bacterial and fungal infection	HPLC proteomic profiling	Gerardo et al. 2010
	Acyrthosiphon pisum	bacterial infection	RT-qPCR	Renoz et al. 2015
	Macrosiphum euphorbiae	wasp parasitism	2-DE/MS proteomic profiling	Nguyen et al. 2008
Host-Plant Mediated Stress				
Defense Related				
	Aulacorthum solani	aphid resistant soybean variety	metabolomic profiling	Sato et al. 2014
	Aphis glycines	Rag1 soybean defense gene	NGS	Bansal et al. 2014
	Aphis glycines	Rag2 soybean defense gene	NGS	Enders et. al. 2014
	Diuraphis noxia	Dn4 wheat defense gene	EST/RT-qPCR	Anathakrishnan et al. 2014
	Macrosiphum euphorbiae	Mi-1.2 tomato defense gene	2-DE/MS proteomic profiling	Francis et al. 2010
Indirect Effects				
	Macrosiphum euphorbiae	drought stressed host	2-DE/MS proteomic profiling	Nguyen et al. 2007
	Macrosiphum euphorbiae	herbivore damage to host	2-DE/MS proteomic profiling	Nguyen et al. 2007
Other Stressors				
	Aulacorthum solani	starvation	metabolomic profiling	Sato et al. 2014
	Aphis glycines	starvation	RT-qPCR	Bansal et al. 2014
	Aphis glycines	starvation	NGS	Enders et al. 2014
	Myzus persicae	host plant switch	2-DE/MS proteomic profiling	Francis et al. 2006
	Myzus persicae	nicotine nutritional stress	microarray	Ramsey et al. 2014
	Macrosiphum euphorbiae	UV radiation	2-DE/MS proteomic profiling	Nguyen et al. 2009

† Expressed sequence tag (EST), High performance liquid chromatography (HPLC), Next generation sequencing (NGS), Reverse transcription quantitative PCR (RT-qPCR,),
Suppression subtractive hybridization (SSH), Two dimentional gel electrophoresis/mass spectromotry (2-DE/MS)

2010). The physiological effects of low temperature and mechanisms enabling aphids to overwinter have been extensively studied (Bale et al. 2007). However, research thus far has only begun to characterize aphid molecular responses to heat stress and the specific genes or pathways enabling aphids to cope with low temperature extremes are currently unknown.

As with many insects, heat shock proteins (HSPs) play an important role in aphid response to both acute and prolonged exposure to elevated temperatures. HSPs are a family of molecular chaperones with varying functions that repair and stabilize damaged or denatured proteins (Sørensen et al. 2003). Transcriptional studies show greater than 15-fold increase in expression of HSP70 at temperatures above 33°C in the pea aphid (Wilson et al. 2006) and soybean aphid (Enders et al. 2014). Although HSPs are often viewed as ubiquitously expressed under heat stress, proteomic profiling indicates HSP70 levels may quickly return to normal when aphids are allowed to recover at benign temperatures (Nguyen et al. 2009). This suggests over-expression and accumulation of aphid HSPs are more likely to be observed directly following sustained heat shock compared to a fluctuating heat stress regime. Additional protective cellular responses involve metabolism of sugar alcohols (polyols) that stabilize cell membranes and proteins under heat stress (Burke et al. 2010, Hendrix and Salvucci 1998). Metabolomic and proteomic profiles of heat stressed *A. pisum* and *M. euphorbiae* also show decreased citric acid cycle activity, consistent with a shift from glycolysis to the pentose phosphate pathway for increased production of sugar alcohols (Burke et al. 2010, Nguyen et al. 2009). Finally, up-regulation of exoskeletal proteins has been observed in heat stressed nymphs (Nguyen et al. 2009) but not adults (Enders et al. 2014). Enhancement of the aphid cuticle is likely important for shielding nymphs from desiccation and other physical stressors during development, but may not occur following the final molt.

Insecticide Exposure

Insecticides are a widely used form of human-mediated stress intended to control aphid populations within agro-ecosystems. These toxins target critical functions, such as transmission of nerve impulses, cellular respiration and lipid biosynthesis, which when disrupted have lethal effects (Nauen et al. 2012, Simon 2011). Several classes of insecticides commonly used on aphids interfere with axonal or synaptic transmission of nerve impulses, causing uncontrollable muscular contractions, lactic acidosis and paralysis (Foster et al. 2007, Simon 2011). Currently, two studies have examined genome-wide transcriptional changes in *M. persicae* and sub-species *M. persicae nicotianae* when exposed to lethal doses of pirimicarb, an anti-cholinesterase insecticide that disrupts synaptic transmission (Cabrera-Brandt et al. 2014, Silva et al. 2012). Among 183 genes up-regulated relative to non-stressful control conditions in a pirimicarb-sensitive *M. persicae* genotype, 49 were involved in primary metabolic processes, 60 in xenobiotic detoxification, 40 in basal cellular processes and four were cuticular proteins (Silva et al. 2012). The transcriptional response of *M. persicae nicotianae* individuals lacking resistance also showed increased expression of several detoxifying enzymes, including glutathione-S transferase and acyl-CoA binding protein, when compared to control conditions (Cabrera-Brandt et al. 2014). In addition, general stress responsive genes associated with restoration of homeostasis were differentially expressed, including heat shock proteins and peptidases that interact with and degrade damaged proteins. Increased expression of several cuticular proteins found in *M. persicae* likely function as a first line of defense against insecticides, by reducing permeability through the aphid exoskeleton (Silva et al. 2012). Overall, transcriptional changes in insecticide stressed aphids are consistent with physiological effects increasing energy demands and mobilization of energy reserves, possibly resulting from neurological impairment and excessive muscle contraction. Activation of genes involved in the various phases of detoxification and cuticular modification is also likely to impose an energetic cost.

Intensive use of insecticides worldwide has resulted in the rapid and frequent evolution of resistance in pest species such as *M. persicae* and *A. gossypii* (Bass et al. 2014, Carletto et al. 2010). Multiple mechanisms exist in aphids that render individuals insensitive to many chemical insecticides, including mutations of the targeted receptor site and heightened activity of detoxification enzymes (Bass et al. 2014, Foster et al. 2007). Functional genomics tools have the power to accelerate discovery and characterization of novel insecticide resistance mechanisms in aphids, which can contribute to improved management of agricultural pest species. One approach is to examine differences in gene regulation and protein expression between insecticide-resistant and susceptible individuals. Transcriptomic and proteomic based comparisons of insecticide-resistant and susceptible cotton aphids (*A. gossypii*) identified candidate genes and metabolic pathways potentially responsible for adaptations to thiamethoxam (Pan et al. 2015) and spirotetramat (Xi et al. 2015). Transcriptomics based studies have also revealed a greater level of complexity underlies adaptation to insecticides than previously thought in aphids. A significant portion of the aphid transcriptome is altered under selective pressure from insecticide stress, with hundreds of genes involving a broad spectrum of pathways affected in resistant genotypes (Pan et al. 2015, Xi et al. 2015). Comparison of transcriptional changes among *M. persicae* genotypes with various resistance mechanisms also demonstrates aphids exhibit considerable plasticity in gene expression (Silva et al. 2012).

Parasites and Pathogens

Aphids encounter a wide range of viral, bacterial, and fungal pathogens in their natural environment capable of causing significant mortality within populations (Hagen and VanDenBosch 1968, Hufbauer 2002, Van Veen et al. 2008). Additional pressure results from parasitoid wasps that use aphids as a host and food source for developing larvae (Hufbauer 2002, Nguyen et al. 2008). To combat biotic stressors aphids utilize behavioral avoidance tactics (e.g., release of alarm phermone, see Chapter 9 "Chemical Ecology of Aphids (Hemiptera: Aphididae)" in this volume) and presumably rely upon both cellular and humoral immune defenses. Aspects of the aphid immune system are covered in detail in Chapter 6 "Aphid Immunity" in this volume. The aphid immune system is considered unusual among insects, lacking many genes previously thought to be critical for recognition, signaling, and destruction of pathogenic microbes (Altincicek et al. 2008, Gerardo et al. 2010). Added complexity stems from the need to mitigate interactions with pathogens, while maintaining mutualistic partnerships with multiple endosymbiotic bacteria. Several studies suggest the aphid immune response is specialized to protect beneficial symbionts, potentially through the reduction of antimicrobial defenses (Altincicek et al. 2008, Renoz et al. 2015). In the pea aphid, ingestion of a facultative endosymbiont, *Serratia symbiotia,* did not trigger a typical immune response, while the pathogen *Serratia marcescens* induced approximately 2-fold higher expression of several genes involved in microbial defense (Renoz et al. 2015). Several studies also provide insight into aspects of general versus specialized immune responses in aphids. Among well-recognized general insect stress responses, heat shock proteins are up-regulated in response to septic wounding and bacterial infection (Altincicek et al. 2008, Gerardo et al. 2010). Changes in the accumulation of proteins involved in basic energy metabolism, detoxification, antioxidant defense, cytoskeletal and cuticular components have also been observed during the initial stages of parasitism of *Macrosiphum euphorbiae* by two wasp species (Nguyen et al. 2008).

Host Plant Mediated Stress

Plant Chemical Defenses

Interaction between aphids and their host plants exemplifies a classic evolutionary arms race. A wealth of knowledge exists on the capability of plants to adjust gene expression and metabolism to produce toxic chemicals, deter feeding, and even recruit predators and parasitoids to prevent damage caused by aphids (de Vos and Jander 2010, Howe and Jander 2008, Smith and Chuang 2014). In contrast, we are only beginning to piece together the molecular changes that occur in aphids exposed to plant defensive stress and understand how counter adaptations arise. Studies thus far show patterns consistent with stress responses resulting from plant production of toxic secondary metabolites and defensive proteins. In a number of species, feeding on plants expressing genes that confer some level of aphid-resistance induces transcriptional responses involved in xenobiotic metabolism. Both the soybean aphid (*A. glycines*) and Russian wheat aphid (*Diuraphis noxia*) show over-expression of enzymes required for detoxification of plant allelochemicals, including many cytochrome P450s, glutathione-S-transferases, and carboxyesterases (Bansal et al. 2014, Sato et al. 2014). In the potato aphid (*M. euphorbiae*), proteomic profiling revealed differential expression of aldehyde dehydrogenase and an E4 esterase in response to feeding on resistant tomato plants carrying the *Mi-1.2* gene (Francis et al. 2010). Altered production of proteases and protease inhibitors (PIs) has also been found in *D. noxia* and *A. glycines* exposed to plant defensive stress (Anathakrishnan et al. 2014, Bansal et al. 2014). Phloem-feeders, including aphids, are insensitive to many plant PIs that target and suppress digestive enzymes of leaf-feeding insects (Rahbé et al. 2003b). However, in several aphid species plant derived serine and cysteine PIs have been shown to inhibit growth and cause mortality not linked to dietary protein digestion (Azzouz et al. 2005, Rahbé et al. 2003a, 2003b). Heighted expression of serine proteases observed in *A. glycines* and *D. noxia* could therefore counteract ingestion of toxic plant serine PIs affecting non-digestive processes. Interestingly, down-regulation of several salivary effector proteins was also observed in soybean aphids feeding on resistant compared to susceptible plants, suggesting a molecular-trade off with other stress responses (e.g., detoxification) or suppression by plant defenses (Bansal et al. 2014). Exposure to soybean plant defenses has also been shown to produce metabolite profiles and induce changes in gene expression similar to those observed in starved aphids (Bansal et al. 2014, Sato et al. 2014). Starvation affects expression of a diverse set of pathways in the soybean aphid, including carbohydrate metabolism, redox regulation, nucleosome assembly, and DNA replication and repair (Enders et al. 2014). Plant production of chemicals that deter aphid feeding may therefore be a potent source of nutritional stress. Other general stress responses to plant defense involve regulation of molecular chaperones (HSPs), energy metabolism, and cytoskeletal proteins (Bansal et al. 2014, Francis et al. 2010).

Recent transcriptomic studies in the soybean aphid indicate different sources of host plant resistance can produce highly variable molecular responses. In contrast to the differential expression of over 900 genes found in response to the *Rag1* soybean defense gene (Bansal et al. 2014), only 12 genes showed altered expression in soybean aphids fed on *Rag2* soybean (Enders et al. 2014). A number of factors could explain varied aphid transcriptional responses to plant defenses, including tolerance being dependent on developmental stage or differences in the mechanistic basis of host plant resistance. Newly hatched nymphs were examined 12 h after feeding on *Rag1* soybean (Bansal et al. 2014), whereas adults were exposed to *Rag2* for 36 h before measuring stress induced transcriptional changes (Enders et al. 2014). It is possible that developing nymphs are more sensitive to plant defenses relative to adults, which may be better able to buffer the physiological effects of stress. Expression differences may also be rapid and transient when aphids encounter plant defenses, suggesting a need for further studies examining the time course of changes in gene expression that occur during aphid-plant interactions.

Indirect Effects of Stress to the Host Plant

Environmental factors can directly impact plant physiological processes, the chemical composition of tissues and nutritional profile of phloem (Farooq et al. 2009, Joern and Mole 2005). Consequently, variation in host plant quality resulting from exposure to abiotic and biotic stressors can indirectly affect aphid performance and abundance (Bale et al. 2007, Nguyen et al. 2007, Tariq et al. 2012). Aphids are particularly sensitive to the nutrient profile of their hosts. For example, the generalist species *M. persicae* exhibits unique expression profiles of proteins involved in several key metabolic pathways depending on host plant (Francis et al. 2006). Drought in particular has serious consequences for plant health, leading to various hypotheses regarding how herbivore populations respond to stress-induced changes in nutrient profiles and metabolism of defensive compounds (Joern and Mole 2005, Larsson 1989, Tariq et al. 2012). In addition, damage caused by competing herbivores (e.g., leaf-chewing insects) can trigger systemic defensive pathways, which has been shown to negatively affect aphid populations (Stout et al. 1997). Alternatively, if prior herbivore damage facilitates increased host plant susceptibility (Karban and Agrawal

2002) or defensive responses are localized to leaf tissue, subsequent aphid population growth may be positively affected or not at all. Proteomic profiling of *M. euphorbiae* on continuously water-stressed potato plants showed modulation of several proteins involved in energy metabolism (e.g., ATP synthase, mitochondrial cytochrome *b*) was correlated with reduced fitness (Nguyen et al. 2007). Decreased abundance of proteins involved in energy metabolism could reflect strategies to conserve energy under stress or reduced metabolic rates resulting from nutrient limitation on drought stressed plants. In the same study, aphids feeding on potato plants defoliated by the beetle *Leptinotarsa decemlineata* did not suffer reductions in growth and survival, but expression profiles showed several proteins involved in neurological processes and cellular communication were down-regulated, suggesting aphid behavior may be altered (Nguyen et al. 2007). Research thus far has primarily focused on aphid population growth and other fitness measures in response to stressed host plants. We currently understand very little about the metabolic changes underlying interactions between aphids and stressed host plants. Integrated studies investigating aphid response to host plants in the context of global climate change are particularly needed, given rising temperatures and increased frequency of drought events will likely alter plant-insect relationships worldwide (Bale et al. 2002, DeLucia et al. 2012, Pritchard et al. 2007).

Additional Molecular Mechanisms

Bacterial Symbioses

There is growing evidence that symbiotic relationships play a prominent role in aphid adaptation to environmental stress. The obligate symbiont *Buchnera aphidicola* performs a vital nutritional role, while several facultative endosymbionts have been shown to influence thermotolerance and resistance to pathogens and parasites (Hansen and Moran 2011, Oliver et al. 2010) (see also Chapter 5 "Bacterial Symbionts of Aphids (Hemiptera: Aphididae)" in this volume). The outcome of aphid-host plant interactions has also been shown to critically depend on microbial associations (Fares et al. 2004, Francis et al. 2010, Tsuchida et al. 2004). However, little is known about the molecular underpinnings of symbiont-mediated responses to stress. In a limited number of cases the molecular basis of symbiont-derived benefits has been characterized. Among pea aphid symbionts, a toxin-encoding bacteriophage that infects *Hamitonella defensa* also provides protection from wasp parasitism (Oliver et al. 2009) and a point mutation in the transcriptional promoter of a *B. aphidicola* heat shock protein contributes to variation in aphid heat tolerance (Dunbar et al. 2007). Symbionts of several insects produce digestive enzymes that counteract and degrade plant defensive compounds (Sugio et al. 2014) and may be involved in suppressing plant defenses against herbivores (Casteel et al. 2012, Hansen and Moran 2014). Interactions between symbionts and the insect immune system have also been demonstrated to provide protection from bacterial and viral infection (Eleftherianos et al. 2013) (see Chapter 6 "Aphid Immunity" in this volume). However, it is unknown whether similar mechanisms function in aphid symbioses.

Transcriptomic and proteomic studies indicate symbionts are highly responsive to the internal environment of the aphid host (Anathakrishnan et al. 2014, Enders et al. 2014, Francis et al. 2010, Nguyen et al. 2009). When challenged by environmental stress, co-regulation of cellular responses can occur, such as the simultaneous up-regulation of aphid and *Buchnera* molecular chaperones (Enders et al. 2014, Nguyen et al. 2008, Wilson et al. 2006). Genes involved in metabolism of peptidoglycan, a component of the bacterial cell wall, have also been shown to be up-regulated in starved aphids (Enders et al. 2014), which could be a defensive response or indicative of symbiont turnover in stressed populations. However, while some endosymbionts provide benefits that allow their aphid hosts to better cope with stress, the *Buchnera* protein GroEL can also trigger plant defensive signaling (Chaudhary et al. 2014). Symbionts thus represent a valuable resource, exploitable for the benefit of both aphids and host plants to manage the effects of stress. Technological advancements enabling simultaneous profiling of aphid, host plant, and symbionts, promise to provide insight into the complexities of multi-trophic interactions under stress.

Epigenetics of Stress Response

Epigenetic mechanisms can regulate gene expression without causing direct changes to the underlying DNA sequence through differential methylation of cytosine residues and chromatin remodeling (Walsh et al. 2010). In a stress biology context, trans-generational stress-induced epigenetic changes could temporarily enhance aphid performance or drive adaptation to chronic stress exposure. Epigenetic changes may be transient, allowing aphid populations to persist during short-term or variable bouts of stress, while long-term stable epigenetic patterns represent an important adaptive process that has yet to be explored. Research has begun to characterize the general epigenetic mechanisms present in aphids and implications for stress response (see Chapter 4 "Epigenetic Control of Polyphenism in Aphids" in this volume). The pea aphid has a functional DNA methylation system that includes homologues of all known vertebrate DNA methyltransferases, the enzymes that add methyl groups to nucleotides (Walsh et al. 2010). Currently, two examples of stress response linked to methylation patterns have been described. In *M. persicae* loss of methylation of some esterase genes is associated with lowered expression and subsequent loss of an insecticide resistant phenotype (Field and Blackman 2003, Hick et al. 1996), suggesting epigenetic control of esterase expression is involved in detoxification of pesticides. Under stress from crowded conditions, pea aphid mothers show greater expression of DNA methyltransferases (Walsh et al. 2010), indicating a possible epigenetic role in determining reproductive strategies in response to stress. However, a number of key knowledge gaps exist, including how epigenetic and phenotypic changes are directly connected and whether epigenetic mechanisms that increase stress tolerance incur additional physiological costs. Investigating the extent to which aphids possess an epigenetic stress memory constitutes an untapped area for further research.

Conclusions

Historically, the molecular basis of response to stress in plants has been intensively studied (see review by Ahuja et al. 2010), while insect herbivores are commonly treated as a source of biotic stress for plants rather than active participants in the interaction (Bilgin et al. 2010). Advancements in functional genomics have facilitated comparative studies and multi-level approaches, deepening our understanding of stress biology within aphid-plant systems (Enders et al. 2014, Francis et al. 2010). Amid the explosive growth of 'omics research the molecular underpinnings of aphid response to a variety of abiotic and biotic stressors has begun to emerge. Figure 1 summarizes the core components of the aphid molecular stress response identified through use of multiple 'omics techniques in the studies outlined in Table 1. Central to coping with all forms of stress is expression of heat shock proteins, modifications of the aphid cuticle and changes to various aspects of primary metabolism and allocation of energy reserves. Regulation of pathways involved in detoxification is shared among molecular responses to insecticides, host plant defenses and parasitism. Altered expression of cell

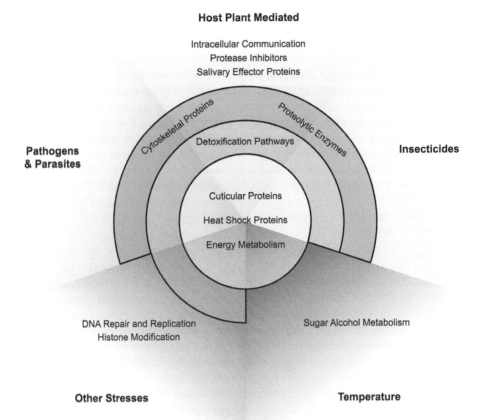

Fig. 1. Core components of the aphid molecular stress response corresponding to five categories of stress.

structural proteins and various classes of proteolytic enzymes is involved in response to host plant mediated stress, pathogens and parasites. While components of a general stress response are evident in aphids, commonly expressed proteins may perform functionally different roles depending on stress type, and the magnitude of response is likely to vary as well. Current research demonstrates aphids exhibit variable and plastic molecular responses to stress, consistent with a profound ability to adapt and persist when challenged by many forms of environmental and anthropogenic stressors. Transcriptional and proteomic profiling have also provided valuable insight into the molecular processes occurring at the plant-aphid interface. Combined use of 'omics technologies will further progress towards a system wide understanding of the effects of multiple interacting stressors on aphid-plant relationships. Finally, unraveling the basis of symbiont-driven stress tolerance and potential epigenetic regulation of stress response are relatively unexplored areas of aphid stress biology.

Acknowledgements

The authors wish to thank Blair Siegfried for his helpful comments and insightful discussions of topics covered in this chapter.

Keywords: aphids, temperature stress, insecticide exposure, parasites, pathogens, plant chemical defenses, bacterial symbiosis, epigenetics

References

Ahuja, I., R.C. de Vos, A.M. Bones and R.D. Hall. 2010. Plant molecular stress responses face climate change. Trends Plant Sci. 15: 664–674.

Altincicek, B., J. Gross and A. Vilcinskas. 2008. Wounding-mediated gene expression and accelerated viviparous reproduction of the pea aphid *Acyrthosiphon pisum*. Insect Mol. Biol. 17: 711–716.

Anathakrishnan, R., D.K. Sinha, M. Murugan, K.Y. Zhu, M.-S. Chen, Y.C. Zhu and M. Smith. 2014. Comparative gut transcriptome analysis reveals differences between virulent and avirulent Russian wheat aphids, *Diuraphis noxia*. Arthropod Plant Interact. 8: 79–88.

Azzouz, H., A. Cherqui, E. Campan, Y. Rahbe, G. Duport, L. Jouanin, L. Kaiser and P. Giordanengo. 2005. Effects of plant protease inhibitors, oryzacystatin I and soybean Bowman–Birk inhibitor, on the aphid *Macrosiphum euphorbiae* (Homoptera, Aphididae) and its parasitoid *Aphelinus abdominalis* (Hymenoptera, Aphelinidae). J. Insect Physiol. 51: 75–86.

Baggerman, G., E. Vierstraete, A. De Loof and L. Schoofs. 2005. Gel-based versus gel-free proteomics: a review. Comb. Chem. High Throughput Screen 8: 669–677.

Bale, J.S., G.J. Masters, I.D. Hodkinson, C. Awmack, T.M. Bezemer, V. Brown, J. Butterfield, A. Buse, J.C. Coulson, J. Farrar, J.G. Good, R. Harrington, S. Hartley, T.H. Jones, R.L. Lindroth, M.C. Press, I. Symrnioudis, A.D. Warr and J.B. Whittaker. 2002. Herbivory in global climate change research: direct effects of rising temperature on insect herbivores. Glob. Change Biol. 8: 1–16.

Bale, J.S., K.L. Ponder and J. Pritchard. 2007. Coping with stress. pp. 287–309. In: H.F. Van Emden and R. Harrington (eds.). Aphids as Crop Pests. CAB International, Wallingford, UK.

Bansal, R., M. Mian, O. Mittapalli and A.P. Michel. 2014. RNA-Seq reveals a xenobiotic stress response in the soybean aphid, *Aphis glycines*, when fed aphid-resistant soybean. BMC Genomics 15: 972. doi: 10.1186/1471-2164-15-972.

Bantscheff, M., S. Lemeer, M.M. Savitski and B. Kuster. 2012. Quantitative mass spectrometry in proteomics: critical review update from 2007 to the present. Anal. Bioanal. Chem. 404: 939–965.

Bass, C., A.M. Puinean, C.T. Zimmer, I. Denholm, L.M. Field, S.P. Foster, O. Gutbrod, R. Nauen, R. Slater and M.S. Williamson. 2014. The evolution of insecticide resistance in the peach potato aphid, *Myzus persicae*. Insect Biochem. Mol. Biol. 51: 41–51.

Bilgin, D.D., J.A. Zavala, J. Zhu, S.T. Clough, D.R. Ort and E.H. DeLuca. 2010. Biotic stress globally downregulates photosynthesis genes. Plant Cell Environ. 33: 1597–1613.

Blackman, R.L. and V.F. Eastop. 2013. Aphids on the World's Plants: An Online Identification and Information Guide.

Blackman, R.L. and V.F. Eastop. 2000. Aphids on the World's Crops. An Identification and Information Guide. John Wiley and Sons, Chichester.

Blackman, R.L. and V.F. Eastop. 2008. Aphids on the World's Herbaceous Plants and Shrubs. John Wiley and Sons, Chichester.

Boerjan, B., D. Cardoen, R. Verdonck, J. Caers and L. Schoofs. 2012. Insect omics research coming of age. Can. J. Zool. 90: 440–455.

Broadbent, L. and M. Hollings. 1951. The influence of heat on some aphids. Ann. Appl. Biol. 38: 577–581.

Burd, J.D. and D.R. Porter. 2006. Biotypic diversity in greenbug (Hemiptera: Aphididae): characterizing new virulence and host associations. J. Econ. Entomol. 99: 959–965.

Burd, J.D., D.R. Porter, G.J. Puterka, S.D. Haley and F.B. Peairs. 2006. Biotypic variation among north American Russian wheat aphid (Homoptera: Aphididae) populations. J. Econ. Entomol. 99: 1862–1866.

Burke, G., O. Fiehn and N. Moran. 2010. Effects of facultative symbionts and heat stress on the metabolome of pea aphids. ISME J. 4: 242–252.

Cabrera-Brandt, M., A.X. Silva, G. Le Trionnaire, D. Tagu and C.C. Figueroa. 2014. Transcriptomic responses of the aphid *Myzus persicae nicotianae* Blackman (Hemiptera: Aphididae) to insecticides: analyses in the single Chilean clone of the tobacco aphid. Chil. J. Agric. Res. 74: 191–199.

Carletto, J., T. Martin, F. Vanlerberghe-Masutti and T. Brévault. 2010. Insecticide resistance traits differ among and within host races in *Aphis gossypii*. Pest Manag. Sci. 66: 301–307.

Casteel, C.L., A.K. Hansen, L.L. Walling and T.D. Paine. 2012. Manipulation of plant defense responses by the tomato psyllid (*Bactericerca cockerelli*) and its associated endosymbiont *candidatus Liberibacter psyllaurous*. PloS One 7: e35191.

Chaudhary, R., H.S. Atamian, Z. Shen, S.P. Briggs and I. Kaloshian. 2014. GroEL from the endosymbiont *Buchnera aphidicola* betrays the aphid by triggering plant defense. Proc. Natl. Acad. Sci. USA 111: 8919–8924.

Chen, M.S. 2008. Inducible direct plant defense against insect herbivores: a review. Insect Sci. 15: 101–114.

de Oliveira, C.F., E.Y. Long and D.L. Finke. 2014. A negative effect of a pathogen on its vector? A plant pathogen increases the vulnerability of its vector to attack by natural enemies. Oecologia 174: 1169–1177.

de Vos, M. and G. Jander. 2010. Volatile communication in plant–aphid interactions. Curr. Opin. Plant Biol. 13: 366–371.

DeLucia, E.H., P.D. Nabity, J.A. Zavala and M.R. Berenbaum. 2012. Climate change: resetting plant-insect interactions. Plant Physiol. 160: 1677–1685.

Donaldson, J.R. and C. Gratton. 2007. Antagonistic effects of soybean viruses on soybean aphid performance. Environ. Entomol. 36: 918–925.

Dunbar, H.E., A.C. Wilson, N.R. Ferguson and N.A. Moran. 2007. Aphid thermal tolerance is governed by a point mutation in bacterial symbionts. PLoS Biol. 5: e96.

Eleftherianos, I., J. Atri, J. Accetta and J.C. Castillo. 2013. Endosymbiotic bacteria in insects: guardians of the immune system? Front. Physiol. 4: 46. doi: 10.3389/fphys.2013.00046.

Enders, L.S., R.D. Bickel, J.A. Brisson, T.M. Heng-Moss, B.D. Siegfried, A.J. Zera and N.J. Miller. 2014. Abiotic and biotic stressors causing equivalent mortality induce highly variable transcriptional responses in the soybean aphid. G3. doi: 10.1534/g3.114.015149.

Fares, M.A., A. Moya and E. Barrio. 2004. GroEL and the maintenance of bacterial endosymbiosis. Trends Genet. 20: 413–416.

Farooq, M., A. Wahid, N. Kobayashi, D. Fujita and S. Basra. 2009. Plant drought stress: effects, mechanisms and management. pp. 153–188. *In*: E. Lichtfouse, M. Navarrrete, P. Debaeke, S. Véronique and C. Alberola (eds.). Sustainable Agriculture. Springer, Netherlands.

Field, L.M. and R.L. Blackman. 2003. Insecticide resistance in the aphid *Myzus persicae* (Sulzer): chromosome location and epigenetic effects on esterase gene expression in clonal lineages. Biol. J. Linnean Soc. 79: 107–113.

Foster, S.P., G. Devine and A.L. Devonshire. 2007. Insecticide resistance. pp. 261–268. *In*: H.F. van Emden and R. Harrington (eds.). Aphids as Crop Pests. CAB International, Wallingford, UK.

Francis, F., P. Gerkens, N. Harmel, G. Mazzucchelli, E. De Pauw and E. Haubruge. 2006. Proteomics in *Myzus persicae*: effect of aphid host plant switch. Insect Biochem. Mol. Biol. 36: 219–227.

Francis, F., F. Guillonneau, P. Leprince, E. De Pauw, E. Haubruge, L. Jia and F. Goggin. 2010. Tritrophic interactions among *Macrosiphum euphorbiae* aphids, their host plants and endosymbionts: investigation by a proteomic approach. J. Insect Physiol. 56: 575–585.

Gerardo, N.M., B. Altincicek, C. Anselme, H. Atamian, S.M. Barribeau, M. de Voss, E.J. Duncan, J.D. Evans, T. Gabaldón, M. Ghanim, A. Heddi, I. Kaloshian, A. Latorre, A. Moya, A. Nakabachi, B.J. Parker, V. Pérez-Brocal, M. Pignatelli, Y. Rahbé, J.S. Ramsey, C.J. Spragg, J. Tamames, D. Tamarit, C. Tamborindeguy, C. Vincent-Monegat and A. Vilcinskas. 2010. Immunity and other defenses in pea aphids, *Acyrthosiphon pisum*. Genome Biol. 11: R21. doi: 10.1186/gb-2010-11-2-r21.

Hagen, K. and R. VanDenBosch. 1968. Impact of pathogens, parasites, and predators on aphids. Annu. Rev. Entomol. 13: 325–384.

Hale, B., J. Bale, J. Pritchard, G. Masters and V. Brown. 2003. Effects of host plant drought stress on the performance of the bird cherry-oat aphid, *Rhopalosiphum padi* (L.): a mechanistic analysis. Ecol. Entomol. 28: 666–677.

Hansen, A.K. and N.A. Moran. 2011. Aphid genome expression reveals host–symbiont cooperation in the production of amino acids. Proc. Nat. Acad. Sci. USA 108: 2849–2854.

Hansen, A.K. and N.A. Moran. 2014. The impact of microbial symbionts on host plant utilization by herbivorous insects. Mol. Ecol. 23: 1473–1496.

Harrington, R., J. Bale and G. Tatchell. 1995. Aphids in a changing climate. pp. 125–155. *In*: R. Harrington and N.E. Stork (eds.). Insects in a Changing Environment. Academic Press, London.

Hendrix, D.L. and M.E. Salvucci. 1998. Polyol metabolism in homopterans at high temperatures: accumulation of mannitol in aphids (Aphididae: Homoptera) and sorbitol in whiteflies (Aleyrodidae: Homoptera). Comp. Biochem. Physiol. A Mol. Integr. Physiol. 120: 487–494.

Hick, C., L. Field and A. Devonshire. 1996. Changes in the methylation of amplified esterase DNA during loss and reselection of insecticide resistance in peach-potato aphids, *Myzus persicae*. Insect Biochem. Mol. Biol. 26: 41–47.

Hill, C., A. Chirumamilla and G. Hartman. 2012. Resistance and virulence in the soybean-*Aphis glycines* interaction. *Euphytica* 186: 635–646.

Hodgson, C. 1991. Dispersal of apterous aphids (Homoptera: Aphididae) from their host plant and its significance. Bull. Entomol. Res. 81: 417–427.

Hoffmann, A.A. and P.A. Parsons. 1991. Evolutionary Genetics and Environmental Stress. Oxford University Press, Oxford.

Hoffmann, A.A. and M.J. Hercus. 2000. Environmental stress as an evolutionary force. BioScience 50: 217–226.

Howe, G.A. and G. Jander. 2008. Plant immunity to insect herbivores. Annu. Rev. Plant Biol. 59: 41–66.

Hufbauer, R. 2002. Aphid population dynamics: does resistance to parasitism influence population size? Ecol. Entomol. 27: 25–32.

Hullé, M., A.C. d'Acier, S. Bankhead-Dronnet and R. Harrington. 2010. Aphids in the face of global changes. C. R. Biol. 333: 497–503.

Hunter, W., P. Dang, M. Bausher, J.X. Chaparro, W. McKendree, R.G. Shatters, Jr., C.L. McKenzie and X.H. Sinsiterra. 2003. Aphid biology: expressed genes from alate *Toxoptera citricida*, the brown citrus aphid. J. Insect Sci. 3: 23.

International Aphid Genomics Consortium IAG. 2010. Genome sequence of the pea aphid *Acyrthosiphon pisum*. PLoS Biol. 8: e1000313.

Joern, A. and S. Mole. 2005. The plant stress hypothesis and variable responses by blue grama grass (*Bouteloua gracilis*) to water, mineral nitrogen, and insect herbivory. J. Chem. Ecol. 31: 2069–2090.

Karban, R. and A.A. Agrawal. 2002. Herbivore offense. Annu. Rev. Ecol. System 33: 641–664.

Karley, A., J. Pitchford, A. Douglas, W. Parker and J. Howardh. 2003. The causes and processes of the mid-summer population crash of the potato aphids *Macrosiphum euphorbiae* and *Myzus persicae* (Hemiptera: Aphididae). Bull. Entomol. Res. 93: 425–438.

Karley, A., W. Parker, J. Pitchford and A. Douglas. 2004. The mid-season crash in aphid populations: why and how does it occur? Ecol. Entomol. 29: 383–388.

Larsson, S. 1989. Stressful times for the plant stress: insect performance hypothesis. Oikos 56: 277–283.

Lee, L., W. Hunter, L. Hunnicutt and P. Dang. 2005. An expressed sequence tag (EST) cDNA library of *Aphis gossypii* alates. American phytopathological society annual meeting. Austin TX: American Phytopathological Society. p. P-579.

Legeai, F., S. Shigenobu, J.P. Gauthier, J. Colbourne, C. Rispe, O. Collin, S. Richards, A.C.C. Wilson and D. Tagu. 2010. AphidBase: a centralized bioinformatic resource for annotation of the pea aphid genome. Insect Mol. Biol. 19(2): 5–12.

Li, Z.-Q., S. Zhang, J.-Y. Luo, C.-Y. Wang, L.-M. Lv, S.-L. Dong and J.-J. Cui. 2013. Ecological adaption analysis of the cotton aphid (*Aphis gossypii*) in different phenotypes by transcriptome comparison. PloS One 8: e83180.

Liu, S., N.P. Chougule, D. Vijayendran and B.C. Bonning. 2012. Deep sequencing of the transcriptomes of soybean aphid and associated endosymbionts. PloS One 7: e45161.

Macel, M., D. Van, M. Nicole and J.J. Keurentjes. 2010. Metabolomics: the chemistry between ecology and genetics. Mol. Ecol. Res. 10: 583–593.

McCornack, B., D. Ragsdale and R. Venette. 2004. Demography of soybean aphid (Homoptera: Aphididae) at summer temperatures. J. Econ. Entomol. 97: 854–861.

Müller, C.B., I.S. Williams and J. Hardie. 2001. The role of nutrition, crowding and interspecific interactions in the development of winged aphids. Ecol. Entomol. 26: 330–340.

Nauen, R., A. Elbert, A. McCaffery, R. Slater and T.C. Sparks. 2012. IRAC: insecticide resistance, and mode of action classification of insecticides. pp. 935–955. *In*: W. Krämer, U. Schirmer, P. Jeschke and M. Witschel (eds.). Modern Crop Protection Compounds, Volumes 1-3, Second Edition. Wiley, Wiley-VCH Verlag GmbH & Co. KGaA, Weinheim, Germany, U.K.

Nguyen, T.T.A., D. Michaud and C. Cloutier. 2007. Proteomic profiling of aphid *Macrosiphum euphorbiae* responses to host-plant-mediated stress induced by defoliation and water deficit. J. Insect Physiol. 53: 601–611.

Nguyen, T.T.A., S. Boudreault, D. Michaud and C. Cloutier. 2008. Proteomes of the aphid *Macrosiphum euphorbiae* in its resistance and susceptibility responses to differently compatible parasitoids. Insect Biochem. Mol. Biol. 38: 730–739.

Nguyen, T.T.A., D. Michaud and C. Cloutier. 2009. A proteomic analysis of the aphid *Macrosiphum euphorbiae* under heat and radiation stress. Insect Biochem. Mol. Biol. 39: 20–30.

Oliver, K.M., P.H. Degnan, M.S. Hunter and N.A. Moran. 2009. Bacteriophages encode factors required for protection in a symbiotic mutualism. Science 325: 992–994.

Oliver, K.M., P.H. Degnan, G.R. Burke and N.A. Moran. 2010. Facultative symbionts in aphids and the horizontal transfer of ecologically important traits. Annu. Rev. Entomol. 55: 247–266.

Pan, Y., T. Peng, X. Gao, L. Zhang, C. Yang, J. Xi, X. Xin, R. Bi and Q. Shang. 2015. Transcriptomic comparison of thiamethoxam-resistance adaptation in resistant and susceptible strains of *Aphis gossypii* Glover. Comp. Biochem. Phys. D: Genomics Proteomics 13: 10–15.

Ponder, K., J. Pritchard, R. Harrington and J. Bale. 2001. Feeding behaviour of the aphid *Rhopalosiphum padi* (Hemiptera: Aphididae) on nitrogen and water-stressed barley (*Hordeum vulgare*) seedlings. Bull. Entomol. Res. 91: 125–130.

Pritchard, J., B. Griffiths and E. Hunt. 2007. Can the plant-mediated impacts on aphids of elevated CO_2 and drought be predicted? Glob. Change Biol. 13: 1616–1629.

Purandare, S.R., B. Tenhumberg and J.A. Brisson. 2014. Comparison of the wing polyphenic response of pea aphids (*Acyrthosiphon pisum*) to crowding and predator cues. Ecol. Entomol. 39: 263–266.

Rahbé, Y., C. Deraison, M. Bonadé-Bottino, C. Girard, C. Nardon and L. Jouanin. 2003a. Effects of the cysteine protease inhibitor oryzacystatin (OC-I) on different aphids and reduced performance of *Myzus persicae* on OC-I expressing transgenic oilseed rape. Plant Sci. 164: 441–450.

Rahbé, Y., E. Ferrasson, H. Rabesona and L. Quillien. 2003b. Toxicity to the pea aphid *Acyrthosiphon pisum* of anti-chymotrypsin isoforms and fragments of Bowman–Birk protease inhibitors from pea seeds. Insect Biochem. Mol. Biol. 33: 299–306.

Ramsey, J.S., A.C. Wilson, M. De Vos, Q. Sun, C. Tamborindeguy, A. Winfield, G. Malloch, D.M. Smith, B. Fenton, S.M. Gray and G. Jander. 2007. Genomic resources for *Myzus persicae*: EST sequencing, SNP identification, and microarray design. BMC Genomics 8: 423. doi: 10.1186/1471-2164-8-423.

Renoz, F., C. Noël, A. Errachid, V. Foray and T. Hance. 2015. Infection dynamic of symbiotic bacteria in the pea aphid *Acyrthosiphon pisum* gut and host immune response at the early steps in the infection process. PloS One 10: e0122099.

Roitberg, B.D., J.H. Myers and B. Frazer. 1979. The influence of predators on the movement of apterous pea aphids between plants. J. Anim. Ecol. 48: 111–122.

Sabater-Muñoz, B., F. Legeai, C. Rispe, J. Bonhomme, P. Dearden, C. Dossat, A. Duclert, J.-P. Gauthier, D. Giblot Ducray, W. Hunter, P. Dang, S. Kambhampati, D. Martinez-Torres, T. Cortes, A. Moya, A. Nakabachi, C. Philippe, N. Prunier-Leterme, Y. Rahbé, J.-C. Simon, D.L. Stern, P. Wincker and

D. Tagu. 2006. Large-scale gene discovery in the pea aphid *Acyrthosiphon pisum* (Hemiptera). Genome Biol. 7: R21. doi: 10.1186/gb-2006-7-3-r21.

Sato, D., M. Sugimoto, H. Akashi, M. Tomita and T. Soga. 2014. Comparative metabolite profiling of foxglove aphids (*Aulacorthum solani* Kaltenbach) on leaves of resistant and susceptible soybean strains. Mol. Biosyst. 10: 909–915.

Schulte, P.M. 2014. What is environmental stress? Insights from fish living in a variable environment. The J. Exp. Biol. 217: 23–34.

Shufran, K., J. Burd, J. Anstead and G. Lushai. 2000. Mitochondrial DNA sequence divergence among greenbug (Homoptera: Aphididae) biotypes: evidence for host-adapted races. Insect Mol. Biol. 9: 179–184.

Sibly, R. and P. Calow. 1989. A life-cycle theory of responses to stress. Biol. J. Linn. Soc. 37: 101–116.

Silva, A.X., G. Jander, H. Samaniego, J.S. Ramsey and C.C. Figueroa. 2012. Insecticide resistance mechanisms in the green peach aphid *Myzus persicae* (Hemiptera: Aphididae) I: a transcriptomic survey. PloS One 7: e36366.

Simon, J.Y. 2011. The Toxicology and Biochemistry of Insecticides. CRC Press, Boca Raton, Florida.

Smith, C.M. and W.P. Chuang. 2014. Plant resistance to aphid feeding: behavioral, physiological, genetic and molecular cues regulate aphid host selection and feeding. Pest Manag. Sci. 70: 528–540.

Sørensen, J.G., T.N. Kristensen and V. Loeschcke. 2003. The evolutionary and ecological role of heat shock proteins. Ecol. Lett. 6: 1025–1037.

Stout, M.J., K.V. Workman, R.M. Bostock and S.S. Duffey. 1997. Specificity of induced resistance in the tomato, *Lycopersicon esculentum*. Oecologia 113: 74–81.

Sugio, A., G. Dubreuil, D. Giron and J.-C. Simon. 2014. Plant–insect interactions under bacterial influence: ecological implications and underlying mechanisms. J. Exp. Bot. doi: 10.1093/jxb/eru435.

Sutherland, O. 1969. The role of crowding in the production of winged forms by two strains of the pea aphid, *Acyrthosiphon pisum*. J. Insect Physiol. 15: 1385–1410.

Tagu, D., N. Prunier-Leterme, F. Legeai, J.P. Gauthier, A. Duclert, B. Sabater-Muñoz, J. Bonhomme and J.C. Simon. 2004. Annotated expressed sequence tags for studies of the regulation of reproductive modes in aphids. Insect Biochem. Mol. Biol. 34: 809–822.

Tariq, M., D.J. Wright, J.T. Rossiter and J.T. Staley. 2012. Aphids in a changing world: testing the plant stress, plant vigour and pulsed stress hypotheses. Agr. Forest Entomol. 14: 177–185.

Tsuchida, T., R. Koga and T. Fukatsu. 2004. Host plant specialization governed by facultative symbiont. Science 303: 1989–1989.

Van Emden, H.F. and R. Harrington. 2007. Aphids as Crop Pests. CABI. doi: 10.1079/9780851998190.0000.

van Ham, R.C., J. Kamerbeek, C. Palacios, C. Rausell, F. Abascal, U. Bastolla, J.M. Fernández, L. Jiménez, M. Postigo, F.J. Silva, J. Tamames, E. Viguera, A. Latorre, A. Valencia, F. Morán and A. Moya. 2003. Reductive genome evolution in *Buchnera aphidicola*. Proc. Natl. Acad. Sci. USA 100: 581–586.

Van Veen, F., C. Müller, J. Pell and H. Godfray. 2008. Food web structure of three guilds of natural enemies: predators, parasitoids and pathogens of aphids. J. Anim. Ecol. 77: 191–200.

Walsh, T.K., J.A. Brisson, H.M. Robertson, K. Gordon, S. Jaubert-Possamai, D. Tagu and O.R. Edwards. 2010. A functional DNA methylation system in the pea aphid, *Acyrthosiphon pisum*. Insect Mol. Biol. 19: 215–228.

Walters, K. and A. Dixon. 1984. The effect of temperature and wind on the flight activity of cereal aphids. Ann. Appl. Biol. 104: 17–26.

Wang, J. and J. Tsai. 2001. Development, survival and reproduction of black citrus aphid, *Toxoptera aurantii* (Hemiptera: Aphididae), as a function of temperature. Bull. Entomol. Res. 91: 477–487.

Wang, Z., M. Gerstein and M. Snyder. 2009. RNA-Seq: a revolutionary tool for transcriptomics. Nature Rev. Genetics 10: 57–63.

Ward, S. and A. Dixon. 1982. Selective resorption of aphid embryos and habitat changes relative to life-span. J. Anim. Ecol. 51: 859–864.

Wilson, A.C., H.E. Dunbar, G.K. Davis, W.B. Hunter, D.L. Stern and N.A. Moran. 2006. A dual-genome microarray for the pea aphid, *Acyrthosiphon pisum*, and its obligate bacterial symbiont, *Buchnera aphidicola*. BMC Genomics 7: 50. doi: 10.1186/1471-2164-7-50.

Xi, J., Y. Pan, Z. Wei, C. Yang, X. Gao, T. Peng, R. Bi, Y. Liu, X. Xin and Q. Shang. 2015. Proteomics-based identification and analysis proteins associated with spirotetramat tolerance in *Aphis gossypii* Glover. Pestic. Biochem. Phys. 119: 47–80.

8

The Effect of Plant Within-Species Variation on Aphid Ecology

Sharon E. Zytynska* and Wolfgang Weisser

Introduction

Research in the area of community genetics has been generating substantial evidence from many different plant-based systems to show that within-species plant genetic variation can have strong and wide-ranging effects on associated communities living on or in the plant, such as herbivores, parasites or mutualists. The majority of community genetics research has been conducted on trees and arthropods, particularly on foundation tree species, i.e., trees that play a pivotal role in defining the structure of the community (Whitham et al. 2012), and through this create locally stable conditions for other species by modulating and stabilizing fundamental ecosystem processes (Ellison et al. 2005). However, there is building evidence that within-species plant variation is also important in non-foundation species, such as short-lived or annual plants, that can influence population dynamics at small-scales and over a single season.

Variation among individuals for ecologically important traits, such as anti-predator defence, disease resistance or competitive ability could have wide-ranging effects on population and community dynamics. And yet, variation within a species is largely ignored in ecological studies despite it being the raw material for natural selection and evolutionary change in a species (Bolnick et al. 2011). In this review we focus

Technische Universität München, Terrestrial Ecology Research Group, Department of Ecology and Ecosystem Management, School of Life Sciences Weihenstephan, Hans-Carl-von-Carlowitz-Platz 2, 85354 Freising, Germany.
* Corresponding author: sharon.zytynska@tum.de

on genetically-based plant variation within a species, as opposed to variation due to phenotypic plasticity (Fordyce 2006). Plant within-species variation can be referred to in a number of ways including plant genotype, cultivar, variety or chemotype. The term genotype is generally restricted to those plants that are genetically identical, e.g., though double haploid breeding or clonal propagation. The terms cultivar or variety indicate that there is potential for some variation to occur within these classifications, albeit substantially less within than between different ones. Chemotype refers to differences in the profile of natural products in plants, often terpenes, and is to some extent genetically based. Plants within a chemotype will also still vary in genotype, and thus chemotyping and genotyping may not necessarily group plants the same way.

Aphids feed on a restricted number of host plants, with many staying on the same host-plant species all year round (autoecious) and approximately 10% of aphid species alternating between taxonomically different plant host species (heteroecious). Aphids disperse either through flight (long-distance dispersal by winged aphids) or walking (short-distance dispersal by unwinged aphids). The abundance and growth-type of the host-plant will therefore influence the effect of plant variation on the aphid, through dispersal limitations; for example, a rare tree or shrub host will require the aphid to have increased dispersal ability to move from host-to-host than would a highly-abundant grassy plant. Indeed, the aphid species themselves will also influence this, depending on their dispersal capabilities. Variation within a host-plant species could influence aphids through direct effects on aphid performance and preference, and indirectly through modification of the interactions between aphids and other species. In this review, we will focus on the direct effect of plant variation on aphids, and explore the indirect effects of plant variation via the ecological communities associated with the aphids, such as aphid natural enemies and aphid mutualists (i.e., ants). We bring together current knowledge on how plant within-species variation can influence aphid population dynamics and discuss potential mechanisms driving these effects to highlight areas that require more attention.

The greatest source of information for the effect of plant variation on aphids comes from work on aphid control (Karban 1992). It has long been known that plant varieties differ in their susceptibility against aphids (and other pest insects) and breeding for resistance is basically the exploitation of plant genetic variation for plant-aphid interactions. Most of our understanding of the (molecular) mechanisms underlying plant variation in susceptibility to aphid attack also comes from work on plant resistance—the last years have seen a variety of mechanisms being unravelled for a number of aphids species (Smith and Chuang 2014). In this review, we will mention this important work as the mechanisms discovered for pest aphid species will very likely also be the mechanisms governing the interactions between non-crop plants and aphids. However, while we use examples from crop systems, our main focus in this chapter is the effect of plant variation on aphid ecology in non-crop systems.

Aphid Performance

Once an aphid has arrived on a plant it will perform a number of pre- and post-stylet insertion behaviours to assess the suitability of the plant for colonisation

(Powell et al. 2006) (see Chapter 11 "Function of Aphid Saliva in Aphid-Plant Interaction" in this volume). Once the host-plant is accepted the aphid will feed for extended periods of time and reproduction will be initiated. Variation within a host-plant species for traits that underlie aphid host-plant acceptance, initiation of reproduction or rate of reproduction can alter aphid population growth rate. Indeed, any variation among aphid genotypes for these factors would also influence aphid performance. Aphid performance can be considered either at the level of an individual aphid or for the aphid colony. Here, we describe an aphid colony as the population of aphids produced from a single founding mother after a number of generations through clonal reproduction, thus assuming every aphid is genetically identical. At the individual aphid level, performance is defined as the number of offspring produced over a given period of time (i.e., fecundity) and will be influenced by the longevity of the aphid. These factors, plus the developmental time from nymph to adult will determine performance of the aphid colony. A high performance of an aphid colony on a host-plant will result from quick initiation of reproduction, a continuous high rate of reproduction, aphid survival for the entire time period (longevity) and fast developmental time of new offspring. When the specific combination of host-plant genotype and aphid genotype determine the performance of the aphid, this means there is an interspecific genotype-by-genotype (GxG) interaction influencing aphid performance. In other words, different aphid genotypes do not respond to host-plant genotypes in the same way, and in turn the host-plant genotypes can vary in their response to different aphid genotypes.

The effect of plant within-species variation (i.e., among genotypes, cultivars, varieties or chemotypes) on aphid performance has been shown in a number of controlled experiments for crop and non-crop systems, including aphids on *Rudbeckia laciniata* (Service 1984); *Lactuca sativa*, lettuce (Reinink et al. 1989); *Dendranthema grandiflora*, chrysanthemum (Bethke et al. 1998); *Triticum aestivum*, wheat (Caillaud et al. 1995, Lamb et al. 2009, Shoffner and Tooker 2013); *Thymus vulgaris*, thyme (Linhart et al. 2005); *Prunus persica*, peach (Sauge et al. 2006); *Hordeum vulgare*, barley (Tétard-Jones et al. 2007, Zytynska and Preziosi 2011); *Fragaria chiloensis*, strawberry (Underwood 2007, Underwood et al. 2011); *Oenothera biennis*, evening primrose (Johnson 2008); *Brassica oleracea*, cabbage (Lamb et al. 2009); *Lagerstroemia fauriei*, crapemyrtle (Herbert et al. 2009); *Elytrigia repens*, quackgrass (Schädler et al. 2010); *Dactylis glomerata*, cocksfoot (Alkhedir et al. 2013); *Medicago truncatula*, medicago (Kanvil et al. 2014); *Schedonorus arundinaceus*, tall fescue (Ryan et al. 2014); and *Solidago altissima*, goldenrod (Utsumi et al. 2011, Williams and Avakian 2015). Field experiments with natural aphid colonisation have also found evidence for variation in aphid performance due to plant genotype or chemotype, either with predator exclusion cages (Utsumi et al. 2011) or without (Krauss et al. 2007, Underwood 2009, Kleine and Müller 2011, Williams and Avakian 2015). For all these studies, aphid performance varied with plant genotype, cultivar or variety. For many of these studies, variation in aphid genotype did not change the effect of plant genotype, such that all aphid genotypes showed similar responses to all plant genotypes. This indicates a common mechanism of effect of plant variation on aphid performance in these species combinations. However, some studies found that different aphid genotypes varied in their response to plant genotype, i.e., one aphid genotype performed better on one particular plant genotype while another performed worse on this same plant genotype. Such GxG interactions between plant and aphid genotypes indicate that there are a number of different mechanisms at work here (Service 1984, Caillaud et al. 1995, Tétard-Jones et al. 2007, Zytynska and Preziosi 2011, Kanvil et al. 2014).

There are many potential mechanisms underlying the different effects of plant within-species variation on aphid performance. Plant genotypes can differ in their defence response to aphids through having resistance genes or stronger induced systemic resistance response to aphid feeding (reviewed by Smith and Clement 2012). For example, there is variation across *M. truncatula* genotypes in the hypersensitive response induced by aphid feeding, which has been linked to a semi-dominant gene *AIN* (*Acyrthosiphon*-induced necrosis) (Klingler et al. 2009). The hypersensitive response results in localised cell-death of the plant cells at the point of pathogen infection and is correlated with host-plant resistance (Klingler et al. 2009). Many aphid resistance genes identified in plants (e.g., *Mi-1.2* or *Vat*) belong to the CC-NBS-LRR sub-family of the NBS-LRR resistant proteins, characterized by nucleotide binding sites (NBS) and leucine-rich repeat (LRR) domains (McHale et al. 2006, Smith and Clement 2012). Further understanding of the molecular basis of aphid resistance will enable the development of insect resistance crop cultivars and, together with IPM (integrated pest management) programs, would benefit the development of sustainable agriculture (Smith and Clement 2012).

Other heritable physiological characteristics of a plant can also influence aphid performance. For example, in evening primrose (*O. biennis*) 49% of variation in aphid performance across 28 genotypes could be explained by leaf water content, trichome density and percentage nitrogen in the leaves (Johnson 2008). Increased leaf water content has a positive effect on aphid performance and, linked to this, Alkhedir et al. (2013) found that an increase in water soluble carbohydrate (WSC) content of plants negatively affecting aphid performance in cocksfoot (*D. glomerata*). This means that aphids were not able to survive on plants with high WSC content (i.e., reduced leaf water content), potentially due to osmotic stress from increased sucrose content of the phloem and as such the aphids cannot process nitrogen and sugars efficiently (Johnson 2008, Alkhedir et al. 2013). Increased leaf nitrogen content usually has a positive effect on aphid population growth rate, for example across milkweed (*Asclepias*) species (Agrawal 2004), but Johnson (2008) found a negative effect across evening primrose plant genotypes, indicating variation in the effect across different species combinations. For many insect species, trichome density is negatively associated with performance by limiting oviposition sites or physically harming the adult insects, particularly when the trichomes are hooked (Levin 1973). However, Johnson (2008) and Agrawal (2004) both found trichome density was positively correlated with aphid population size. Aphids are small and so the beneficial effect is thought to occur by protecting early instars from predators, as they can hide more efficiently among the trichomes. Predation on early instars can decrease population stability and thus impact the future growth rate and persistence of the aphid population (Tenhumberg 2010). The mechanisms behind the effect of chemical differences between plants on aphid performance are less clear, although many presumably act as general defence chemicals. For example, goldenrod plants containing β-pinene tend to support larger aphid populations (Williams and Avakian 2015), whereas tansy plants with more camphor support smaller aphid populations (Kleine and Müller 2011) and thyme plants with more linalool also have fewer aphids (Linhart et al. 2005).

Variation in aphid performance across different plants within a host-species can directly influence aphid population dynamics, and any effect will likely be enhanced as plant within-species diversity increases. Non-additive effects of plant genotypic variation on aphid population dynamics have been shown in strawberry-aphid and goldenrod-aphid systems (Underwood 2009, Utsumi et al. 2011). A non-additive effect means that the effect of plant genotype in a diverse community cannot be simply explained by the sum of the independent effects of each genotype in isolation. Such non-additive effects are thought to be driven by source-sink dynamics and short-distance movement from plant to plant by unwinged aphids, followed by rapid clonal reproduction (Van Zandt and Mopper 1998, Utsumi et al. 2011). Aphids do not distribute randomly across plant genotypes but rather exhibit preference for particular host-plants and genotypes within host-plants (Zytynska and Preziosi 2011). Variation in aphid performance on different plant genotypes would lead to unequal aphid abundances on different host plants across the population and could be exasperated by aphid preference. The effect of aphid host-genotype preference could further explain the dynamics of aphid populations among host-plant populations that are not easily understood through performance aspects alone.

Aphid Preference

As an aphid arrives onto a new plant it will make the decision to stay and reproduce, or to leave and find a new host plant. When this decision is non-random based on plant genotypes, and consistent across them, then there is aphid preference. Thus, the definition of aphid preference is here given as 'non-random associations between host-plant genotypes and aphids', such that aphids will typically colonise the preferred host-plant when provided with a choice.

Aphid preference is partially down to active aphid choice, which can be shown by the movement of aphids towards a preferred host plant and the settling behaviour after a short period of time. As such, many studies of aphid preference use controlled experimental conditions either with leaf-discs or potted plants to assess the amount of time spent or the number of offspring deposited on particular plants. Aphid preference is also influenced by the propensity to leave a host plant, with movement to a different host requiring the aphid to decide to leave the current one. Thus, there may be some 'threshold of suitability' for a plant and only below this would the aphid leave to find a new host. If dispersal is limited, the aphids only leave if the host is really unsuitable whereas if dispersal is not limited the aphids are more likely to leave more suitable plants to find a better one. The propensity to leave a plant will thus be influenced by a number of factors, including the distance to the next host-plant, the presence of competitors or predation risk. The majority of preference studies use unwinged aphids to understand local intrapopulation dispersal between host-plant genotypes, as opposed to long-range dispersal by winged aphids. The propensity to leave a host-plant is therefore also host-plant species dependent; dispersal to another host individual by unwinged aphids on trees would be more limited than for aphids on common grasses that grow in much higher densities. A study on aphid movement in strawberries showed that unwinged aphids may move multiple times from one host-genotype to another, in a population, over the course of a few days (Underwood et al. 2011). It is likely that while they are on each host, they also deposit a batch of offspring before moving onto another plant, either to maximise resources for the offspring or in order to search for a more preferred genotype.

Variation in aphid preference among plant genotypes has been demonstrated in various systems, including aphids on *H. vulgare*, barley (Pettersson et al. 1999, Ninkovic et al. 2002, Zytynska and Preziosi 2011, 2013), *L. fauriei*, crapemyrtle (Herbert et al. 2009), *F. chiloensis*, strawberry (Underwood et al. 2011), and *T. aestivum*, wheat (De Zutter et al. 2012). Aphid preference for different barley genotypes was found to also vary among aphid genotypes, with different aphid genotypes choosing different barley genotypes, i.e., a GxG interaction influencing aphid preference (Zytynska and Preziosi 2011). It is really only when reproduction occurs that plant acceptance can be confirmed since aphids cannot survive on all plant species (Powell et al. 2006). However, within a plant species aphids can survive on the majority of the plant genotypes and thus reproduction could be initiated on many of the plants they encounter. Aphids are considered specialised insects as they feed on a restricted number of host-plant species. This specialisation is a spectrum within itself, with some aphid species feeding on more host-plants than others. Within-plant variation can further expand this spectrum. Consider two aphid species highly specialised to feeding on only one host-plant species, one aphid may readily accept all genotypes within this species whereas another will further specialise to accept only a subset. In this case, specialisation of an aphid extends further than just the host-plant species to also include host-plant genotypes. Discrimination among host-plant genotypes occurs as not all aphids respond to the same plant traits as cues for the initiation of feeding and reproduction. For example, De Zutter et al. (2012) found that two aphid species, on barley, differed in the time taken to make a choice of host-genotype across plant growth stages. One aphid species (*Sitobion avenae*) preferred the barley ears and made a much quicker decision when presented with the ears of different genotypes than when presented with the leaves. The other (*Metopolophium dirhodum*) preferred to colonise the leaves and made a quicker decision of host-genotype when presented with leaf material than with barley ears. Such variation can therefore support the co-occurrence of different aphid species on a single host and further provide information regarding potential mechanisms driving plant genotype preferences.

Under the preference-performance hypothesis (PPH) one may assume that aphid preference and performance would be highly correlated. However, evidence does not support this assumption for aphids across and within plant species (Gripenberg et al. 2010, Underwood et al. 2011, Zytynska and Preziosi 2011). A meta-analysis by Gripenberg et al. (2010), on a range of insect taxa (including Diptera, Coleoptera and Hemiptera) across different host-plant species, found the strongest link between preference and performance in those species that have intermediate level of diet specialisation, i.e., within a plant family, as opposed to monophagus (within a plant genus) or polyphagus (multiple plant families) species. Within-species, Herbert et al. (2009) found an association between plant suitability and preference, but only in crapemyrtle genotypes that were a result of hybridisation with an unsuitable host species. Across six barley genotypes, only one aphid genotype out of four tested showed a positive correlation between aphid performance and preference (Zytynska and Preziosi 2011). The lack of correlation between preference and performance for aphids across host-plant genotypes suggests that the traits involved in aphid performance and aphid choice are not tightly linked. Thus, there is little selective pressure driving an association between preference and performance and aphids do not actively seek out those plants that confer a high performance.

Within a host-plant species, all plants could be potential hosts so the choice of plant is no longer the difference between survival and extinction (as might be across plant species), but rather the difference between slow or fast population growth rates. The cues evolved by the aphids to distinguish host to non-host plant species may therefore be obsolete to discriminate between high or low performance plant genotypes, and the cues associated with host-preference may not (yet) be strongly linked to performance traits. Over time, in a highly specialised system, co-evolutionary processes between a plant and its aphids would be expected to lead to a stronger positive association between preference and performance. In this case, any preference among host-plant genotypes would be enhanced by the increased performance on preferred plants and decreased performance on non-preferred plants (Fig. 1a). No correlation between preference and performance does not mean that the aphids exhibit no active choice, but that it would be possible to observe aphids showing active choice towards one genotype and against another for which the performance is the same (e.g., Zytynska and Preziosi 2011). Any aphid on those plants would experience the same performance but the population size would be increased for the host-plant genotype the aphids chose to colonise (Fig. 1b). An unlikely scenario would be that aphid preference and performance is negatively correlated, such that aphids actively choose the plants that confer the lowest performance; this would theoretically act to standardise the aphid population size across plants (Fig. 1c). Over time, with high performance on preferred plants and subsequent reduction in quality of these plants, the preference of aphids

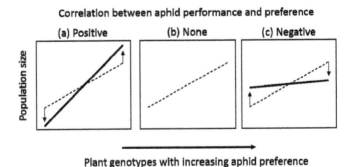

Fig. 1. The effect of preference (for particular genotypes) and performance (fitness differences on different genotypes) on aphid population sizes across host-plant genotypes. (a) Positive correlation between aphid preference and performance would act to enhance preference effects with aphids preferentially settling and reproducing on the most suitable plants, (b) no correlation would lead to a greater influence of aphid preference than performance on population sizes across plant genotypes, and (c) a negative correlation could theoretically standardise the aphid population size across host-plant genotypes, as few aphids would settle on suitable genotypes while more would settle on those conferring low performance.

may change. In this case, source-sink dynamics from preferred to less preferred hosts could also standardise the aphid population size across host-plant genotypes.

Challenges in Investigating Plant Variation Effects on Aphid Performance and Preference

Aphid performance is usually measured under controlled conditions, either in a growth chamber or a greenhouse, where abiotic factors (e.g., temperature or rainfall) can be controlled and other herbivores or predators excluded. This method provides a good measure of aphid performance on a plant but can still include a number of variables, such as intraspecific competition between aphids that are not controlled for (but see Bethke et al. 1998, Underwood et al. 2011). As a performance experiment continues, the density of aphids on a plant increases that can, in turn, decrease plant quality directly feeding back to the aphid to reduce performance. In some aphid species, increased density can induce the production of winged morphs through crowding effects (Müller et al. 2001, Mehrparvar et al. 2013). Winged morphs often exhibit reduced fecundity and longer developmental times as they are more adapted to dispersal than reproduction (Braendle et al. 2006), thus if the duration of the experiment is too long this may affect robust measuring of aphid performance. When an aphid experiences crowding, it is the next generation of aphids that develop winged morphs. Therefore, the environment of the aphid prior to any performance experiment must be maintained at low densities to avoid this. The age at which the initial aphid is introduced to the experimental plant would also influence performance estimation, for example in an extreme case where an older adult is introduced fecundity would be reduced due to prior reproduction. Generally for performance experiments, an adult aphid is allowed to reproduce on a plant and then the offspring and adult removed leaving only one or two nymphs. These are allowed to develop and reproduce, which as well as controlling for aphid age also controls for maternal effects and can be repeated to control for grand-maternal effects (Mousseau and Dingle 1991).

Aphid preference is a complex measure and often the settling behaviour after a short-period of time on leaf discs or using olfactometers is used to assess preference, i.e., by comparing the number of aphids on one plant to the other (Herbert et al. 2009, De Zutter et al. 2012). Alternatively, the movement across living plants can be assessed using plants grown together (Underwood et al. 2011, Zytynska and Preziosi 2011). Both of these methods have advantages and disadvantages. Leaf disc methods allow for greater movement of the aphids between treatments and thus purely measures aphid active choice, but the choice is also reliant on cues in the plant that are still detectable once it has been cut. Olfactometers are particularly useful to study volatile effects, and thus focus on pre-settling choice behaviour of aphids but rarely consider the propensity to leave a host since they are conducted over short periods of time (hours as opposed to days). Performance assays with living plants measure the combined effect of pre-settling choice behaviour with the decision to stay on a plant or leave as the experiment can be continued for a longer period of time. Here, the effect of performance and preference can be assessed. Therefore, the methodology should be chosen based on the question to be answered and a combination of olfactometer and living plants may provide the most information on aphid preference/performance across plant genotypes.

Indirect Effects of Plant Within-Species Variation on Aphids

The interaction between plants and aphids can be modified by a third species, through either an interaction chain or interaction modification (Wootton 1994). A change in the environment (biotic or abiotic) can lead to phenotypic plasticity in a plant genotype's response to aphids (Fordyce 2006). We focus here on the effect of a third species on the effect of within-species plant variation on plant-aphid interactions.

Plant Variation Effects on Aphid-Ant Interactions

Many aphids are tended by ants, which harvest the honeydew produced by the aphid and in return provide protection against natural enemies (see Chapter 10 "Aphid Honeydew: Rubbish or Signaler" in this volume). Ant-attendance among aphid species can vary from obligate to opportunistic, depending on the system, with up to a third of aphid species classified as obligate myrmecophiles in Central Europe (Stadler and Dixon 2005). The rate at which ants attend aphids can vary due to host-plant species, aphid colony isolation and the production of dispersal morphs through the season, with greater attendance on woody-plant species and reduced attendance on isolated or highly mobile aphid species (Stadler and Dixon 2005). Variation within a plant species for such traits could also influence the interaction between aphids and ants.

   The cottonwood system in North America is a well-studied system in community genetics, with parental and hybrid trees influencing many associated ecological communities (Whitham et al. 2006). In this system, ant colony abundance on hybrid tree genotypes was much lower than on narrowleaf parental genotypes and led to the distribution of an obligate myrmecophile aphid, *Chaitophorus populicola*, being restricted to only 21% of their potential habitat space (Wimp and Whitham 2001). Thus, while the aphids could survive on all trees, they were restricted to parental genotypes due to their interaction with the ants, via an interaction chain. For ant-tended aphids (*Aphis asclepiadis*) on milkweed, plant genotype has been shown to alter the ant-aphid interaction from mutualistic (20 of 32 genotypes) to antagonistic (seen for 12 of 32 genotypes) (Mooney and Agrawal 2008). The authors suggest this is due to heritable variation in the phloem sap through an interaction modification by ants. Mooney and Agrawal (2008) speculate that interaction modifications are required for selection to occur from indirect interactions, for example in their results if the ants select plants due to genetically-based variation in phloem sap quality or quantity this would drive changes in ant recruitment across plant genotypes. Whereas, if the plant only influenced the ants through changes in aphid density (e.g., Wimp and Whitham 2001) then selection on plant traits would be unchanged by ants. Plant genotype does, however, not always influence the mutualistic interactions among ants and aphids, as found in a study on 28 evening primrose genotypes (Johnson 2008). Plant genotype was found to influence ant abundance both directly, and indirectly via aphid density;

however, there was no interaction between plant genotype and ant abundance on aphid performance (Johnson 2008). Similarly, plant genotypic diversity increased aphid abundance but did not influence the aphid-ant interactions on the shrubby host-plant *Baccharis salicifolia* (Moreira and Mooney 2013).

Any effect of plant variation on the aphid-ant interaction requires variation in the plant for ecologically important traits involved in this interaction, e.g., phloem sap quality. Another important trait could be the production of extra-floral nectar (EFN) by the host-plant, which is often produced to attract predatory arthropods to a plant. Extra-floral nectar abundance and composition can vary within a plant species (Heil et al. 2000) and EFN production can influence ant-aphid interactions by both increasing and decreasing ant-attendance of aphids depending on the system (Rudgers and Gardener 2004, Martinez et al. 2011). Aphids are not considered to function as EFNs do (Offenberg 2000) but the sugar content of the nectar can be higher than that of the aphid honeydew (Katayama et al. 2013). Thus, in some systems the ants will preferentially obtain sugar from the EFN, and protein by predating on the aphids (Buckley 1987). Aphids do not seem to induce changes in nectar production (Engel et al. 2001, Katayama et al. 2013), although leaf damage to *Vicia faba* plants can strongly increase EFN production, suggesting some mechanism to attract predatory ants to a plant with high herbivory (Mondor and Addicott 2003). Recent work suggests that at the beginning of the season ants are more likely to feed on plant EFN but shift to aphids as their populations increase (Katayama et al. 2013); this is explained through EFN secretion levels not changing through the season whereas the amount of honeydew produced increases as an aphid population increases. If some host-plant genotypes produced higher quality EFN they would be more likely to attract ants at the beginning of the season. Then, when myrmecophilous aphids arrive to the host plant they may be immediately attended by the ants and experience higher survival on this plant genotype than another where ants are less abundant. In this way ants could influence aphid population dynamics through preferences based on variation in plant EFN production.

Plant Variation Effects on Aphid-natural Enemy Interactions

Natural enemies of aphids can drive strong top-down effects on aphids whilst the plant drives bottom-up effects leading to aphids being 'wedged between bottom-up and top-down processes' (*sensu* Stadler (2004)). Plant within-species variation can also indirectly influence aphids through mediating top-down effects of natural enemies. For example, the effect of natural enemy presence on aphid abundance varied among different thyme chemotypes (Linhart et al. 2005) and bird predation of galling aphids varied among different genotypes of *Populus angustifolia*, with one tree genotype being preferentially visited by bird predators (Smith et al. 2011). Such effects could also vary with host-plant patch size, as increasing patch size was found to be associated with increasing parasitoid density in a wheat-cereal aphid system (Hamback et al. 2007).

In general, aphid density explains the majority of variation in abundances among natural enemies with higher aphid density leading to higher natural enemy abundance. However, not all the variation is explained through cascading effects of plant variation on natural enemies via the aphids. There are also direct effects of plant variation on natural enemies. For example, variation in the proportion of mummified *Rhopalosiphum padi* aphids on quackgrass (*E. repens*) genotypes was found to be partly explained by direct effects of the plant on the parasitoids, in addition to the bottom-up effect of aphid density (Schädler et al. 2010). In a cabbage-aphid system, there was some evidence for plant mediated top-down effects albeit weaker than the direct effect of plant within-species variation on aphid performance and preference (Kos et al. 2011). On *Lolium perenne* plants, the effect of cultivar diversity was only significant for the primary and secondary parasitoid wasp trophic levels, indicating that plant-mediated top-down effects are stronger in this system than direct effects of plant cultivar diversity on the aphids (Jones et al. 2011). In a barley-aphid system, parasitoid wasp body size (a proxy for fitness) was influenced by an interaction between barley and aphid genotype, showing that for particular combinations of genotypes the emerging parasitoids were on average larger than for other combinations (Zytynska et al. 2010). With non-random associations between aphid and plant genotypes (Zytynska and Preziosi 2011) this could lead to changes in the parasitoid and aphid population dynamics in a diverse community. The presence of a predator can also influence aphid host-plant preference, with aphids more likely to colonise a less-preferred host-plant genotype due to previous infestation of the preferred genotype by ladybirds (Wilson and Leather 2012). Indeed, it has been shown that pea aphids will produce more winged offspring when feeding on plants that have tracks left by ladybirds, while other aphid species do not show such responses (Dixon and Agarwala 1999).

Plant Variation Effects on Aphid Interactions with other Organisms

Whilst ants and natural enemies have strong effects on aphid populations, other less obvious species groups may also affect aphid-plant interactions, e.g., microorganisms, earthworms and other plants. Microorganisms in the soil, in particular beneficial rhizobacteria or mycorrhiza, can change the interaction between a plant and its aphids (Pineda et al. 2013). Genetic interactions between barley and aphids can be mediated by the addition of the plant-growth-promoting-rhizobacteria (PGPR) *Pseudomonas aeruginosa* (Tétard-Jones et al. 2007). Here, the final population size of the aphid genotypes was dependent on both the aphid genotype, the barley genotype and the presence of the rhizobacteria. In other words, the effect of the aphid-plant GxG was dependent on whether *P. aeruginosa* was present among the roots of the plant. In this system, five plant QTLs have been identified that are associated with the rhizobacteria-aphid effect, indicating there is a genetic-basis for this effect and is the first step to identifying the genes involved (Tétard-Jones et al. 2012). The microbial community of a plant may prime the plant through induced resistance responses to be more resistant to aphid attack (Pineda et al. 2013) and, in turn, aphids can manipulate the plant defensive systems in order to avoid the negative impacts of plant defences (Will et al. 2013). Variation among plant genotypes for aphid resistance (Delp et al. 2009) provides material for plant genotypes to mediate the effect of microorganisms on aphids.

Earthworms are known decomposer ecosystem engineers (Jones et al. 1994) and enhance plant performance by increasing nitrogen availability in the soil, which can further benefit aphids through an interaction chain. However, earthworms have been found to differentially affect the influence of host-plant genotype on two aphid species (Singh et al. 2014). An interaction between plant variety and earthworm treatment influenced the performance of *Aphis fabae* aphids but not of *Acyrthosiphon pisum* aphids in the same experiment. Other belowground effects on plant-aphid interactions can include other plants. In a barley-aphid system, the presence (and population origin) of a hemi-parasitic plant changed the distribution of aphids among host-plants (Zytynska et al. 2014). The effect on aphid distribution was dependent on a four-way interaction between plant genotype, aphid genotype, the competing aphid genotype and the population origin of the hemi-parasitic plant. Since no aphids were ever observed on the parasitic plant itself, it is assumed the interaction occurred indirectly through the host-plant. Volatiles from other plants within the same species can also influence aphid settling behaviour (Ninkovic et al. 2002), they may also act to attract natural enemies and further influence the aphid population dynamics (Paré and Tumlinson 1999). Thus, in a diverse community there can be many community members that alter aphid spatial distribution and, in turn, aphid population dynamics. It is likely that plant variation affects many more three or more-way interactions between plants, aphids, and other organisms. The few experiments mentioned above have only started to unravel the complexity of these interactions.

Aphid Population Genetics

Plant variation may also influence aphid population genetics, e.g., if aphid genotypes differ in their preference for, or performance on particular plant genotypes this could creating fine-scale genetic structuring of the population. The identity of the host-plant species can further influence the effect of plant variation on aphid population genetic structure through creating isolated patches via dispersal limitations. In other words, greater differentiation between aphid populations would be expected on more spatially separated long-lived trees than for a common grass species that grows more continuously over a landscape. In Israel a distance of 150 m between *Pistacia palaestina* trees was sufficient to create genetic differentiation among host-alternating galling aphids (*Baizongia pistaciae*) (Martinez et al. 2005). This means that each patch must have consisted of one host-tree individual plus the secondary grassy host, such that the descendent aphids from one tree returned to the same tree the next season. However, a follow-up study found no differentiation suggesting that variation in long-range dispersal capabilities of the aphids across populations may further influence the dynamics in this system (Ben-Shlomo and Inbar 2012). In two non-host-alternating species, *Macrosiphoniella tanacetaria* and *Metopeurum fuscoviride*, specialising on tansy plants (*Tanacetum vulgare*) genetic differentiation could be observed at the level of the individual plant in a single site (i.e., within a few meters of one another) (Loxdale et al. 2011). While Martinez et al. (2005) and Loxdale et al. (2011) both found that several aphid clones occupied each host individual, it was unknown if there was any association between plant genotype and aphid genotypes.

Herbivores may adapt to genetically-based traits of their hosts following the adaptive deme formation hypothesis, i.e., sedentary, specialist insect herbivores should form localized demes adapted to the host-individual (Edmunds and Alstad 1978). A meta-analysis by Van Zandt and Mopper (1998) showed this was particularly important for parthenogenic species such as aphids due to the potential for rapid evolution (Van Zandt and Mopper 1998). In aphids, such rapid evolution was studied using two clonal genotypes of *Myzus persicae* and showed that evolving populations grew faster compared to non-evolving populations in the presence of competitors and predators (Turcotte et al. 2011). In this study, plant variation was controlled but differential effects of preference and performance across the aphid genotype clones for different plant genotypes could further alter aphid population genetic structuring. For example, the presence of another aphid genotype was found to alter aphid preference across barley genotypes, leading some aphid genotypes to colonize a less-preferred plant genotype when in a genotypically diverse (aphid and plant) population (Zytynska and Preziosi 2013). Genetic differentiation will also be enhanced through limited gene flow across sub-populations, exasperated by limited dispersal, habitat patchiness and host-plant longevity (Mopper 1996, Van Zandt and Mopper 1998). The adaptation of the non-migratory galling aphid *Kaltenbachiella japonica* to the budburst phenologies of the host-trees (*Ulmus davidiana* var. *japonica*) was studied in Japan (Komatsu and Akimoto 1995). They found a strong association between tree budburst phenology and aphid egg-hatching rate using controlled reciprocal mating experiments, suggesting a strong selective pressure acting on the aphids at the level of the individual host-tree. Ants may also inhibit the production of winged aphids (Braendle et al. 2006), further enhancing genetic differentiation among sub-populations through acting to limit aphid dispersal.

Most temperate aphid species undergo sexual reproduction over winter, and so the genetic diversity of a population will be highest at the beginning of the season when the overwintering eggs hatch. Over the season, due to a variety of selection pressures, this genetic diversity would likely decrease (Loxdale et al. 2011). Plant within-species variation may also drive aphid population genetic structuring through influences on aphid performance, such that a plant genotype that confers high performance for an aphid would provide the best chance for survival of the aphid genotypes it is hosting. If similar aphid genotypes actively seek out similar host-plant genotypes (host preference) this would further enhance any effect. This is consistent with the genetic-similarity rule, which has predominantly been considered in studies on trees where more similar tree genotypes have been often found to host more similar associated communities of arthropods, fungi and plants (Bangert et al. 2006, Whitham et al. 2012). Such community genetic studies consider variation at a 'community of species' level, whereas in a plant-aphid system that community would refer to our aphid populations as a 'community of genotypes' associated with different plant genotypes. The effect of plant variation on aphid population genetics is still rarely studied but we show that variation in preference, performance and competition among aphid genotypes could all be mediated by plant variation to further influence aphid population genetic structure.

Detecting Effects of Plant Variation on Aphid Ecology

Comparing aphid preference and performance on plants of differing genotypes can unravel strong effects of plant variation, but for this to be ecologically meaningful requires sampling over the appropriate spatial scale. At the plant species level, the pea aphid exhibits strong genetic divergence among races associated with different plant species (Via 1991, Ferrari et al. 2006, Peccoud et al. 2009). The variation between host-plant species is obviously much greater than the variation within a species, and to create such differentiation it must also be associated with an important ecological trait. Therefore, the amount of variation for ecologically important traits within a host-plant species will also determine the strength of differentiation across different plant genotypes (Hughes et al. 2008). In practice this means that the particular genotypes chosen to assess aphid performance or preference will influence the final results. In many studies, plant genotypes are collected from a large geographical area or chosen for strong phenotypic variation in order to maximise the differences seen. This is a reasonable method to use to first determine if host-plant genotypic variation can have an effect on aphid performance or preference. However, by maximising the variation in greenhouse and common garden experiments the effect of plant variation on aphid ecology might be being over-estimated and other factors overlooked. A meta-analysis by Tack et al. (2012) found that spatial mismatch (i.e., genotypes collected from a wide geographical area but grown in a single site for a common garden experiment) was correlated with the relative amount of variation explained by host-plant genotype. In particular, species-specific abundances were overestimated when the spatial scale did not match (Tack et al. 2012), confirming that the effect of plant variation on aphid performance may be overestimated if the plant variation is greater than would be expected at the scale of the aphid population. Such high levels of variation are expected to be found in crop-species that are bred for particular traits and so inferences on the effect of plant variation on aphid ecology using agricultural crops must be interpreted as such. Thus, to fully understand the extent to which this could influence aphid dynamics on non-crop plants there needs to be a focus on naturally occurring variation at the scale that encompasses the observed aphid populations. Such studies would then be able to more accurately estimate the effect size of plant within-species variation on the aphid population.

Conclusions

Plant variation has long been ignored in aphid ecology and its study has often been restricted to studies of plant resistance to aphids, which has unravelled a number of mechanisms of plant defence and aphid adaptation to this defence. We have shown that plant variation can have a multitude of effects on aphids also in natural systems, either by direct effects of the plant on aphids, or indirect, by affecting the interaction between plant and aphid through other organisms such as microorganisms, natural enemies or mutualists. The study of the ecological consequences of within-species plant variation is, however, still in its infancy and future work is likely to unravel a number of further exciting effects on aphid ecology and genetics. Aphid populations in a diverse community experience many direct and indirect interactions that act to influence the population dynamics. Plant within-species variation can directly influence the aphids through changing aphid preference and performance, but importantly can also mediate other interactions that an aphid experiences. Figure 2 shows a conceptual framework to illustrate how plant variation will affect aphid ecology. The multitude of interactions will vary depending on the particular structure of the community in which

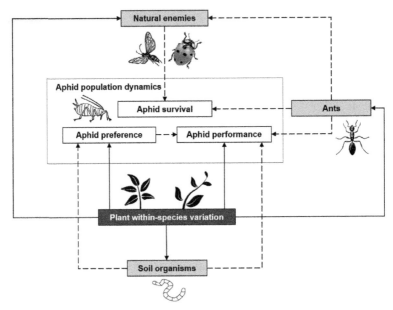

Fig. 2. Plant within-species variation can have direct effects (bold lines) on aphid performance and preference, and indirect effects (dashed lines) on aphid performance and survival through mediating interactions between aphids and natural enemies or ants. Soil organisms can also influence aphids through the host-plant.

the aphid resides, and the structure of the community may itself vary between locations. This means that the resulting population dynamics of an aphid population across the season are context dependent on the structure of the associated community in which it resides (Fig. 2). Further work should therefore strive to elucidate not only the effect of plant variation on aphid ecology, but also the context-dependency of these effects.

Keywords: plant variation effects, aphid performance, aphid preference, aphid-ant interactions, effects on aphid-natural enemies, population genetics, aphid ecology

References

Agrawal, A.A. 2004. Plant defense and density dependence in the population growth of herbivores. Amer. Nat. 164: 113–120.

Alkhedir, H., P. Karlovsky and S. Vidal. 2013. Relationship between water soluble carbohydrate content, aphid endosymbionts and clonal performance of *Sitobion avenae* on cocksfoot cultivars. PLoS One 8: e54327.

Bangert, R.K., R.J. Turek, B. Rehill, G.M. Wimp, J.A. Schweitzer and G.J. Allan. 2006. A genetic similarity rule determines arthropod community structure. Mol. Ecol. 15: 1379–1391.

Ben-Shlomo, R. and M. Inbar. 2012. Patch size of gall-forming aphids: deme formation revisited. Popul. Ecol. 54: 135–144.

Bethke, J.A., R.A. Redak and U.K. Schuch. 1998. Melon aphid performance on chrysanthemum as mediated by cultivar, and differential levels of fertilization and irrigation. Entomol. Exp. Appl. 88: 41–47.

Bolnick, D.I., P. Amarasekare, M.S. Araújo, R. Bürger, J.M. Levine and M. Novak. 2011. Why intraspecific trait variation matters in community ecology. Trends Ecol. Evol. 26: 183–192.

Braendle, C., G.K. Davis, J.A. Brisson and D.L. Stern. 2006. Wing dimorphism in aphids. Heredity 97: 192–199.

Buckley, R. 1987. Ant-plant-homopteran interactions. Adv. Ecol. Res. 16: 53–85.

Caillaud, C.M., C.A. Dedryver, J.P. Dipietro, J.C. Simon, F. Fima and B. Chaubet. 1995. Clonal variability in the response of *Sitobion avenae* (Homoptera, Aphididae) to resistant and susceptible wheat. Bull. Entomol. Res. 85: 189–195.

De Zutter, N., K. Audenaert, G. Haesaert and G. Smagghe. 2012. Preference of cereal aphids for different varieties of winter wheat. Arthropod Plant Interact. 6: 345–350.

Delp, G., T. Gradin, I. Åhman and L.M.V. Jonsson. 2009. Microarray analysis of the interaction between the aphid *Rhopalosiphum padi* and host plants reveals both differences and similarities between susceptible and partially resistant barley lines. Mol. Genet. Genomics 281: 233–248.

Dixon, A. and B. Agarwala. 1999. Ladybird-induced life–history changes in aphids. Proc. R. Soc. Lond. B Biol. Sci. 266: 1549–1553.

Edmunds, G.F. and D.N. Alstad. 1978. Coevolution in insect herbivores and conifers. Science 199: 941–945.

Ellison, A.M., M.S. Bank, B.D. Clinton, E.A. Colburn, K. Elliott and C.R. Ford. 2005. Loss of foundation species: consequences for the structure and dynamics of forested ecosystems. Front. Ecol. Environ. 3: 479–486.

Engel, V., M.K. Fischer, F.L. Wackers and W. Volkl. 2001. Interactions between extrafloral nectaries, aphids and ants: are there competition effects between plant and homopteran sugar sources? Oecologia 129: 577–584.

Ferrari, J., H.C.J. Godfray, A.S. Faulconbridge, K. Prior and S. Via. 2006. Population differentiation and genetic variation in host choice among pea aphids from eight host plant genera. Evolution 60: 1574–1584.

Fordyce, J.A. 2006. The evolutionary consequences of ecological interactions mediated through phenotypic plasticity. J. Exp. Biol. 209: 2377–2383.

Gripenberg, S., P.J. Mayhew, M. Parnell and T. Roslin. 2010. A meta-analysis of preference–performance relationships in phytophagous insects. Ecol. Lett. 13: 383–393.

Hamback, P.A., M. Vogt, T. Tscharntke, C. Thies and G. Englund. 2007. Top-down and bottom-up effects on the spatiotemporal dynamics of cereal aphids: testing scaling theory for local density. Oikos 116: 1995–2006.

Heil, M., B. Fiala, B. Baumann and K. Linsenmair. 2000. Temporal, spatial and biotic variations in extrafloral nectar secretion by *Macaranga tanarius*. Funct. Ecol. 14: 749–757.

Herbert, J.J., R. Mizell and H. McAuslane. 2009. Host preference of the Crapemyrtle aphid (Hemiptera: Aphididae) and host suitability of Crapemyrtle cultivars. Environ. Entomol. 38: 1155–1160.

Hughes, A.R., B.D. Inouye, M.T.J. Johnson, N. Underwood and M. Vellend. 2008. Ecological consequences of genetic diversity. Ecol. Lett. 11: 609–623.

Johnson, M.T.J. 2008. Bottom-up effects of plant genotype on aphids, ants, and predators. Ecology 89: 145–154.

Jones, C.G., J.H. Lawton and M. Shachak. 1994. Organisms as ecosystem engineers. Oikos 69(3): 373–386.

Jones, T.S., E. Allan, S.A. Härri, J. Krauss, C.B. Müller and F. van Veen. 2011. Effects of genetic diversity of grass on insect species diversity at higher trophic levels are not due to cascading diversity effects. Oikos 120: 1031–1036.

Kanvil, S., G. Powell and C. Turnbull. 2014. Pea aphid biotype performance on diverse Medicago host genotypes indicates highly specific virulence and resistance functions. Bull. Entomol. Res. 104: 689–701.

Karban, R. 1992. Plant variation: its effects on populations of herbivorous insects. pp. 195–215. *In*: R.S. Fritz and E.L. Simms (eds.). Plant Resistance to Herbivores and Pathogens: Ecology, Evolution, and Genetics. University of Chicago Press, Chicago.

Katayama, N., D.H. Hembry, M.K. Hojo and N. Suzuki. 2013. Why do ants shift their foraging from extrafloral nectar to aphid honeydew? Ecol. Res. 28: 919–926.

Kleine, S. and C. Müller. 2011. Intraspecific plant chemical diversity and its relation to herbivory. Oecologia 166: 175–186.

Klingler, J.P., R.M. Nair, O.R. Edwards and K.B. Singh. 2009. A single gene, AIN, in *Medicago truncatula* mediates a hypersensitive response to both bluegreen aphid and pea aphid, but confers resistance only to bluegreen aphid. J. Exp. Bot. 60(14): 4115–4127.

Komatsu, T. and S. Akimoto. 1995. Genetic differentiation as a result of adaptation to the phenologies of individual host trees in the galling aphid *Kaltenbachiella japonica*. Ecol. Entomol. 20: 33–42.

Kos, M., C. Broekgaarden, P. Kabouw, K. Oude Lenferink, E.H. Poelman and L.E. Vet. 2011. Relative importance of plant-mediated bottom-up and top-down forces on herbivore abundance on *Brassica oleracea*. Funct. Ecol. 25: 1113–1124.

Krauss, J., S.A. Härri, L. Bush, R. Husi, L. Bigler and S.A. Power. 2007. Effects of fertilizer, fungal endophytes and plant cultivar on the performance of insect herbivores and their natural enemies. Funct. Ecol. 21: 107–116.

Lamb, R.J., P.A. MacKay and S.M. Migui. 2009. Measuring the performance of aphids: fecundity versus biomass. Can. Entomol. 141: 401–405.

Levin, D.A. 1973. The role of trichomes in plant defense. Q. Rev. Biol. 48(1): 3–15.

Linhart, Y.B., K. Keefover-Ring, K.A. Mooney, B. Breland and J.D. Thompson. 2005. A chemical polymorphism in a multitrophic setting: thyme monoterpene composition and food web structure. Amer. Nat. 166: 517–529.

Loxdale, H.D., G. Schöfl, K.R. Wiesner, F.N. Nyabuga, D.G. Heckel and W.W. Weisser. 2011. Stay at home aphids: comparative spatial and seasonal metapopulation structure and dynamics of two specialist tansy aphid species studied using microsatellite markers. Biol. J. Linnean Soc. 104: 838–865.

Martinez, J.J.I., O. Mokady and D. Wool. 2005. Patch size and patch quality of gall-inducing aphids in a mosaic landscape in Israel. Landsc. Ecol. 20: 1013–1024.

Martinez, J.J.I., M. Cohen and N. Mgocheki. 2011. The response of an aphid tending ant to artificial extra-floral nectaries on different host plants. Arthropod Plant Interact. 5: 185–192.

McHale, L., X. Tan, P. Koehl and R. Michelmore. 2006. Plant NBS-LRR proteins: adaptable guards. Genome Biol. 7: 212. doi:10.1186/gb-2006-7-4-212.

Mehrparvar, M., S.E. Zytynska and W.W. Weisser. 2013. Multiple cues for winged morph production in an aphid metacommunity. PLoS ONE 8: e58323.

Mondor, E.B. and J.F. Addicott. 2003. Conspicuous extra-floral nectaries are inducible in *Vicia faba*. Ecol. Lett. 6: 495–497.

Mooney, K.A. and A.A. Agrawal. 2008. Plant genotype shapes ant-aphid interactions: implications for community structure and indirect plant defense. Amer. Nat. 171(6): E195–E205. doi: 10.1086/587758.

Mopper, S. 1996. Adaptive genetic structure in phytophagous insect populations. Trends Ecol. Evol. 11: 235–238.

Moreira, X. and K.A. Mooney. 2013. Influence of plant genetic diversity on interactions between higher trophic levels. Biol. Lett. 9(3): 20130133. doi: 10.1098/rsbl.2013.0133.

Mousseau, T.A. and H. Dingle. 1991. Maternal effects in insect life histories. Annu. Rev. Entomol. 36: 511–534.

Müller, C.B., I.S. Williams and J. Hardie. 2001. The role of nutrition, crowding and interspecific interactions in the development of winged aphids. Ecol. Entomol. 26: 330–340.

Ninkovic, V., U. Olsson and J. Pettersson. 2002. Mixing barley cultivars affects aphid host plant acceptance in field experiments. Entomol. Exp. Appl. 102: 177–182.

Offenberg, J. 2000. Correlated evolution of the association between aphids and ants and the association between aphids and plants with extrafloral nectaries. Oikos 91: 146–152.

Paré, P.W. and J.H. Tumlinson. 1999. Plant volatiles as a defense against insect herbivores. Plant Physiol. 121: 325–332.

Peccoud, J., A. Ollivier, M. Plantegenest and J.-C. Simon. 2009. A continuum of genetic divergence from sympatric host races to species in the pea aphid complex. Proc. Nat. Acad. Sci. 106: 7495–7500.

Pettersson, J., V. Ninkovic and E. Ahmed. 1999. Volatiles from different barley cultivars affect aphid acceptance of neighbouring plants. Acta Agric. Scand., Sect. B Plant Soil Sci. 49: 152–157.

Pineda, A., M. Dicke, C.M.J. Pieterse and M.J. Pozo. 2013. Beneficial microbes in a changing environment: are they always helping plants to deal with insects? Funct. Ecol. 27: 574–586.

Powell, G., C.R. Tosh and J. Hardie. 2006. Host plant selection by aphids: behavioral, evolutionary, and applied perspectives. Annu. Rev. Entomol. 51: 309–330.

Reinink, K., F.L. Dieleman, J. Jansen and A.M. Montenarie. 1989. Interactions between plant and aphid genotypes in resistance of lettuce to *Myzus persicae* and *Macrosiphum euphorbiae*. Euphytica 43: 215–222.

Rudgers, J.A. and M.C. Gardener. 2004. Extrafloral nectar as a resource mediating multispecies interactions. Ecology 85: 1495–1502.

Ryan, G.D., L. Emiljanowicz, S.A. Haerri and J.A. Newman. 2014. Aphid and host-plant genotype×genotype interactions under elevated CO_2. Ecol. Entomol. 39: 309–315.

Sauge, M.-H., F. Mus, J.-P. Lacroze, T. Pascal, J. Kervella and J.-L. Poessel. 2006. Genotypic variation in induced resistance and induced susceptibility in the peach-*Myzus persicae* aphid system. Oikos 113: 305–313.

Schädler, M., R. Brandl and A. Kempel. 2010. Host plant genotype determines bottom up effects in an aphid parasitoid predator system. Entomol. Exp. Appl. 135: 162–169.

Service, P. 1984. Genotypic interactions in an aphid-host plant relationship: *Uroleucon rudbeckiae* and *Rudbeckia laciniata*. Oecologia 61: 271–276.

Shoffner, A.V. and J.F. Tooker. 2013. The potential of genotypically diverse cultivar mixtures to moderate aphid populations in wheat (*Triticum aestivum* L.). Arthropod Plant Interact. 7: 33–43.

Singh, A., J. Braun, E. Decker, S. Hans, A. Wagner and W. Weisser. 2014. Plant genetic variation mediates an indirect ecological effect between belowground earthworms and aboveground aphids. BMC Ecol. 14: 25. doi:10.1186/s12898-014-0025-5.

Smith, C.M. and S.L. Clement. 2012. Molecular bases of plant resistance to arthropods. Annu. Rev. Entomol. 57: 309–328.

Smith, C.M. and W.P. Chuang. 2014. Plant resistance to aphid feeding: behavioral, physiological, genetic and molecular cues regulate aphid host selection and feeding. Pest Manag. Sci. 70: 528–540.

Smith, D.S., J.K. Bailey, S.M. Shuster and T.G. Whitham. 2011. A geographic mosaic of trophic interactions and selection: trees, aphids and birds. J. Evol. Biol. 24: 422–429.

Stadler, B. 2004. Wedged between bottom-up and top-down processes: aphids on tansy. Ecol. Entomol. 29: 106–116.

Stadler, B. and A.F.G. Dixon. 2005. Ecology and evolution of aphid-ant interactions. Ann. Rev. Ecol. Evol. Syst. 36: 345–372.

Tack, A.J.M., M.T.J. Johnson and T. Roslin. 2012. Sizing up community genetics: it's a matter of scale. Oikos 121: 481–488.

Tenhumberg, B. 2010. Ignoring population structure can lead to erroneous predictions of future population size. Nat. Educ. Knowl. 3: 2.

Tétard-Jones, C., M.A. Kertesz, P. Gallois and R.F. Preziosi. 2007. Genotype-by-genotype interactions modified by a third species in a plant-insect system. Amer. Nat. 170: 492–499.

Tétard-Jones, C., M.A. Kertesz and R.F. Preziosi. 2012. Identification of plant quantitative trait loci modulating a rhizobacteria-aphid indirect effect. PloS One 7: e41524.

Turcotte, M.M., D.N. Reznick and J.D. Hare. 2011. The impact of rapid evolution on population dynamics in the wild: experimental test of eco-evolutionary dynamics. Ecol. Lett. 14: 1084–1092.

Underwood, N. 2007. Variation in and correlation between intrinsic rate of increase and carrying capacity. Amer. Nat. 169: 136–141.

Underwood, N. 2009. Effect of genetic variance in plant quality on the population dynamics of a herbivorous insect. J. Animal Ecol. 78: 839–847.

Underwood, N., S. Halpern and C. Klein. 2011. Effect of host-plant genotype and neighboring plants on strawberry aphid movement in the greenhouse and field. Am. Midl. Nat. 165: 38–49.

Utsumi, S., Y. Ando, T.P. Craig and T. Ohgushi. 2011. Plant genotypic diversity increases population size of a herbivorous insect. Proc. R. Soc. B, Biol. Sci. 278: 3108–3115.

Van Zandt, P.A. and S. Mopper. 1998. A meta-analysis of adaptive deme formation in phytophagous insect populations. Amer. Nat. 152: 595–604.

Via, S. 1991. Specialized host plant performance of pea aphid clones is not altered by experience. Ecology 72: 1420–1427.

Whitham, T.G., J.K. Bailey, J.A. Schweitzer, S.M. Shuster, R.K. Bangert and C.J. LeRoy. 2006. A framework for community and ecosystem genetics: from genes to ecosystems. Nat. Rev. Genet. 7: 510–523.

Whitham, T.G., C.A. Gehring, L.J. Lamit, T. Wojtowicz, L.M. Evans and A.R. Keith. 2012. Community specificity: life and afterlife effects of genes. Trends Plant Sci. 17: 271–281.

Will, T., A.C.U. Furch and M.R. Zimmermann. 2013. How phloem-feeding insects face the challenge of phloem-located defenses. Front. Plant Sci. 4: 336. doi: 10.3389/fpls.2013.00336.

Williams, R.S. and M.A. Avakian. 2015. Colonization of *Solidago altissima* by the specialist aphid *Uroleucon nigrotuberculatum*: effects of genetic identity and leaf chemistry. J. Chem. Ecol. 41: 129–138.

Wilson, M.R. and S.R. Leather. 2012. The effect of past natural enemy activity on host-plant preference of two aphid species. Entomol. Exp. Appl. 144: 216–222.

Wimp, G.M. and T.G. Whitham. 2001. Biodiversity consequences of predation and host plant hybridization on an aphid-ant mutualism. Ecology 82: 440–452.

Wootton, J.T. 1994. The nature and consequences of indirect effects in ecological communities. Annu. Rev. Ecol. Syst. 25: 443–466.

Zytynska, S.E. and R.F. Preziosi. 2011. Genetic interactions influence host preference and performance in a plant-insect system. Evol. Ecol. 25: 1321–1333.

Zytynska, S.E. and R.F. Preziosi. 2013. Host preference of plant genotypes is altered by intraspecific competition in a phytophagous insect. Arthropod Plant Interact. 7: 349–357.

Zytynska, S.E., S. Fleming, C. Tétard-Jones, M.A. Kertesz and R.F. Preziosi. 2010. Community genetic interactions mediate indirect ecological effects between a parasitoid wasp and rhizobacteria. Ecology 91: 1563–1568.

Zytynska, S.E., L. Franz, B. Hurst, A. Johnson, R.F. Preziosi and J. Rowntree. 2014. Host-plant genotypic diversity and community genetic interactions mediate aphid spatial distribution. Ecol. Evol. 4: 121–131.

9

Chemical Ecology of Aphids (Hemiptera: Aphididae)

Antoine Boullis and *François J. Verheggen**

Aphid Olfaction

Aphids are among the most important agricultural pests in the temperate regions of the north hemisphere (Dixon 1998), and are currently used as models for evolutionary and functional biological research. Because of their negative impact on agricultural production, aphids were among the first organisms to be studied for their semiochemistry. Semiochemicals are chemical substances that carry a message from one living organism to another for the purpose of communication. The identification of chemical cues that guide individuals according to their needs is now a major discipline, termed chemical ecology, studied in a wide range of organisms, especially insects (Thompson et al. 1999). The semiochemicals emitted and/or perceived by aphids have been extensively documented over the last 40 yr, with the first sex and alarm pheromones being identified in the early 1970s (Bowers et al. 1972, Marsh 1972).

The chemical signals involved in communication are usually separated according to whether they affect the behavior of individuals belonging to the same species as that of the emitter or not. Pheromones are semiochemicals that are used within a species, while allelochemicals act between individuals from different species. Sometimes, the same substance may be used in both intra- and inter-specific communication, such as the aphid alarm pheromone, which acts as an alarm signal within an aphid colony and as a kairomone (allelochemical) when it is eavesdropped by a predator (e.g., Nakatuma 1991).

Functional and evolutionary entomology, Gembloux Agro-bio Tech (University of Liège), Passage des déportés 2, 5030 Gembloux, Belgium.
* Corresponding author: fverheggen@ulg.ac.be

The perception of semiochemicals in aphids is mediated by receptors situated on the antennae. These sensory organs, called sensilla, are classified in different categories depending on their position on the antenna, their physiological role, or their structure. Aphid sensilla were first studied at the start of the 20th century by Flögel (1905), and were fully described by Bromley et al. (1979, 1980). In aphids, the semiochemicals are perceived by sensory structures called rhinaria. Anatomically, they are circular openings covered by a membrane on the antennae of aphids (Torre-Bueno and Tulloch 1962), and contain receptor neurons that allow aphids to detect volatile compounds. Rhinaria are classified in two main groups: primary and secondary rhinaria.

Primary rhinaria are mainly sensitive organs constituted by an assembly of several sensillum types and are involved in host plant location (Yan and Visser 1982). They are located on the 5th and 6th segments of the antenna (Zhang and Zhang 2000). For example, distal (DPR) and proximal (PPR) primary rhinaria allow all morphs and lifestages of *Aphis fabae* Scopoli to detect (E)-2-hexenal, a common volatile of their host-plants (Park and Hardie 2004) that is not detected by secondary rhinaria. PPR are usually associated with the perception of host and non-host volatile chemicals. DPR are probably involved in the perception of the alarm pheromone (Wohlers and Tjallingii 1983, Pickett et al. 1992), which is transported throughout the sensillar lymph by odorant-binding proteins (OBPs) (Vandermoten et al. 2011, Sun et al. 2012).

Secondary rhinaria are made of sensilla placoidea that contain neurons with receptors that are sensitive to the aphid sex pheromone constituents (Marsh 1975, Dawson et al. 1987a). These rhinaria are present in larger numbers on the antennae of male aphids compared to female aphids (Hardie et al. 1994a). In males, secondary rhinaria are distributed from the third to the fifth segment, while in females they are concentrated on the third antennal segment (Park and Hardie 2002). In addition to this sexual dimorphism, differences also occur in terms of number of the secondary rhinaria between alate and wingless individuals (Pickett et al. 1992, Park and Hardie 2002), with this variation differing also among species (Shambaugh et al. 1978). In asexual morphs that do not behaviorally respond to sexual stimuli, these structures are sensitive to plant volatiles and alarm pheromone (Pickett et al. 1992, Hardie et al. 1994a). Contrarywise, their high prevalence on antennae of alate individuals suggests that they are implicated in sex pheromone detection (Pickett et al. 1992). However, they would also be implicated in the spacing and aggregation phenomena of individuals and in the possible recognition of alarm pheromone and plant volatiles (Hardie et al. 1994a).

The chemical ecology of aphids is increasingly being studied with the aim of providing alternatives to insecticide applications (Dewhirst et al. 2010). In this chapter, we present recent state of the art information about aphid chemical ecology, including the role of semiochemicals in intraspecific aphid relations, in addition to between aphids and their host plants, and between aphids and their natural enemies. The potential application of semiochemicals to protect crops against aphids is also discussed.

Aphid Pheromones

Within an insect colony, communication mostly occurs through the use of chemicals emitted and perceived by individuals (Richard and Hunt 2013). Compared to solitary insects, gregarious aphids are more likely to communicate amongst themselves by using different kinds of signals, particularly chemicals. While it is commonly accepted that pheromones may be classified into six behaviorally functional groups (Birch 1974, Tillman et al. 1999), aphids tend to rely on four of these groups; specifically, sex (Pettersson 1970a, Marsh 1972), alarm (Kislow and Edwards 1972), aggregation, and spacing (Pettersson et al. 1995) pheromones, which are classified according to the physiological and behavioral modifications induced in the receivers.

Sex Pheromone

Aphid species that are recognized as important pests for agriculture belong to the Aphidinae subfamily (Pickett and Glinwood 2007). Aphidinae have complex life cycles, including different reproductive phases and host plant alternation: these species are called heteroecious. In early spring, females (called fundatrices) hatch on the winter host plant and begin to feed and reproduce asexually, producing asexual wingless and winged females. Winged females spread to find and feed on summer hosts plants, generally consisting of grass species. While feeding on summer host plants, females continue to reproduce by parthenogenesis for several generations, allowing aphid populations to grow fast. After this phase of rapid and asexual reproduction, the second reproductive phase begins in early autumn (characterized by a decrease in day length and temperature), leading to the production of alate males and females (Moran 1992). Alate females, called gynoparae, then migrate on the primary (winter) hosts to produce oviparae, which are wingless females able to produce sex pheromones from scent plaques situated on their hind tibiae that attract males and promote sexual reproduction (Pettersson 1970a, Marsh 1972). After mating, females lay overwintering eggs (Fig. 1) (see also Chapter 2 "The Ontogenesis of the Pea Aphid *Acyrthosiphon pisum*" in this volume).

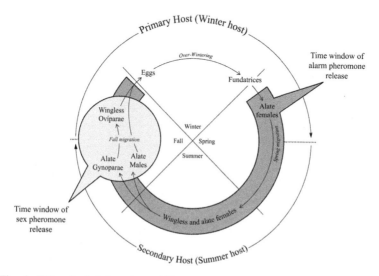

Fig. 1. Theorical lifecycle of a heteroecious aphid species and associated periods of pheromone production.

The role of sex pheromones in Aphidinae was first described by Pettersson (1970a) for the *Schizaphis* genus, followed by additional species (Marsh 1972), via olfactometric assays. These assays showed that males are attracted by odors released by oviparae. Subsequent research focused on the chemical identification of this sex pheromone and evaluating potential differences that occur among aphid species (Hardie et al. 1999). In the majority of Aphidinae species, the sex pheromone is composed of two main chemical compounds; namely, (4a*S*,7*S*,7a*R*)-nepetalactone (**I** - Fig. 2), and (1*R*,4a*S*,7*S*,7a*R*)-nepetalactol (**II** - Fig. 2). Both compounds were first identified in the vetch aphid *Megoura viciae* Buckton by single cell recording (SCR) coupled with gas chromatography (GC) (Dawson et al. 1987a). In other aphid species considered economically-important, such as *Acyrthosiphon pisum* Harris, *Myzus persicae* Sulzer, and *A. fabae*, the same two semiochemicals were also identified (Dawson et al. 1990) with species-specific ratios of these compounds, as common in insects (Webster 2012, Pickett et al. 2013). Additional stereoisomers were identified by Campbell et al. (1990) as components of the sex pheromone of the damson-hop aphid, *Phorodon humuli* Schrank: (1*R*,4a*R*,7*S*,7a*S*)-nepetalactol and (1*S*,4a*R*,7*S*,7a*S*)-nepetalactol (**III**, **IV** respectively - Fig. 2). Moreover, field trap and olfactometry assays have shown that individuals from *P. humuli* do not respond to nepetalactone (**I**) and show little behavioral responses toward nepetalactol (**II**) compared to the species-specific pheromone (Campbell et al. 2003). The different compounds of the

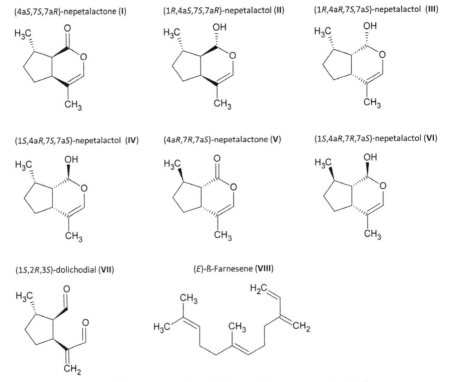

Fig. 2. Chiral representation of different aphid pheromone compounds.

sex pheromone are perceived by different sensitive cells situated onto the secondary rhinaria (Dawson et al. 1990, Lilley et al. 1994/95).

The sexual pheromonal blend identified in the majority of aphid species is generally composed of one of the nepetalactone enantiomers (**I** and/or **V** - Fig. 2) and/or one of the nepetalactol enantiomers (**II** and/or **VI** - Fig. 2) (Stewart-Jones et al. 2007) (Table 1). The stereochemistry of the aphid sex pheromone is, therefore, a major element that explains the species-specific attraction capacity of the pheromonal blend (Hardie et al. 1997, Campbell et al. 2003). Some synergisms were also identified with specific host-plant volatiles (Hardie et al. 1994b, Pope et al. 2007). This phenomenon has been suggested in the blackcurrant aphid, *Cryptomyzus galeopsidis* Kaltenbach. For instance, the females of various *Cryptomyzus* species produce a sex pheromone with an identical ratio of (**I**) and (**II**) (i.e., 1:30, respectively), with males from *C. galeopsidis* being able to distinguish conspecific odors from sibling species in choice experiments (Guldemond et al. 1993). Moreover, *C. galeopsidis* males are also able to differentiate synthetic odors (1:30 natural mimicked ratio) from oviparae conspecific females releasing their own sex pheromone. These results indicate that additional stimuli may be involved in the mechanisms of specific recognition (Lilley and Hardie 1996).

In some species, the quantity and enantiomeric composition (lactone:lactol) of the sex pheromone may be modulated, as observed in *M. viciae*, which releases higher quantities of pheromones about one week after the beginning of its adult lifestage (Hardie et al. 1990). Similarly, *Dysaphis plantaginea* Passerini oviparae release the greatest quantity of sex pheromone on the seventh day after the last molting (Stewart-Jones et al. 2007). The subsequent decline in sex pheromone production may be the result of senescence in the sex pheromone-producing glandular tissues. Photoperiod also influences pattern in the emission of sex pheromone levels, with greater amounts being released during the photophase, as observed in *Aphis spiraecola* Patch (Jeon et al. 2003) and *D. plantaginea* (Stewart-Jones et al. 2007). These changes may be implicated in a species differentiation mechanism that improves isolation between sympatric aphid species.

Although the ratio of nepetalactone (**I** and/or **V**) and nepetalactol (**II** and/or **VI**) is species-specific, it may be qualitatively modified over time, leading to cases of interspecific activity, as documented by laboratory and field trappings studies (Dawson et al. 1990, Hardie et al. 1990, Boo et al. 2000). In the case of sympatric species that share the same host plant during the same period (e.g., *D. plantaginea* and *Rhopalosiphum insertum* Walker on primary host apple trees *Malus* spp.—Table 1), differences in their ratios should be sufficiently important to permit aphids from each species to avoid mistakes in specific recognition and to prevent the effect of temporal variation in their pheromone release. Indeed, *R. insertum* and *D. plantaginea* release greater amounts of nepetalactol and nepetalactone, respectively, in their sexual pheromones, conferring a significant difference between the two species pheromones (Table 1). However, other criteria might enhance this differentiation between the two pheromonal blends.

In addition to nepetalactone (**I** and **V**) and nepetalactol (**II** and **VI**) isomers, Dewhirst et al. (2008) identified additional semiochemicals released in the headspace of oviparae of the rosy apple aphid (*D. plantaginea*): phenylacetonitrile and

Table 1. Ratios of the sex pheromone components (nepetalactone I–V and nepetalactol II–VI) released by oviparae females of several aphid species.

Common name	Species name	Ratio lactone : lactol	Reference
Greenbug	*Schizaphis graminum* Rondani	1 : 8	Dawson et al. 1988
Damson hop aphid	*Phorodon humuli*	0 : 1 [w]	Campbell et al. 1990
Vetch aphid	*Megoura viciae*	5 : 1 to 12 : 1 [*]	Hardie et al. 1990
Pea aphid	*Acyrthosiphon pisum*	1 : 1	Dawson et al. 1990
Black-bean aphid	*Aphis fabae*	29 : 1	Dawson et al. 1990
Peach-potato aphid	*Myzus persicae*	1 : 1.5	Dawson et al. 1990
Black-berry cereal aphid	*Sitobion fragariae* Walker	1 : 0	Hardie et al. 1992
Currant aphid	*Cryptomyzus spp.*	1 : 30	Guldemond et al. 1993
Bird cherry-oat aphid	*Rhopalosiphum padi*	0 : 1	Hardie et al. 1994a
Grain aphid	*Sitobion avenae* F.	1 : 0	Lilley et al. 1994/95
Cabbage aphid	*Brevicoryne brassicae*	1 : 0	Gabrys et al. 1997
Peach aphid	*Tuberocephalus momonis* Matsumura	4 : 1	Boo et al. 2000
Spiraea aphid	*Aphis spiraecola*	(2 : 1) [x] – (6 : 1 to 8 : 1) [y] [*]	Jeon et al. 2003
Potato aphid	*Macrosiphum euphorbiae* Thomas	1 : 2 to 1 : 4 [*]	Goldansaz et al. 2004
Soybean aphid	*Aphis glycines*	2 : 1	Zhu et al. 2006
Lettuce aphid	*Nasonovia ribis-nigri* Moseley	1.5 : 1	Dewhirst 2007
Peach aphid	*Ovatus insitus* Walker	2 : 1	Dewhirst 2007
Apple grass aphid	*Rhopalosiphum insertum*	1 : 21	Dewhirst 2007
Rosy apple aphid	*Dysaphis plantaginea*	1 : 3.7	Stewart-Jones et al. 2007
		1 : 4	Dewhirst et al. 2008
Mealy plum aphid	*Hyalopterus pruni* Geoffrey	(2.5 : 1 – 3.4 : 1) [z]	Symmes et al. 2012
Leaf-curl plum aphid	*Brachycaudus helichrysi* Kaltenbach	2.6 : 1	Symmes et al. 2012
Tea aphid	*Toxoptera aurantii* Fonscolombe	4.3 : 1 – 4.9 : 1	Han et al. 2014

[w] Diastereisomers **III** and **IV** of nepetalactol, [x] Field-collected oviparae, [y] Lab-reared oviparae, [z] Differences between strains, [*] Changes linked with aphid age

(1*S*,2*R*,3*S*)-dolichodial (**VII** - Fig. 2). Both chemicals elicited electrophysiological responses from male antennae, with (1*S*,2*R*,3*S*)-dolichodial increasing the attractive potential of the "classic" pheromonal blend (a 1:4 ratio of nepetalactone (**I**) and nepetalactol (**II**)) during bioassays. Moreover, other unpublished studies (see Dewhirst et al. 2008) have shown that this third pheromonal compound may also be present in the sex pheromones of other aphid species, such as *Cryptomyzus maudamanti* Guldemond or *Cryptomyzus ribis* L. Thus, (1*S*,2*R*,3*S*)-dolichodial may play a predominant role in the species-specific composition of the sex pheromones in *Cryptomyzus* species.

Alarm Pheromone

Another important functional group of aphid semiochemicals is alarm signals. Alarm pheromones are widespread among insects (Verheggen et al. 2010), and are considered to be the second functional group of insect pheromone in terms of production, after sex pheromones (Barbier 1982). As the majority of insect pheromones, alarm signals are composed of a range of chemical molecules, with their activity depending on the composition, the ratio of different compounds, and/or the stereoisomerism of the prevailing semiochemicals (Wadhams 1990). However, despite this range of possibilities, alarm pheromones tend to be less specialized than other types of pheromones (Blum 1985).

Aphids are defenseless soft-bodied insects. However, they secrete liquid droplets from two cornicles situated on the upper posterior surface of their abdomen in response to predation or other threats (Fig. 3). These droplets contain triglycerides that may serve as mechanical defenses by gluing mouthparts of natural enemies (Callow et al. 1973,

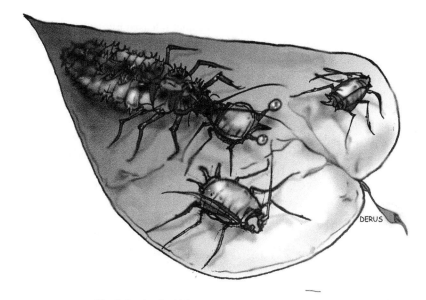

Fig. 3. Predated aphid emitting droplets from cornicles.

Minks and Harrewijn 1987). In addition, the droplets release a repellent odor that alerts conspecifics to the threat (Kislow and Edwards 1972). The nature of the behavioral response varies within and among species (Montgomery and Nault 1977, 1978) as well as with the predation stimulus and costs of escape. Following the perception of the alarm signal, individuals in the vicinity may cease feeding, escape from the release point or, in some cases, drop from the leaves of the host plant (Pickett et al. 1992). Moreover, this alarm pheromone may stimulate an increase in the production of winged individuals that are specialized for dispersal (Kunert et al. 2005, Hatano et al. 2010). This wing polyphenism induction may also be enhanced when physical contacts between aphids within a colony are combined with the chemical signal ("pseudo-crowding hypothesis"—Sloggett and Weisser 2002, Kunert and Weisser 2005).

The chemical characterization of this pheromone was first highlighted for several aphid species by Bowers and his colleagues in 1972 (Bowers et al. 1972). They found the main component of this alarm pheromone to be the sesquiterpene (E)-7,11-dimethyl-3-methylene-1,6,10-dodecatriene, $C_{15}H_{24}$ (also known as (E)-β-Farnesene, Eβf or *trans*-β-Farnesene) (**VIII** - Fig. 2). This molecule is considered as the only or the main active component of the alarm pheromone of most Aphidinae species. For instance, Francis et al. (2005a) published a list of 16 aphid species that use Eβf as the only, or the main, component of the alarm pheromone. Four other species have been added to this list: *Rhopalosiphum maidis* Fitch, *Aphis glycines* Matsumura, *A. spiraecola*, and *Brachycaudus persicae* Passerini (F.J. Verheggen, unpublished data). In addition, in alfalfa aphids (i.e., genus *Therioaphis*), sesquiterpene germacrene A, which is considered as a biogenic precursor of many sesquiterpenes, plays a role in alarm communication (Bowers et al. 1977); however, this chemical does not induce escape behavior by other species. The pheromone composition of some Aphidinae species (e.g., *M. viciae*) includes additional chemicals, such as α-pinene, β-pinene, and limonene (Pickett and Griffiths 1980, Francis et al. 2005a).

It has long been considered that each droplet released by the cornicles contain Eβf (Mondor et al. 2000, Schwartzberg et al. 2008). A recent study showed that only 30–60% of droplets emitted by *A. pisum* contained Eβf (depending on the predator attacking), even though all individuals contained Eβf (Joachim et al. 2013). This result contradicts the concept that droplet production by aphids is correlated with alarm signal emission.

Benefits and costs related to the emission of aphid alarm pheromones have been frequently discussed since its original identification in the early 1970s. Cornicles release a liquid that may glue the mouthparts of the attacking predator, providing direct benefit to the releaser. However, the escape of the emitter occurs only in around 10% of attacks (Dixon 1958, Edwards 1966), indicating that the function of cornicle secretions may be more significant for colony defense than as a direct defensive mechanism for individual aphids (Wu et al. 2010, Vandermoten et al. 2012). Within a colony, aphids are generally clone-mates. For instance, the inclusive fitness theory (Hamilton 1964) suggests that alarm signals deployed during predation events may have substantial inclusive fitness benefits for the aphid producing them, as the nearby individuals who benefit share exactly the same genotype (Mondor and Roitberg 2004). As demonstrated by Robertson et al. (1995), *A. pisum* individuals are more likely to emit alarm pheromone when situated near to clonal conspecifics compared to heteroclonal conspecifics or other aphid species. $E\beta f$ production per individual is also relatively more important in pre-reproductive instars compared to adults (Mondor et al. 2000, Schwartzberg et al. 2008), which may be because larval instars are more likely to find related clones in close proximity to them compared to reproductive adults, which are more likely to disperse.

Moreover, costs related with pheromone production are not negligible in terms of physiology and ecology. Because plant sap is generally poor in lipids, and because triglycerides are the main components of droplets emitted via the cornicles, it is generally costly for aphids to produce these droplets (Vandermoten et al. 2012). Indeed, because the *de novo* production of $E\beta f$ is related to the biosynthetic pathways of juvenile hormones (Gut et al. 1987, Vandermoten et al. 2012), its synthesis may impact individual development and offspring production. Mondor and Roitberg (2003) confirmed this hypothesis by showing that the emission of alarm pheromone in pre-reproductive *A. pisum* instars has a negative impact on their reproduction. Moreover, to avoid the useless production of this alarm signal, its emission is related to their social environment, with higher quantities being produced in individuals situated in larger colonies compared to smaller ones (Verheggen et al. 2009). $E\beta f$ release does not appear to be contagious, whereby perceiving aphids do not release additional $E\beta f$ upon the perception of the conspecific alarm signal. This finding supports the reduction of the costs associated with the production of this semiochemical (Hatano et al. 2008a, Verheggen et al. 2008a). Thus, to avoid the overproduction of $E\beta f$, physical contact may replace the role of chemical signaling in high-density colonies.

Host plants are also able to produce and emit $E\beta f$ in the volatile blends of both intact and damaged plants (Gibson and Pickett 1983, Agelopoulos et al. 2000, Webster et al. 2008a). In the absence of predation pressure, the release of $E\beta f$ in their constitutive volatile emissions might be advantageous for plants to reduce herbivory pressure in, at least, three different ways (Kunert et al. 2010): (i) plants might directly avoid the settling of aphids by repelling them via the continuous release of their alarm signal; (ii) plants might disrupt the feeding episodes of aphids, reducing the direct fitness of aphids; or (iii) plants might influence the production of winged offspring, reducing herbivory pressure. Moreover, plants might also habituate aphids to this semiochemical and make them less reactive to short- or long-term release events of $E\beta f$ (Petrescu et al. 2001, de Vos et al. 2010), or might enhance the attraction of natural enemies (see below).

To counteract the defense strategies established by host plants, aphids have become sensitive to variations in $E\beta f$ release events (frequency of release), rather than to the intensity of $E\beta f$ release events (Kunert et al. 2005). The time profile of $E\beta f$ emission by a predated *A. pisum* is characterized by a rapid increase, directly after predatory attack (within the first 15 min of the aphid being attacked), followed by a continuous release at lower concentrations for about 1 hr (Schwartzberg et al. 2008, Joachim et al. 2013). Neither the predator species (Chrysopidae and Coccinellidae) nor the developmental stage (adult or larvae) affects the $E\beta f$ emission profile (Joachim et al. 2013, A. Boullis, unpublished data). Thus, in an ecological context, the dynamics of alarm pheromone release may be more important than the quantity of pheromone being released.

Other Pheromones

In addition to sex and alarm pheromones, aphids are also able to respond to, at least, two other functional groups of semiochemicals; namely, aggregation and spacing pheromones (Pettersson et al. 1995). These pheromones inform conspecifics about host-plant quality and the density of conspecifics in the vicinity (Way 1973), and may influence the attraction and aversion of conspecifics.

As found for several animal species, one of the benefits of forming aggregates is to decrease the predation pressure exerted by natural enemies (Allee 1931, Turchin and Kareiva 1989). Aggregation phenomena tend to be most common in heteroecious aphid species because of their need to change host-plant during seasons, with aggregations possibly improving the efficiency of alarm signals during predation events. Several field and laboratory studies have demonstrated the capacity of aphids to find areas where conspecifics are already feeding on viable plants. Pettersson (1993) used olfactometer bioassays to show that the gynoparae of *Rhopalosiphum padi* L. are attracted by other conspecific gynoparae odors. This observation was confirmed by field trap experiments and observations of aphids on winter host-plants. These data inform about the eventual presence of an aggregation pheromone emitted by the gynoparae of *R. padi*, even though the chemical characterization of this pheromone has yet to be elucidated. The aggregation of conspecific gynoparae on the winter host might also benefit aggregations of newborn oviparae, in parallel to increasing the perception of sex pheromone by males, due to the greater concentrations being released, which would promote mating. Moreover, even if the components of the sex pheromone (4aS,7S,7aR)-nepetalactone (**I**) and (1R,4aS,7S,7aR)-nepetalactol (**II**) are important for male attraction (e.g., Boo et al. 2000), both olfactory bioassays and electrophysiological analyses have shown that gynopae are also lured by these compounds (Campbell et al. 1990, 2003, Park et al. 2000, Pope et al. 2004, Zhu et al. 2006, Dewhirst et al. 2008, Symmes et al. 2012, Han et al. 2014), especially when considering species-specific ratios. Thus, because gynoparae do not have any direct fitness benefit by responding to a sex pheromone emitted by oviparae, the two sex pheromone compounds might serve as aggregation stimuli that improve host plant location. Thus, the secondary rhinaria on gynoparae antennae might function to perceive these sex pheromone compounds.

As for aggregation pheromones, aphid spacing pheromones (also called avoidance or epideictic pheromones) have received very limited research focus. Spacing pheromones regulate aphid population density to maintain the optimal quality of the food source because, as explained by Way (1973), a surplus in conspecific density may induce a decrease in host-plant quality. Pettersson and colleagues are one of the few groups that worked on aphid aggregation and spacing pheromones (Pettersson and Stephansson 1991, Pettersson et al. 1995), with no recent works on this subject being available. Apterae *R. padi* individuals placed on oat leaves were observed to be repelled from other apterous individuals in bioassays (Pettersson et al. 1995). Moreover, repellent behavior was stronger with increasing aphid density. According to Pettersson's hypotheses, this repellent behavior may be due to the cumulative release of semiochemicals responsible for the spacing behavior of apterae, which increases with aphid density. Pettersson's second hypothesis was that apterous females produce these semiochemicals only when the population density reaches a certain threshold. In both cases, a common strategy would be involved to obtain the optimal population density, to retain the advantage of both aggregation and spacing behavior. High-density populations might also lead to an increase in the production of winged individuals, through both physical contact and the chemical perception of volatiles. However, because this polyphenism is also observed in the absence of alarm pheromone emission, a spacing pheromone might exist. To substantiate these findings, Quiroz et al. (1997) conducted gas chromatography analyses on the volatiles collected in the headspace of apterae *R. padi*, and identified three semiochemicals, the quantities of which were related to the density of the aphid colony; these chemicals were 6-methyl-5-hepten-2-one, 6-methyl-5-hepten-2-ol, and 2-tridecanone. All three chemicals acted as repellents when presented individually or in a blend to apterae aphids in bioassays. However, the origin of these compounds remains unclear, because of the lack of complementary research on this topic. Indeed, these semiochemicals might be emitted by a large aphid colony or by the host plant under high predation pressure.

In addition to using pheromones (intraspecific communication), aphids communicate with their environment by allelochemicals (interspecific communication). For example, plant semiochemicals inform aphids about the location and quality of their host-plant and act as synergists for aphid pheromones. Moreover, because of their feeding behavior, aphids may modify the chemical composition of the emitted plant volatiles to inform other aphid individuals about the plant health; however, such modifications may also inform aphid natural enemies about the presence of their herbivore preys or hosts.

Aphids and Plant Semiochemicals

Volatile chemicals released by plants play a major role in host-plant recognition and location by aphids (Fraenkel 1959) (see Chapter 7 "Aphid Molecular Stress Biology" and Chapter 8 "The Effect of Plant Within-Species Variation on Aphid Ecology" in this volume). During their life cycle, heteroecious aphid species go through two major foraging stages related to their summer and winter hosts. To locate adequate host plants, aphids rely on the composition of the chemical blends, which allows aphids to differentiate between host and non-host plants (Webster 2012, Döring 2014). The gynoparae of many aphid species are sensitive to the odorant cues released by their primary host plant (winter host) on which they are reproducing (Powell and Hardie 2001). Many studies have shown that plant volatiles, particularly those of the host plants, induce electrophysiological (Bromley and Anderson 1982, Yan and Visser 1982, Dawson et al. 1987b, Wadhams 1990) and behavioral responses (Alikhan 1960, Pettersson 1970b, 1973, 1993) in different aphid species.

Aphids are able to discriminate a chemical blend released by a host plant from that of a non-host one based on blend composition. Plants odor blends are also composed of different chemicals, which vary depending on their taxonomic (Knudsen et al. 1993), developmental (Holopainen et al. 2010), and physiological status (Jiménez-Martínez et al. 2004a). Some of these compounds are generally present in a wide range of plant groups, including short-chain alcohol and aldehydes (usually called green leaf volatiles). In contrast, other compounds are more specific, being exclusively released by a plant order, family, genus, or species. These taxonomically-specific plant compounds generally induce behavioral responses in herbivores that feed on the associated plants. For example, Brassicaceae release volatiles compounds belonging to the isothiocyanates group (Fahey et al. 2001), which is a group of volatiles considered to be taxonomically specific to this plant family. Dilawari and Atwal (1989) demonstrated the ability of the turnip aphid *Lipaphis erysimi* Kaltenbach to respond behaviorally to isothiocyanates, and, thus, locate its brassica host. In 1991, Nottingham and colleagues highlighted the presence of neural cells in the proximal primary rhinaria (PPR) that are sensitive to isothiocyanates in *Brevicoryne brassicae* L. and *L. erysimi*. In both species, the perception of isothiocyanates induces attraction towards the odor source. In addition, the authors found that cells from the PPR of black bean aphid *A. fabae* are also sensitive to isothiocyanates; however, the host range of this aphid species is larger and does not include cruciferous plants (Isaacs et al. 1993). The functional reason for this physiological adaptation may be to inform aphids about the presence of a non-host plant, which is reinforced by the observed repellent behavior of isothiocyanates on *A. fabae* (Nottingham et al. 1991, Isaacs et al. 1993). Other examples involve the presence of taxonomically-specific compounds in aphid host plant location. The sulfur-containing volatiles from plants of the *Allium* genus act as attractants for the onion aphid *Neotoxoptera formosana* Takahashi (Hori 2007). Another example is the presence of α-agarofuran (considered as a taxonomically-specific compound) in the odors of *Nothofagus* trees (southern beech), which informs generalist (which have a host range that includes several *Nothofagus* species) and specialist (specialized to a single *Nothofagus* species) aphid species from the genus *Neuquenaphis* about the presence of a potential host plant (Quiroz et al. 1999, Russell et al. 2004).

In addition to their ability to determine whether a plant belongs to their host range according to taxonomically specific volatile compounds, aphids might also be able to recognize their host plant based on ubiquitous chemicals. Webster et al. (2008a) showed that the blend emitted by healthy broad bean *Vicia faba* L. serves as an attractant for the black bean aphid *A. fabae*. Furthermore, none of the 15 chemical components of this blend are specific to the host plant *V. faba* (Knudsen et al. 1993). When every compound of the constitutive blend is presented individually in olfactometric bioassays, only one compound, (Z)-3-hexen-1-ol (presented in high concentrations), is a significant attractant for winged *A. fabae* individuals when compared to clean-air; yet, this compound is still less attractive compared to the entire blend (Webster et al. 2008b, 2010). The other compounds either elicited no response or induced repellent behavior. Presented as a blend, these volatile compounds may be integrated as host cues, leading to aphid attraction/arrestment toward the odor source (Webster et al. 2010).

Plant pathogens may cause significant changes to the volatile blend emission of their host to enhance the attraction of aphid vectors. Indeed, many aphid species are important vectors of plant viruses (Katis et al. 2007), which rely on insects to spread to new hosts. Thus, non-viruliferous aphids are more receptive to infected plants compared to uninfected plants (Eigenbrode et al. 2002, Jiménez-Martínez et al. 2004a, Srinivasan et al. 2006, Medina-Ortega et al. 2009). The *Barley yellow dwarf virus* (BYDV) and the *Potato leaf roll virus* (PLRV) are highly dependent on *R. padi* and *M. persicae* for dispersal, respectively. These viruses modify the volatile profile of the infected host plants, which increases the frequency of interactions with their vectors. For instance, the infestation of *Solanum tuberosum* L. with PLRV induces an increase in the released quantities of at least 11 volatiles, directly affecting the foraging behavior of *M. persicae* (Eigenbrode et al. 2002). *Triticum aestivum* L. infected with BYDV also releases greater quantities of volatiles compared to healthy plants, leading to higher frequencies of visitation by *R. padi* (Jiménez-Martínez et al. 2004b). This close interaction between viruses and aphids might be considered to be a case of mutualism, in which both partners increase their fitness. Specifically, the virus gains higher transmission rates, while the infected plant provides better conditions for aphid growth and reproduction compared to uninfected plants (e.g., Araya and Foster 1987, Castle and Berger 1993).

Insect feeding usually leads to modifications in the volatile blend emissions of a plant, with aphids being no exception (de Vos and Jander 2010). The qualitative and quantitative composition of a plant volatile blend varies according to the herbivore type of feeding, such as sucking or chewing the plant, as observed for turnip plants (*Brassica rapa* L.) under attack by aphids (*M. persicae*) or by chewing lepidopteran larvae (*Heliothis virescens* F.) (Verheggen et al. 2013). These differences likely reflect divergence in the mode of feeding by the pests and in insect-derived biochemical elicitors. Chewing insects cause significant damage to plant foliage and strongly activate wound-signaling pathways, whereas phloem-feeding aphids cause relatively little injury to the host plant. These herbivore-induced odors may have anti-attractive effects on foraging herbivores, including aphids (Pettersson et al. 1995, Verheggen et al. 2013). Indeed, although glucosinolate-derived compounds, such as isothiocyanates, serve as attractants for specialist aphid species (see above), they are considered to be a toxic component for several generalist aphid species. Thus, plants may increase the volatile emission of these compounds as defensive strategies (Verheggen et al. 2013). As another example, wheat plants under infestation by *R. padi* produce additional volatile cues (including short-chained alcohols and ketones) that make the plant less attractive for subsequent aphids (Pettersson et al. 1995). However, in some cases aphids do not induce, or induce very little changes in the chemical emissions of plants on which they are feeding. For instance, only one compound, (Z)-3-hexenol, is emitted in greater quantity by *V. faba* when infested with *A. pisum* (Schwartzberg et al. 2011). Furthermore, the modifications in the volatile profile emitted by *V. faba* under infestation by *Spodotpera exigua* Hübner are reduced when the plants are co-infested by *A. pisum*, showing that certain herbivore-induced plant volatiles are inhibited by the pea aphid.

Host plant volatiles may also have synergistic effects with aphid semiochemicals, essentially with their sex pheromone. Despite the contradictory results obtained by field trappings and laboratory bioassays (Park et al. 2000, Pope et al. 2007), it has been demonstrated that the attractive effect of oviparae sex pheromone on conspecific males is reinforced by the addition of the volatile blend released by their primary host plant, as demonstrated for *P. humuli* during laboratory assays and field trials (Campbell et al. 1990). Aphid-induced plant volatiles also act as synergists with aphid pheromones (de Vos and Jander 2010). Feeding oviparae of the rosy apple aphid *D. plantaginea* induce an increase in the emission of four compounds in *Malus* spp. (van Tol et al. 2009). When combined, these four chemicals did not attract males in field trials;

however, more *D. plantaginea* males were caught when they were presented with combinations of the **I** and **II** components of the sex pheromone.

Aphid-Associated Volatiles and Interactions with Natural Enemies

Although semiochemicals provide a powerful way for organisms to communicate and coordinate their behaviors, they also provide opportunities for other organisms to intercept and exploit signals intended for conspecifics, as reviewed by Hatano et al. (2008b). Predators and parasitoids frequently exhibit innate responses to chemical cues that are reliably associated with hosts and preys, with abundant evidence that profitable chemical cues may be learnt (Du et al. 1997, Mölck et al. 2000) (Fig. 4). Aphid natural enemies locate their preys or hosts by relying on three groups of volatile chemicals: (i) aphid-produced semiochemicals (not exclusively their pheromones) (e.g., Leroy et al. 2011); (ii) volatile chemicals released by non infested plants that are potentially used as food source by their aphid hosts (e.g., Pareja et al. 2007); and (iii) plant semiochemicals released during aphid infestation (e.g., Guerrieri et al. 1993). This phenomenon is particularly true for the aphid parasitoids that mainly belong to two families of Hymenoptera: Braconidae and Aphelinidae. Although less specialized than parasitic wasps, predators are also able to distinguish damaged plants from intact ones, but do not use constitutive blends of plants for long range foraging (e.g., Verheggen et al. 2008b, Oliveira and Pareja 2014).

Consequently, there are numerous examples of natural enemies having learned or evolved to use the semiochemicals associated with the presence of their prey as foraging cues (Vet and Dicke 1992, Paré and Tumlinson 1999). This phenomenon has led to the potential uses of aphid semiochemicals within integrated management strategies being explored to manipulate the behavior of aphid natural enemies

Fig. 4. Aphid natural enemies are attracted by aphid products and aphid-induced plant volatiles.

(see section below). For example, the host location behavior of several aphid-parasitoid wasps has been extensively studied in a biological control context. Naïve *Aphidius ervi* Haliday females are able to discriminate aphid-infested plants from uninfested ones (Du et al. 1998). In dual choice assays, *Diaeretiella rapae* M'Intosh is attracted to volatiles from *Arabidopsis thaliana* (L.) Heynh infested with *M. persicae* over volatiles from undamaged *A. thaliana* (Girling et al. 2006). The parasitoid shows no interest in the volatiles released by aphids alone, their honeydew, or mechanically damaged plants, indicating an interaction between the plant and the aphid leading to the production of a different volatile blend, which *D. rapae* learns and responds to.

Informing natural enemies about the presence of infesting aphids by modifying the volatile headspace composition in an aphid-infested plant is an indirect strategy allowing plants to counter infestation events. The attraction of lady beetles (Sarmento et al. 2008) and hoverflies (Verheggen et al. 2008c) to aphid-infested tomato plants has been clearly demonstrated. Brassicaceae also direct aphid predators to their prey by releasing isothiocyanates. These glucosinolate-derived compounds attract specialist aphid species, but also inform the parasitoid *D. rapae* about the presence of their hosts (Reed et al. 1985). Here, again, as suggested previously for aphids and their alarm pheromone, plants must face an ecological trade-off between the attraction of specialist aphids (e.g., *L. erysimi*, *B. brassicae*) and beneficial insects (as predators or parasitoids).

In addition to their ability to locate preys/hosts by using aphid-induced plant volatiles, natural enemies also have the capacity to detect aphid-produced volatile chemicals directly during foraging behavior (Hatano et al. 2008b). Several predators and parasitoids are sensitive to molecules used in aphid intra-specific communication, including the components of the sex and alarm pheromones. Parasitoid females from the genus *Praon* are highly attracted by (4a*S*,7*S*,7a*R*)-nepetalactone (**I**) in field trials (Lilley et al. 1994), and also by (1*R*,4a*S*,7*S*,7a*R*)-nepetalactol (**II**), but less significantly (Hardie et al. 1991, 1994c). In field trapping assays, the sex pheromone of the cabbage aphid *B. brassicae* (only composed of the compound nepetalactone I - see Table 1) attracts the parasitoid *D. rapae* (Gabrys et al. 1997). In laboratory behavioral assays, female *Aphidius* parasitoids are also attracted by sex pheromone components. Specifically, *A. ervi* and *Aphidius eadyi* Starý are sensitive to (4a*S*,7*S*,7a*R*)-nepetalactone (**I**) and (1*R*,4a*S*,7*S*,7a*R*)-nepetalactol (**II**), either combined or presented alone (Glinwood et al. 1999a,b), while *Aphidius colemani* Viereck is sensitive to nepetalactone (**I** and/or **V**) (Ameixa and Kindlmann 2012).

As for parasitoids, predators also have the ability to detect their aphid prey via the perception of their sex pheromone components. In both behavioral laboratory tests and in field trials, several lacewing species from the genera *Chrysopa* and *Chrysoperla* are attracted to aphid sex pheromone compounds. The antennae of *Chrysoperla carnea* Stephens, *Chrysopa oculata* Say, and *Chrysopa cognata* McLachlan produce strong electrophysiological depolarization when exposed to (4a*S*,7*S*,7a*R*)-nepetalactone (**I**) and (1*R*,4a*S*,7*S*,7a*R*)-nepetalactol (**II**) (Boo et al. 1998, Zhu et al. 2005). *C. oculata* and *C. cognata* are also attracted by field traps releasing these two components of the aphid sex pheromone (Boo et al. 1998, Zhu et al. 2005). In comparison, the common green lacewing *C. carnea* shows a lower aptitude at being attracted by these compounds, but has a good perception of these chemical signals (Zhu et al. 2005, Koczor et al.

2010). Other species from the *Chrysopa* genus, including *Chrysopa formosa* Brauer and *Chrysopa pallens* Rambur, also exhibit an ability to detect aphid sex pheromone in the field, with a higher capture frequency when both components are used (Koczor et al. 2010). Lady beetles may also rely on the perception of the aphid sex pheromone in their foraging behavior. For instance, *Z,E*-nepetalactone elicits increased searching behavior by the multicolored Asian lady beetle *Harmonia axyridis* Pallas during both wind tunnel and field assays (Leroy et al. 2012a).

Because sex pheromones are only released by oviparae females and for only a short time window (autumn for majority of species), natural enemies should rely on other aphid-related cues during prey searching behavior. About 50 y ago, it was first suggested that aphid honeydew has kairomonal properties and may be used in the foraging behavior of natural enemies (Bombosch and Volk 1966). Many studies have since reported that several natural enemies from various feeding-guilds elicit behavioral responses to aphid honeydew, including parasitoid wasps, coccinellids, chrysopids, and syrphid flies (e.g., Budenberg and Powell 1992, Han and Chen 2002). Honeydew-associated volatiles elicit electrophysiological responses (Han and Chen 2002) and allow natural enemies to find their food source (Ide et al. 2007) or to find an oviposition site (Leroy et al. 2011). Honeydew, and associated volatiles, guide lady beetles adults and larvae toward their aphid prey (Han and Chen 2002, Ide et al. 2007, Leroy et al. 2012b, Purandare and Tenhumberg 2012), with similar observations being made for lacewing (Han and Chen 2002) and hoverfly larvae (Leroy et al. 2014). In contrast to their larval instars, adult hoverflies feed on flower pollen, rather than aphids (Larson et al. 2001), and use the volatiles from honeydew to find a food source for their offspring, namely the oviposition site (Leroy et al. 2011, 2014). Leroy et al. (2011) showed that several of these volatiles are released from *A. pisum* honeydew because of a bacterium that has been isolated and identified as *Staphylococcus sciuri* (Bacilles: Staphylococcaceae). This bacterium has previously been found in the gut of aphid laboratory strains, and has also been identified in the honeydew of other aphid natural species. To determine whether this bacterium is the source of honeydew volatiles, the authors reinoculated sterilized honeydew with *S. sciuri*. Of the volatiles that were identified in the crude honeydew, nine were found in the reinoculated sample, but not in the sterile honeydew, indicating that they are produced by *S. sciuri*. In wind tunnel assays, plants sprayed with either a solution of *S. sciuri* or with the bacterium-produced volatiles stimulated hoverfly *Episyrphus balteatus* (DeGeer) attraction and oviposition to a similar extent as aphid infestation.

The alarm pheromone, particularly its main component (E)-β-Farnesene which is normally used in intra-specific aphid communication, also has kairomonal properties for several groups of natural enemies (Vandermoten et al. 2012). Whatever the source of the odor (emitted by plant or aphid), studies have reported that hoverfly adults and larvae locate their prey by using this compound (Francis et al. 2005b, Harmel et al. 2007, Verheggen et al. 2008b,c). Many lady beetle species also rely on *E*βf to locate aphid populations in their vicinity, including *H. axyridis* (Verheggen et al. 2007, Leroy et al. 2012a), *Adalia bipunctata* L. (Francis et al. 2004), *Coccinella septempunctata* L. (Nakatuma 1991, Al Abassi et al. 2000), and *Hippodamia convergens* Guérin-Méneville (Acar et al. 2001). Attraction to this sesquiterpene has also been demonstrated for the common green lacewing *C. carnea* (Zhu et al. 2005) and parasitoids, such as *D. rapae* (Beale et al. 2006, Moayeri et al. 2014) and *A. ervi* (Du et al. 1998). In contrast, the lacewing species *C. cognata* did not present any behavioral response to this sesquiterpene in both electrophysiological and olfactometric tests (Boo et al. 1998). However, many of the experiments providing evidence supporting the attractiveness of the aphid alarm pheromone to aphid natural enemies were performed under variable *E*βf concentrations; for instance, concentrations higher than those typically emitted by aphids under enemy attack were sometimes used (Vandermoten et al. 2012). Thus, while there is strong evidence that the aphid alarm pheromone acts as a foraging cue for many aphid predators and parasitoids when presented in high concentrations, the use of realistic amounts of *E*βf does not increase the foraging success of predators, at least in laboratory and field bioassays (Joachim et al. 2014). Predators did not find their aphid prey more often, nor did they reside for a longer time on the plant when natural doses of *E*βf where added. Additional studies conducted under realistic field conditions would help to clarify the exact role of such cues (e.g., in short- and long range host location) and their ecological significance.

Aphid Semiochemicals in Integrated Management

Studying insect semiochemicals allows intriguing questions to be answered regarding their basic biology, physiology, and ecology. Moreover, the semiochemical identification of important insect pests for agriculture also has applied implications for their management. Insect pheromones have already been successfully deployed, such as the use of lepidopteran sex pheromones to monitor, trap, and disrupt the mating of these pests (Carde and Minks 1995, Copping 2001, Witzgall et al. 2010). Semiochemicals produced or used by aphids could also be implemented in pest management programs. Because of their intraspecific and interspecific activities, the molecules implicated in aphid communication have both a direct and indirect effect on the aphid infestation of crops and orchards. For instance, mating disruption and repelling behavior induction represent strategies that directly affect population development, while the attraction of aphid natural enemies represents an indirect means of controlling infestation.

(4a*S*,7*S*,7a*R*)-Nepetalactone (**I**) and (1*R*,4a*S*,7*S*,7a*R*)-nepetalactol (**II**) are the two components of the sex pheromone that are present in a wide range of aphid species, and so represent good candidates for a semiochemical-based control strategy against aphids. (4a*S*,7*S*,7a*R*)-Nepetalactone (**I**) may be obtained from fresh catmint *Nepeta cataria* L. (Birkett and Pickett 2003); yet, despite the relatively low production cost (≈1€/gr), the use of these two components remains limited. The isomerism of these molecules is important for eliciting the behavioral responses of aphids and natural enemies (Hardie et al. 1997). However, based on several published field-trapping bioassays, the efficiency of nepetalactone (**I**) and nepetalactol (**II**) has been confirmed for trapping aphid males (Gabrys et al. 1997, Hardie et al. 1997, Campbell et al. 2003) and gynoparae (Hardie et al. 1996, Boo et al. 2000, Symmes et al. 2012). Because most aphid species that use sex pheromones have complex and specific life cycles, the time window available to exploit the effects of aphid sex pheromone on mating is generally short (Fig. 1), and usually occurs in orchards, in the presence of the primary host for heteroecious aphid species (Moran 1992). Thus, the use of sex pheromone compounds could mainly be used in the context of mating disruption or for the mass trapping of aphid species economically affecting orchards yields. During the autumn season, when several aphid species migrate from crops to their winter hosts, the use of sex pheromone might cause an imbalance in the mating process, reducing egg production and the number of newly hatched individuals.

Following the identification of $E\beta f$ as the key component of the aphid alarm pheromone of most Aphidinae species, the opportunities for using this semiochemical to repel aphids have been discussed (Bowers et al. 1972, Griffiths and Pickett 1980). Because $E\beta f$ affects the dispersal of aphid populations, the potential use of this aphid pheromone as a direct control mechanism has been explored in a number of studies. Field experiments confirmed the dispersal behavior of aphids subjected to their alarm pheromone in 41 species (Xiangyu et al. 2002). However, once $E\beta f$ concentrations decrease, aphids tend to reinfest their host plants (Calabrese and Sorensen 1978). Methods to apply $E\beta f$ in crop fields have been diversified. The use of synthetic $E\beta f$ has long been considered to be difficult, due to the highly unstable properties of this molecule (Bruce et al. 2005). Extracting $E\beta f$ from the essential oils of *Matricaria chamomilla* L. has been proposed as a cheap way to obtain this aphid pheromone in sufficient quantities to facilitate field application (Heuskin et al. 2009) and would present better resistance in contact with air compared to the chemically synthesized molecule. Through the use of semiochemical slow-release devices (Heuskin et al. 2011), $E\beta f$ may also be released in the field for a longer period of time. Several plant species release $E\beta f$ in their volatile headspace (Mostafavi et al. 1996, Harmel et al. 2007, Webster et al. 2008a). It has been hypothesized that this compound might increase plant protection against aphids by (i) deterring aphids from settling, (ii) reducing aphid performance due to the frequent interruption of feeding, or (iii) inducing the production of more winged offspring (Kunert et al. 2010, Gibson and Pickett 1983). However, aphid behavioral responses toward $E\beta f$ released by plants appear to be strongly influenced by other common plant volatiles, including (-)-β-caryophyllene, which acts as an inhibitor of alarm pheromone activity (Dawson et al. 1984, Mostafavi et al. 1996). Recently, new methods have been developed that modify plant genomes, allowing plants to continuously release $E\beta f$ (Pickett et al. 2013). Kunert et al. (2010) highlighted that $E\beta f$ emissions by transgenic $E\beta f$-emitting *A. thaliana* provide limited protection against the aphid *M. persicae*. Although no metabolic cost of $E\beta f$ synthesis was demonstrated, there was no evidence of any direct benefits to the plants. For instance, $E\beta f$ emission did not appear to repel *M. persicae*, reduce its reproduction, or increase the proportion of winged offspring. The authors suggested two possible explanations (i) habituation to the $E\beta f$ of aphids reared on these constitutively $E\beta f$-producing plants, or (ii) that aphids might only react to $E\beta f$ emitted in pulses (i.e., mimicking $E\beta f$ emission by aphids). In contrast, under laboratory conditions, *M. persicae* aphids were repelled by genetically modified *A. thaliana* that released $E\beta f$ (Beale et al. 2006). These results were later supported by other laboratory assays with a transgenic elite wheat variety, which induced greater escape behavior than *Triticum* wild type (Pickett et al. 2013). This technique might also have promising applications in integrated pest management via the kairomone function of $E\beta f$ as a cue for aphid natural enemies (Vandermoten et al. 2012). $E\beta f$ emission by transgenic plants enhances the level of aphid control by enhancing the recruitment of aphid parasitoids

(Beale et al. 2006, Pickett et al. 2013). de Vos et al. (2010), using the same aphid-plant system as Beale et al. (2006) (i.e., *M. persicae–A. thaliana*), observed that habituated *M. persicae* aphids failed to respond to aphid-emitted Eβf, and suffered higher levels of predation by the coccinelid predator *H. convergens*. Further studies are needed to determine whether these effects may be effectively exploited to enhance the biological control of aphid pests in commercial agricultural settings (Vandermoten et al. 2012).

Because aphids are also sensitive to volatiles released by plants during their host location inducing attractant/arrestment or repellent behaviors (Webster 2012), these molecules might also be included in integrated pest management approaches. Isothiocyanates are considered to be a repellent and are toxic for many generalist aphid species (Fahey et al. 2001, Verheggen et al. 2013). These chemicals could be used to repel generalist aphids in non-Brassicaceae crops. Attractant blends that aphids use to locate host plants could also be implemented to draw aphids outside the growing areas. However, plant semiochemicals may have a different effect on winged and wingless individuals, males and females, and different species, with the activity of these chemicals varying depending on whether they are presented alone or as a complete blend (Webster et al. 2008b, Webster 2012).

Using semiochemicals to attract aphid natural enemies to an infested area also represents a promising approach. Field trappings with aphid sex pheromones significantly attract female parasitoids (e.g., Hardie et al. 1991, 1994c). However, the use of semiochemicals in the context of aphid control by attracting parasitoids is subject to several constraints, including the optimization of synchrony between host and natural enemy populations, which is generally difficult (Powell et al. 1998). The alarm pheromone also induces the positive response of natural enemies in field bioassays. Cui et al. (2012a) showed that more ladybeetles and parasitoids were found in areas treated with Eβf compared to control areas, with aphid density being significantly reduced in treated areas. Moreover, the continuous release of Eβf by plants in crops is not repellent for aphids, but may induce the habituation of aphids to the alarm signal, making them less reactive to aphid signaling when an attack occurs (de Vos et al. 2010).

Honeydew also has some properties that could be implemented to control aphids by the amplification of foraging and ovipositing behaviors of several natural enemies. Leroy et al. (2011) used wind tunnel, greenhouse and field assays to test the effect of the volatiles produced by a bacteria previously identified in honeydew on *E. balteatus*. Sterile honeydew had no effect, but crude and reinoculated honeydew attracted hoverflies to the treated plants. Out of the separate bacterially derived honeydew volatiles, 3-methylbutanoic acid was an attractant, while 3-methyl-2-butenal acid and 2-methylbutanoic acid were both attractants and oviposition stimulators.

Coupling the use of semiochemicals with insecticides could be one solution to control aphids efficiently with reduced doses of chemicals. For example, spraying insecticides during autumn when oviparae arrive in orchards (i.e., on primary hosts) significantly reduces infestation the following spring (Kehrli and Wyss 2001). Thus, the use of sex pheromone to detect the arrival of aphids might facilitate targeted treatment, and could be a good way of reducing the use of insecticides. Moreover, a recent study highlighted that the release of Eβf improves the efficiency of subsequent insecticide treatments (Cui et al. 2012b), probably because of an increase in aphid exposure to phytosanitary products.

Conclusions

Aphid-related semiochemicals include a broad range of molecules that are able to synergize with each other, permitting a wide variety of signaling. Signals are used to inform aphids about the physiological status, location of a food source, threat, and mating opportunities. These signals may be directly dedicated to targeted organisms, but might also be eavesdropped by natural enemies.

Over the last 40 y, since the first aphid pheromone was identified, detailed studies have been conducted to enhance our knowledge about aphid chemical ecology, leading to the identification and evaluation of several semiochemicals of interest in terms of aphid control. However, their use is still poorly developed, despite the availability of several biotechnologies, including genetically modified plants, efficient semiochemical dispensers, and pheromone traps. Push-pull strategies are currently in development, which would allow pest repulsion outside the crops via repellent signals inside crops and attractant ones outside crops, generating mass trappings. These strategies might be applied by combining different groups of molecules, such as alarm and spacing pheromones as repellents and sex and aggregation pheromones as attractants. Plant volatiles could be used in both attracting and repelling blends, depending on the aphid species being, the plant group they are feeding on, or the natural enemies required to control the targeted species.

Keywords: aphids, olfaction, volatiles, aphid pheromones, semiochemicals, integrated management

References

Acar, E.B., J.C. Medina, M.L. Lee and G.M. Booth. 2001. Olfactory behavior of convergent lady beetles (Coleoptera: Coccinellidae) to alarm pheromone of green peach aphid (Hemiptera: Aphidiidae). Can. Entomol. 133: 389–397.

Agelopoulos, N.G., K. Chamberlain and J.A. Pickett. 2000. Factors affecting volatile emissions of intact potato plants, *Solanum tuberosum*: variability of quantities and stability of ratios. J. Chem. Ecol. 26: 497–511.

Al Abassi, S., M.A. Birkett, J. Pettersson, J.A. Pickett, L.J. Wadhams and C.M. Woodcock. 2000. Response of the seven-spot ladybird to an aphid alarm pheromone and an alarm pheromone inhibitor is mediated by paired olfactory cells. J. Chem. Ecol. 26: 1765–1771.

Alikhan, M.A. 1960. The experimental study of the chemotactic basis of host-specificity in a phytophagous insect, *Aphis fabae* Scop. (Aphididae: Homoptera). Ann. Univ. Mariae Curie-Sklodowska Sect. C 15: 117–159.

Allee, W.C. 1931. Animal Aggregations, a Study in General Sociology. The University of Chicago Press, Chicago, Illinois.

Ameixa, O.M.C.C. and P. Kindlmann. 2012. Effect of synthetic and plant-extracted aphid pheromones on the behaviour of *Aphidius colemani*. J. Appl. Entomol. 136: 292–301.

Araya, J.E. and J.E. Foster. 1987. Laboratory study on the effects of barley yellow dwarf virus on the life cycle of *Rhopalosiphum padi* (L.). J. Plant Dis. Prot. 94: 578–583.

Barbier, M. 1982. Les phéromones. Aspects biochimiques et biologiques. Masson, Paris, France.

Beale, M.H., M.A. Birkett, T.J.A. Bruce, K. Chamberlain, L.M. Field, A.K. Huttly, J.L. Martin, R. Parker, A.L. Phillips, J.A. Pickett, I.M. Prosser, P.R. Shewry, L.E. Smart, L.J. Wadhams, C.M. Woodcock and

Y.H. Zhang. 2006. Aphid alarm pheromone produced by transgenic plants affects aphid and parasitoid behavior. Proc. Nat. Acad. Sci. USA 103: 10509–10513.

Birch, M.C. 1974. Introduction. pp. 1–7. *In*: M.C. Birch (ed.). Pheromones. North Holland, Amsterdam.

Birkett, M.A. and J.A. Pickett. 2003. Aphid sex pheromones: from discovery to commercial production. Phytochemistry 62: 651–656.

Blum, M.S. 1985. Alarm pheromones. pp. 193–224. *In*: G.A. Kerkut and L.I. Gilbert (eds.). Comprehensive Insect Physiology, Biochemistry and Pharmacology. Pergamon Press, Oxford, UK.

Bombosch, S. and S.T. Volk. 1966. Selection of the oviposition site by *Syrphus corollae* F. pp. 117–119. *In*: I. Hodek (ed.). Proceedings of the Symposium on the Ecology of Aphidophagous Insects. Academia, Prague.

Boo, K.S., I.B. Chung, K.S. Han, J.A. Pickett and L.J. Wadhams. 1998. Response of the lacewing *Chrysopa cognata* to pheromones of its aphid prey. J. Chem. Ecol. 24: 631–643.

Boo, K.S., M.Y. Choi, I.B. Chung, V.F. Eastop, J.A. Pickett, L.J. Wadhams and C.M. Woodcock. 2000. Sex pheromone of the peach aphid, *Tuberocephalus momonis*, and optimal blends for trapping males and females in the field. J. Chem. Ecol. 26: 601–609.

Bowers, W.S., L.R. Nault, R. Webb and S.R. Dutky. 1972. Aphid alarm pheromone: isolation, identification, synthesis. Science 177: 1121–1122.

Bowers, W.S., C. Nishino, M.E. Montgomery, L.R. Nault and M.W. Nielson. 1977. Sesquiterpene progenitor, Germacrene A: an alarm pheromone in aphids. Science 196: 680–681.

Bromley, A.K. and M. Anderson. 1982. An electrophysiological study of olfaction in the aphid *Nasanovia ribis-nigri*. Entomol. Exp. Appl. 32: 101–110.

Bromley, A.K., J.A. Dunn and M. Anderson. 1979. Ultrastructure of the antennal sensilla of aphids. 1. Coeloconic and placoid sensilla. Cell Tissue Res. 203: 427–442.

Bromley, A.K., J.A. Dunn and M. Anderson. 1980. Ultrastructure of the antennal sensilla of aphids. 2. Trichoid, chordotonal and campaniform sensilla. Cell Tissue Res. 205: 493–511.

Bruce, T.J.A., M.A. Birkett, J. Blande, A.M. Hooper, J.L. Martin, B. Khambay, I. Prosser, L.E. Smart and L.J. Wadhams. 2005. Response of economically important aphids to components of *Hemizygia petiolata* essential oil. Pest Manag. Sci. 61: 1115–1121.

Budenberg, W.J. and B. Powell. 1992. The role of honeydew as an oviposition stimulant for two species of syrphids. Entomol. Exp. Appl. 64: 57–61.

Calabrese, E.J. and A.J. Sorensen. 1978. Dispersal and recolonization by *Myzus persicae* following aphid alarm pheromone exposure. Ann. Entomol. Soc. Am. 71: 181–182.

Callow, R.K., A.R. Greenway and D.C. Griffiths. 1973. Chemistry of the secretion from the cornicles of various species of aphids. J. Insect Physiol. 19: 737–748.

Campbell, C.A.M., G.W. Dawson, D.C. Griffiths, J. Pettersson, J.A. Pickett, L.J. Wadhams and C.M. Woodcock. 1990. Sex attractant pheromone of damson-hop aphid *Phorodon humuli* (Homoptera, aphididae). J. Chem. Ecol. 16: 3455–3465.

Campbell, C.A.M., F.J. Cook, J.A. Pickett, T.W. Popoe, L.J. Wadhams and C.M. Woodcock. 2003. Responses of the aphids *Phorodon humuli* and *Rhopalosiphum padi* to sex pheromone stereochemistry in the field. J. Chem. Ecol. 29: 2225–2234.

Carde, R.T. and A.K. Minks. 1995. Control of moth pests by mating disruption: successes and constraints. Annu. Rev. Entomol. 40: 559–585.

Castle, S.J. and P.H. Berger. 1993. Rates of growth and increase of *Myzus persicae* on virus-infected potatoes according to type of virus-vector relationship. Entomol. Exp. Appl. 69: 51–60.

Copping, L.G. 2001. The Biopesticide Manual: A World Compendium of Naturally Occuring Biopesticides. British Crop Protection Council, Farnham, UK.

Cui, L.L., F. Francis, S. Heuskin, G. Lognay, Y.J. Liu, J. Dong, J. Chen, X. Song and Y. Liu. 2012a. The functional significance of E-β-farnesene: does it influence the populations of aphid natural enemies in the fields? Biol. Control 60: 108–112.

Cui, L.L., J. Dong, F. Francis, Y.J. Liu, S. Heuskin, G. Lognay, J.L. Chen, C. Bragard, J.F. Tooker and Y. Liu. 2012b. E-β-farnesene synergizes the influence of an insecticide to improve control of cabbage aphids in China. Crop Prot. 35: 91–96.

Dawson, G.W., D.C. Griffiths, J.A. Pickett, M.C. Smith and C.M. Woodcock. 1984. Natural inhibition of the aphid alarm pheromone. Entomol. Exp. Appl. 36: 197–199.

Dawson, G.W., D.C. Griffiths, N.F. Janes, A. Mudd, J.A. Pickett, L.J. Wadhams and C.M. Woodcock. 1987a. Identification of an aphid sex pheromone. Nature 325: 614–616.

Dawson, G.W., D.C. Griffiths, J.A. Pickett, L.J. Wadhams and C.M. Woodcock. 1987b. Plant-derived synergists of alarm pheromone from turnip aphid, *Lipaphis (Hyadaphis) erysimi* (Homoptera, Aphididae). J. Chem. Ecol. 13: 1663–1671.

Dawson, G.W., D.C. Griffiths, L.A. Merritt, A. Mudd, J.A. Pickett, L.J. Wadhams and C.M. Woodcock. 1988. The sex pheromone of the greenbug, *Schizaphis graminum*. Entomol. Exp. Appl. 48: 91–93.

Dawson, G.W., D.C. Griffiths, L.A. Merritt, A. Mudd, J.A. Pickett, L.J. Wadhams and C.M. Woodcock. 1990. Aphid semiochemicals—a review, and recent advances on the sex pheromone. J. Chem. Ecol. 16: 3019–3030.

de Vos, M. and G. Jander. 2010. Volatile communication in plant-aphid interactions. Curr. Opin. Plant Biol. 13: 366–371.

de Vos, M., W.Y. Cheng, H.E. Summers, R.A. Raguso and G. Jander. 2010. Alarm pheromone habituation in *Myzus persicae* has fitness consequences and causes extensive gene expression changes. Proc. Nat. Acad. Sci. USA 107: 14673–14678.

Dewhirst, S.Y. 2007. Aspects of aphid chemical ecology: sex pheromones and induced plant defences. Ph.D. Thesis, Imperial College, London, UK.

Dewhirst, S.Y., M.A. Birkett, J.D. Fitzgerald, A. Stewart-Jones, L.J. Wadhams, C.M. Woodcock, J. Hardie and J.A. Pickett. 2008. Dolichodial: a new aphid sex pheromone component? J. Chem. Ecol. 34: 1575–1583.

Dewhirst, S.Y., J.A. Pickett and J. Hardie. 2010. Chapter 22: Aphid Pheromones. Vitamins and Hormones 83: 551–574.

Dilawari, V.K. and A.S. Atwal. 1989. Response of mustard aphid, *Lipaphis erysimi* (Kalt.) to allylisothiocyanate. J. Insect Sci. 2: 103–108.

Dixon, A.F.G. 1958. The escape responses shown by certain aphids to the presence of the coccinellid *Adalia decempunctata* (L.). Trans. R. Entomol. Soc. Lond. 110: 319–334.

Dixon, A.F.G. 1998. Aphid Ecology, 2nd ed. Chapman and Hall, London.

Döring, T. 2014. How aphids find their host plants, and how they don't. Ann. Appl. Biol. 165: 3–26.

Du, Y., G.M. Poppy, W. Powell and L.J. Wadhams. 1997. Chemically mediated associative learning in the host foraging behavior of the aphid parasitoid *Aphidius ervi* (Hymenoptera: Braconidae). J. Insect Behav. 10: 509–522.

Du, Y., G.M. Poppy, W. Powell, J.A. Pickett, L.J. Wadhams and C.M. Woodcock. 1998. Identification of semiochemicals released during aphid feeding that attract parasitoid *Aphidius ervi*. J. Chem. Ecol. 24: 1355–1368.

Edwards, J.S. 1966. Defence by smear—supercooling in cornicle wax of aphids. Nature 211: 73–74.

Eigenbrode, S.D., H. Ding, P. Shiel and P.H. Berger. 2002. Volatiles from potato plants infected with potato leafroll virus attract and arrest the virus vector, *Myzus persicae* (Homoptera: Aphididae). Proc. R. Soc. Lond. B Biol. Sci. 269: 455–460.

Fahey, J.W., A.T. Zalcmann and P. Talalay. 2001. The chemical diversity and distribution of glucosinolates and isothiocyanates among plants. Phytochemistry 56: 5–51.

Flögel, J.H.L. 1905. Monographie der Johannisbeeren-Blattlaus, *Aphis ribis* L. Z. wiss. Insekten-biologie 1: 49–63.

Fraenkel, G.S. 1959. The raison d'être of secondary plant substances. Science 129: 1466–1470.

Francis, F., G. Lognay and E. Haubruge. 2004. Olfactory responses to aphid and host plant volatile releases: (E)-β-farnesene an effective kairomone for the predator *Adalia bipunctata*. J. Chem. Ecol. 30: 741–755.

Francis, F., S. Vandermoten, F.J. Verheggen, G. Lognay and E. Haubruge. 2005a. Is the (E)-β-farnesene only volatile terpenoid in aphids? J. Appl. Entomol. 129: 6–11.

Francis, F., T. Martin, G. Lognay and E. Haubruge. 2005b. Role of (E)-beta-farnesene in systematic aphid prey location by *Episyrphus balteatus* larvae (Diptera: Syrphidae). Eur. J. Entomol. 102: 431–436.

Gabrys, B.J., H.J. Gadomski, Z. Klukowski, J.A. Pickett, G.T. Sobota, L.J. Wadhams and C.M. Woodcock. 1997. Sex pheromone of cabbage aphid *Brevicoryne brassicae*: identification and field trapping of male aphids and parasitoids. J. Chem. Ecol. 23: 1881–1890.

Gibson, R.W. and J.A. Pickett. 1983. Wild potato repels aphids by release of aphid alarm pheromone. Nature 302: 608–609.

Girling, R.D., M. Hassall, J.D. Turner and G.M. Poppy. 2006. Behavioural responses of the aphid parasitoid *Diaeretiella rapae* to volatiles from *Arabidopsis thaliana* induced by *Myzus persicae*. Entomol. Exp. Appl. 120: 1–9.

Glinwood, R.T., Y.J. Du and W. Powell. 1999a. Responses to aphid sex pheromones by the pea aphid parasitoids *Aphidius ervi* and *Aphidius eadyi*. Entomol. Exp. Appl. 92: 227–232.

Glinwood, R.T., Y.J. Du, D.W.M. Smiley and W. Powell. 1999b. Comparative responses of parasitoids to synthetic and plant-extracted nepetalactone component of aphid sex pheromones. J. Chem. Ecol. 25: 1481–1488.

Goldansaz, S.H., S. Dewhirst, M.A. Birkett, A.M. Hooper, D.W.M. Smiley, J.A. Pickett, L.J. Wadhams and J.N. McNeil. 2004. Identification of two sex pheromone components of the potato aphid, *Macrosiphum euphorbiae* (Thomas). J. Chem. Ecol. 30: 819–834.

Griffiths, D.C. and J.A. Pickett. 1980. A potential application of aphid alarm pheromones. Entomol. Exp. Appl. 27: 199–201.

Guerrieri, E., F. Pennacchio and E. Tremblay. 1993. Flight behaviour of the aphid parasitoid *Aphidius ervi* (Hymenoptera: Braconidae) in response to plant and host volatiles. Eur. J. Entomol. 90: 415–421.

Guldemond, J.A., A.F.G. Dixon, J.A. Pickett, L.J. Wadhams and C.M. Woodcock. 1993. Specificity of sex pheromones, the role of host plant odour in the olfactory attraction of males, and mate recognition in the aphid *Cryptomyzus*. Physiol. Entomol. 18: 137–143.

Gut, J., P. Harrewijn, A.M. van Oosten and B. van Rheenen. 1987. Additional functions of alarm pheromones in development processes of aphids. Meded. Fac. Landbouwwet. Rijksuniv. Gent. 52: 371–378.

Hamilton, W.D. 1964. The genetical evolution of social behaviour. J. Theor. Biol. 7: 1–16.

Han, B. and Z. Chen. 2002. Behavioral and electrophysiological responses of natural enemies to synomones from tea shoots and kairomones from tea aphids, *Toxoptera aurantii*. J. Chem. Ecol. 28: 2203–2219.

Han, B.Y., M.X. Wang, Y.C. Zheng, Y.Q. Niu, C. Pan, L. Cui, K.R. Chauhan and Q.H. Zhang. 2014. Sex pheromone of the tea aphid, *Toxoptera aurantii* (Boyer de Fonscolombe) (Hemiptera: Aphididae). Chemoecology 24: 179–187.

Hardie, J., M. Holyoak, J. Nicholas, S.F. Nottingham, J.A. Pickett, L.J. Wadhams and C.M. Woodcock. 1990. Aphid sex pheromone components: Age dependent release by females and species-specific male response. Chemoecology 1: 63–68.

Hardie, J., S.F. Nottingham, W. Powell and L.J. Wadhams. 1991. Synthetic aphid sex-pheromone lures female parasitoids. Entomol. Exp. Appl. 61: 97–99.

Hardie, J., S.F. Nottingham, G.W. Dawson, R. Harrington, J.A. Pickett and L.J. Wadhams. 1992. Attraction of field-flying aphid males to synthetic sex pheromone. Chemoecology 3: 113–117.

Hardie, J., J.H. Visser and P.G.M. Piron. 1994a. Perception of volatiles associated with sex and food by different adult forms of the black bean aphid, *Aphis fabae*. Physiol. Entomol. 19: 278–284.

Hardie, J., J.R. Storer, S.F. Nottingham, L. Peace, R. Harrington, L.A. Merritt, L.J. Wadhams and D.K. Wood. 1994b. The interaction of sex pheromone and plant volatiles for field attraction of male bird-cherry aphid, *Rhopalosiphum padi*. Brighton Crop Prot. Conf.—Pests and Diseases 3: 1223–1230.

Hardie, J., A.J. Hick, C. Holler, J. Mann, L. Merritt, S.F. Nottingham, W. Powell, L.J. Wadhams, J. Witthinrich and A.F. Wright. 1994c. The responses of *Praon* spp. parasitoids to aphid sex-pheromone components in the field. Entomol. Exp. Appl. 71: 95–99.

Hardie, J., J.R. Storer, F.J. Cook, C.A.M. Campbell, L.J. Wadhams, R. Lilley and L. Peace. 1996. Sex pheromone and visual trap interactions in mate location strategies and aggregation by host-alternating aphids in the field. Physiol. Entomol. 21: 97–106.

Hardie, J., L. Peace, J.A. Pickett, D.W.M. Smiley, J.R. Storer and L.J. Wadhams. 1997. Sex pheromone stereochemistry and purity affect field catches of male aphids. J. Chem. Ecol. 23: 2547–2554.

Hardie, J., J.A. Pickett, E.M. Pow and D.W.M. Smiley. 1999. Aphids. pp. 250–277. *In*: R.J. Hardie and A.K. Minks (eds.). Pheromones of Non-lepidopteran Insects Associated with Agricultural Plants. CAB International, Wallingford, UK.

Harmel, N., R. Almohamad, M.L. Fauconnier, P. Du Jardin, F.J. Verheggen, M. Marlier, E. Haubruge and F. Francis. 2007. Role of terpenes from aphid-infested potato on searching and oviposition behavior of *Episyrphus balteatus*. Insect Sci. 14: 57–63.

Hatano, E., G. Kunert, S. Bartram, W. Boland, J. Gershenzon and W.W. Weisser. 2008a. Do aphid colonies amplify their emission of alarm pheromone? J. Chem. Ecol. 34: 1149–1152.

Hatano, E., G. Kunert, J.P. Michaud and W.W. Weisser. 2008b. Chemical cues mediating aphid location by natural enemies. Eur. J. Entomol. 105: 797–806.

Hatano, E., G. Kunert and W.W. Weisser. 2010. Aphid wing induction and ecological costs of alarm pheromone emission under field conditions. PloS one 5: e11188.

Heuskin, S., B. Godin, P. Leroy, Q. Capella, J.P. Wathelet, F.J. Verheggen, E. Haubruge and G. Lognay. 2009. Fast gas chromatography characterisation of purified semiochemicals from essential oils of *Matricaria chamomilla* L. (Asteraceae) and *Nepeta Cataria* L. (Lamiaceae). J. Chromatogr. A 1216: 2768–2775.

Heuskin, S., F.J. Verheggen, E. Haubruge, J.P. Wathelet and G. Lognay. 2011. The use of semiochemical slow-release devices in integrated pest management strategies. Biotechnol. Agron. Soc. Environ. 15: 459–470.

Holopainen, J.K., J. Heijari, E. Oksanen and G.A. Alessio. 2010. Leaf volatile emissions of *Betula pendula* during autumn coloration and leaf fall. J. Chem. Ecol. 36: 1068–1075.

Hori, M. 2007. Onion aphid (*Neotoxoptera formosana*) attractants, in the headspace of *Allium fistulosum* and *A. tuberosum* leaves. J. Appl. Entomol. 131: 8–12.

Ide, T., N. Suzuki and N. Katayama. 2007. The use of honeydew in foraging for aphids by larvae of the ladybird beetle, *Coccinella septempunctata* L. (Coleoptera: Coccinellidae). Ecol. Entomol. 32: 455–460.

Isaacs, R., J. Hardie, A.J. Hick, B.J. Pye, L.E. Smart, L.J. Wadhams and C.M. Woodcock. 1993. Behavioral responses of *Aphis fabae* to isothiocyanates in the laboratory and field. Pestic. Sci. 39: 349–355.

Jeon, H., K.S. Han and K.S. Boo. 2003. Sex pheromone of *Aphis spiraecola* (Homoptera: Aphididae): composition and circadian rhythm in release. J. Asia Pacific Entomol. 6: 159–165.

Jiménez-Martínez, E.S., N.A. Bosque-Pérez, P.H. Berger and R.S. Zemetra. 2004a. Life history of the bird cherry-oat aphid, *Rhopalosiphum padi* (Homoptera: Aphididae), on transgenic and untransformed wheat challenged with *Barley yellow dwarf virus*. J. Econom. Entomol. 97: 203–212.

Jiménez-Martínez, E.S., N.A. Bosque-Pérez, P.H. Berger, R.S. Zemetra, H. Ding and S.D. Eigenbrode. 2004b. Volatile cues influence the response of *Rhopalosiphum padi* (Homoptera: Aphididae) to Barley yellow *dwarf virus*-infected transgenic and untransformed wheat. Environ. Entomol. 33: 1207–1216.

Joachim, C., E. Hatano, A. David, M. Kunert, C. Linse and W.W. Weisser. 2013. Modulation of aphid alarm pheromone emission of pea aphid prey by predators. J. Chem. Ecol. 39: 773–782.

Joachim, C., I. Vosteen and W.W. Weisser. 2014. The aphid alarm pheromone (E)-β-farnesene does not act as a cue for predators searching on a plant. Chemoecology. doi : 10.1007/s00049-014-0176-z.

Katis, N.I., J.A. Tsitsipis, M. Stevens and G. Powell. 2007. Transmission of plant viruses. pp. 353–390. *In*: H.F. Van Emden and R. Harrington (eds.). Aphids as Crop Pests. CAB International, Wallingford, UK.

Kehrli, P. and E. Wyss. 2001. Effects of augmentative releases of the coccinellid, *Adalia bipunctata*, and of insecticide treatments in autumn on the spring population of aphids of the genus *Dysaphis* in apple orchards. Entomol. Exp. Appl. 99: 245–252.

Kislow, C. and L.J. Edwards. 1972. Repellent odour in aphids. Nature 235: 108–109.

Knudsen, J.T., L. Tollsten and L.G. Bergstrom. 1993. Floral scents—a checklist of volatile compounds isolated by headspace techniques. Phytochemistry 33: 253–280.

Koczor, S., F. Szentkiralyi, M.A. Birkett, J.A. Pickett, E. Voigt and M. Toth. 2010. Attraction of *Chrysoperla carnea* complex and *Chrysopa* spp. Lacewings (Neuroptera: Chrysopidae) to aphid sex pheromone components and a synthetic blend of floral compounds in Hungary. Pest Manag. Sci. 66: 1374–1379.

Kunert, G. and W.W. Weisser. 2005. The importance of antennae for pea aphid wing induction in the presence of natural enemies. Bull. Entomol. Res. 95: 125–131.

Kunert, G., S. Otto, U.S.R. Röse, J. Gershenzon and W.W. Weisser. 2005. Alarm pheromone mediates production of winged dispersal morphs in aphids. Ecol. Lett. 8: 596–603.

Kunert, G., C. Reinhold and J. Gershenzon. 2010. Constitutive emission of the aphid alarm pheromone, (*E*)-beta-farnesene, from plants does not serve as a direct defense against aphids. BMC Ecol. 10: 23. doi: 10.1186/1472-6785-10-23.

Larson, B.M.H., P.G. Kevan and D.W. Inouye. 2001. Flies and flowers: taxonomic diversity of anthophiles and pollinators. Can. Entomol. 133: 439–465.

Leroy, P.D., A. Sabri, S. Heuskin, P. Thonart, G. Lognay, F.J. Verheggen, F. Francis, Y. Brostaux, G.W. Felton and E. Haubruge. 2011. Microorganisms from aphid honeydew attract and enhance the efficacy of natural enemies. Nat. Commun. 2: 348. doi: 10.1038/ncomms1347.

Leroy, P.D., T. Schillings, J. Farmakidis, S. Heuskin, G. Lognay, F.J. Verheggen, Y. Brostaux, E. Haubruge and F. Francis. 2012a. Testing semiochemicals from aphid, plant and conspecific: attraction of *Harmonia axyridis*. Insect Sci. 19: 372–382.

Leroy, P.D., S. Heuskin, A. Sabri, F.J. Verheggen, J. Farmakidis, G. Lognay, P. Thonart, J.P. Wathelet, Y. Brostaux and E. Haubruge. 2012b. Honeydew volatile emission acts as a kairomonal message for the Asian lady beetle *Harmonia axyridis* (Coleoptera: Coccinellidae). Insect Sci. 19: 498–506.

Leroy, P.D., R. Almohamad, S. Attia, Q. Capella, F.J. Verheggen, E. Haubruge and F. Francis. 2014. Aphid honeydew: an arrestant and a contact kairomone for *Episyrphus balteatus* (Diptera: Syrphidae) larvae and adults. Eur. J. Entomol. 111: 1–6.

Lilley, R. and J. Hardie. 1996. Cereal aphid responses to sex pheromones and host-plant odours in the laboratory. Physiol. Entomol. 21: 304–308.

Lilley, R., J. Hardie and L.J. Wadhams. 1994. Field manipulation of *Praon* populations using semiochemicals. Norw. J. Agric. Sci. Suppl. 16: 221–226.

Lilley, R., J. Hardie, L.A. Merritt, J.A. Pickett, L.J. Wadhams and C.M. Woodcock. 1994/95. The sex pheromone of the grain aphid, *Sitobion avenae* (Fab.) (Homoptera, Aphididae). Chemoecology 5/6: 43–46.

Marsh, D. 1972. Sex pheromone in the aphid *Megoura viciae*. Nature 238: 31–32.

Marsh, D. 1975. Responses of male aphids to the female sex pheromone in *Megoura viciae* Buckton. J. Entomol. Ser. A 50: 43–64.

Medina-Ortega, K.J., N.A. Bosque-Perez, E. Ngumbi, E.S. Jiménez-Martinez and S.D. Eigenbrode. 2009. *Rhopalosiphum padi* (Hemiptera Aphididae) responses to volatile cues from barley yellow dwarf virus-infected wheat. Environ. Entomol. 38: 836–845.

Minks, A.K. and P. Harrewijn. 1987. Aphids, their Biology, Natural Enemies and Control, Vol. A. Elsevier, Amsterdam, Netherlands.

Moayeri, H.R., A. Rasekh and A. Enkegaard. 2014. Influence of cornicle droplet secretions of the cabbage aphid, *Brevicoryne brassicae*, on parasitism behavior of naive and experienced *Diaeretiella rapae*. Insect Sci. 21: 56–64.

Mölck, G., H. Pinn and U. Wyss. 2000. Manipulation of plant odour preference by learning in the aphid parasitoid *Aphelinus abdominalis* (Hymenoptera: Aphelinidae). Eur. J. Entomol. 97: 533–538.

Mondor, E.B. and B.D. Roitberg. 2003. Age-dependent fitness costs of alarm signaling in aphids. Can. J. Zool. 81: 757–762.

Mondor, E.B. and B.D. Roitberg. 2004. Inclusive fitness benefits of scent-marking predators. Proc. R. Soc. Lond. B Biol. Sci. 271: S341–S343.

Mondor, E.B., D.S. Baird, K.N. Slessor and B.D. Roitberg. 2000. Ontogeny of alarm pheromone secretion in pea aphid, *Acyrthosiphon pisum*. J. Chem. Ecol. 26: 2875–2882.

Montgomery, M.E. and L.R. Nault. 1977. Comparative response of aphids to alarm pheromone, (E)-β-farnesene. Entomol. Exp. Appl. 22: 236–242.

Montgomery, M.E. and L.R. Nault. 1978. Effects of age and wing polymorphism on sensitivity of *Myzus persicae* to alarm pheromone. Ann. Entomol. Soc. Am. 71: 788–790.

Moran, N.A. 1992. The evolution of aphid life cycles. Ann. Rev. Entomol. 37: 321–348.

Mostafavi, R., J.A. Henning, J. Gardea-Torresday and I.M. Ray. 1996. Variation in aphid alarm pheromone content among glandular and eglandular-haired *Medicago* accessions. J. Chem. Ecol. 22: 1629–1638.

Nakatuma, K. 1991. Aphid alarm pheromone component, (E)-beta-farnesene, and local search by a predatory lady beetle, *Coccinella septempunctata* Bruckii mulsant (Coleoptera, Coccinellidae). Appl. Entomol. Zool. 26: 1–7.

Nottingham, S.F., J. Hardie, G.W. Dawson, A.J. Hick, J.A. Pickett, L.J. Wadhams and C.M. Woodcock. 1991. Behavioral and electrophysiological responses of aphids to host and non-host plant volatiles. J. Chem. Ecol. 17: 1231–1242.

Oliveira, M.S. and M. Pareja. 2014. Attraction of a ladybird to sweet pepper damaged by two aphid species simultaneously or sequentially. Arthropod-Plant Interact. 8: 547–555.

Paré, P.W. and J.H. Tumlinson. 1999. Plant volatiles as a defense against insect herbivores. Plant Physiol. 121: 325–331.

Pareja, M., M.C.B. Moraes, S.J. Clark, M.A. Birkett and W. Powell. 2007. Response of the aphid parasitoid *Aphidius funebris* to volatiles from undamaged and aphid-infested *Centaurea nigra*. J. Chem. Ecol. 33: 695–710.

Park, K.C. and J. Hardie. 2002. Functional specialisation and polyphenism in aphid olfactory sensilla. J. Insect Physiol. 48: 527–535.

Park, K.C. and J. Hardie. 2004. Electrophysiological characterisation of olfactory sensilla in the black bean aphid, *Aphis fabae*. J. Insect Physiol. 50: 647–655.

Park, K.C., D. Elias, B. Donato and J. Hardie. 2000. Electroantennogram and behavioural responses of different forms of the bird cherry-oat aphid, *Rhopalosiphum padi*, to sex pheromone and a plant volatile. J. Insect Physiol. 46: 597–604.

Petrescu, A.S., E.B. Mondor and B.D. Roitberg. 2001. Subversion of alarm communication: do plants habituate aphids to their own alarm signals? Can. J. Zool. 79: 737–740.

Pettersson, J. 1970a. An aphid sex attractant. Entomol. Scand. 1: 63–73.

Pettersson, J. 1970b. Studies on *Rhopalosiphum padi* (L.). Laboratory studies on olfactometric responses to winter host *Prunus padus* L. LantbrHogsk. Annlr. 36: 381–399.

Pettersson, J. 1973. Olfactory reactions of *Brevicoryne brassicae* (L.) (Hom.: Aph.). Swed. J. Agric. Res. 3: 95–103.

Pettersson, J. 1993. Odour stimuli affecting autumn migration of *Rhopalosiphum padi* (L.) (Hemiptera: Homoptera). Ann. Appl. Biol. 122: 417–425.

Pettersson, J. and D. Stephansson. 1991. Odor communication in two brassica feeding aphid species (Homoptera: Aphidinea: Aphididae). Entomol. Gen. 16: 241–247.

Pettersson, J., A. Quiroz, D. Stephansson and H.M. Niemeyer. 1995. Odor communication of *Rhopalosiphum padi* on grasses. Entomol. Exp. Appl. 76: 325–328.

Pickett, J.A. and D.C. Griffiths. 1980. Composition of aphid alarm pheromones. J. Chem. Ecol. 6: 349–360.

Pickett, J.A. and R.T. Glinwood. 2007. Chemical ecology. pp. 235–260. *In*: H.F. van Emden and R. Harrington (eds.). Aphids as Crop Pests. CAB International, Wallingford, UK.

Pickett, J.A., L.J. Wadhams and C.M. Woodcock. 1992. The chemical ecology of aphids. Annu. Rev. Entomol. 37: 69–90.

Pickett, J.A., R.K. Allemann and M.A. Birkett. 2013. The semiochemistry of aphids. Nat. Prod. Rep. 30: 1277–1283.

Pope, T.W., C.A.M. Campbell, J. Hardie and L.J. Wadhams. 2004. Electroantennogram responses of the three migratory forms of the damson-hop aphid, *Phorodon humuli*, to aphid pheromones and plant volatiles. J. Insect Physiol. 50: 1083–1092.

Pope, T.W., C.A.M. Campbell, J. Hardie, J.A. Pickett and L.J. Wadhams. 2007. Interactions between host-plant volatiles and the sex pheromones of the bird cherry-oat aphid, *Rhopalosiphum padi* and the damson-hop aphid, *Phorodon humuli*. J. Chem. Ecol. 33: 157–165.

Powell, G. and J. Hardie. 2001. The chemical ecology of aphid host alternation: how do return migrants find the primary host plant? Appl. Entomol. Zool. 36: 259–267.

Powell, W., F. Pennacchio, G.M. Poppy and E. Tremblay. 1998. Strategies involved in the location of hosts by the parasitoid *Aphidius ervi* Haliday (Hymenoptera: Braconidae: Aphidiidae). Biol. Control 11: 104–112.

Purandare, S.R. and B. Tenhumberg. 2012. Influence of aphid honeydew on the foraging behavior of *Hippodamia convergens* larvae (Coleoptera: Coccinellidae). Ecol. Entomol. 37: 184–192.

Quiroz, A., J. Pettersson, J.A. Pickett, L.J. Wadhams and H.M. Niemeyer. 1997. Semiochemicals mediating spacing behavior of bird cherry-oat aphid *Rhopalosiphum padi* feeding on cereals. J. Chem. Ecol. 23: 2599–2607.

Quiroz, A., E. Fuentes-Contreras, C.C. Ramirez, G.B. Russell and H.M. Niemeyer. 1999. Host plant chemicals and distribution of *Neuquenaphis* on *Nothofagus*. J. Chem. Ecol. 25: 1043–1054.

Reed, H.C., S.H. Tan, K. Haapenen, M. Killmon, D.K. Reed and N.C. Elliot. 1985. Olfactory responses of the parasitoid *Diaeretiella rapae* (Hymenoptera: Aphididae) to odour of plants, aphids, and plant-aphid complexes. J. Chem. Ecol. 21: 407–418.

Richard, F.J. and J.H. Hunt. 2013. Intracolony chemical communication in social insects. Insect Soc. 60: 275–291.

Robertson, I., B. Roitberg, I. Williamson and S. Senger. 1995. Contextual chemical ecology: an evolutionary approach to the chemical ecology of insects. Am. Entomol. 41: 237–239.

Russell, G.B., E.H. Faundez and H.M. Niemeyer. 2004. Selection of *Nothofagus* host trees by the aphids *Neuquenaphis staryi* and *Neuquenaphis edwardsi*. J. Chem. Ecol. 30: 2231–2241.

Sarmento, R.A., F. de Lemos, C.R. Dias, A. Pallini and M. Venzon. 2008. Infoquímicos induzidos por herbivoria mediando a comunicação entre plantas de tomate e o predador *Cycloneda sanguinea* (Coleoptera: Coccinellidae). [Herbivore-Induced infochemicals mediating communication between tomato plants and the predator *Cycloneda sanguinea* (Coleoptera: Coccinellidae)]. Ceres 55: 439–444.

Schwartzberg, E.G., G. Kunert, C. Stephan, A. David, U.S.R. Röse, J. Gershenzon, W. Boland and W.W. Weisser. 2008. Real-time analysis of alarm pheromone emission by the pea aphid (*Acyrthosiphon pisum*) under predation. J. Chem. Ecol. 34: 76–81.

Schwartzberg, E.G., K. Böröczky and J.H. Tumlinson. 2011. Pea aphids, *Acyrthosiphon pisum*, suppress induced plant volatiles in broad bean, *Vicia Faba*. J. Chem. Ecol. 37: 1055–1062.

Shambaugh, G.F., J.L. Frazier, A.E.M. Castell and L.B. Coon. 1978. Antennal sensilla of seventeen aphid species (Homoptera: Aphidinae). Int. J. Insect Morphol. Embryol. 7: 389–404.

Sloggett, J.J. and W.W. Weisser. 2002. Parasitoids induce production of the dispersal morph in the pea aphid, *Acyrthosiphon pisum*. Oikos 98: 323–333.

Srinivasan, R., J.M. Alvarez, S.D. Eigenbrode and N.A. Bosque-Pérez. 2006. Influence of hairy nightshade *Solanum sarrachoides* (Sendtner) and Potato leafroll virus (Luteoviridae: Poleovirus) on the host preference of *Myzus persicae* (Sulzer) (Homoptera: Aphididae). Environ. Entomol. 35: 546–553.

Stewart-Jones, A., S.Y. Dewhirst, L. Durrant, J.D. Fitzgerald, J. Hardie, A.M. Hooper, J.A. Pickett and G.M. Poppy. 2007. Structure, ratios and patterns of release in the sex pheromone of an aphid, *Dysaphis plantaginea*. J. Exp. Biol. 210: 4335–4344.

Sun, Y.F., F. De Biasio, H.L. Qiao, I. Iovinella, S.X. Yang, Y. Ling, L. Riviello, D. Battaglia, P. Falabella, X.L. Yang and P. Pelosi. 2012. Two odorant-binding proteins mediate the behavioural response of aphids to the alarm pheromone (E)-β-farnesene and structural analogues. PLoS one 7: e32759.

Symmes, E.J., S.Y. Dewhirst, M.A. Birkett, C.A.M. Campbell, K. Chamberlain, J.A. Pickett and F.G. Zalom. 2012. The sex pheromones of mealy plum (*Hyalopterus pruni*) and leaf-curl plum (*Brachycaudus helichrysi*) aphids: identification and field trapping of male and gynoparous aphids in prune orchards. J. Chem. Ecol. 38: 576–583.

Thompson, D.R., L.J. Gut and J.W. Jenkins. 1999. Pheromones for insect control: strategies and successes. pp. 385–412. *In*: F.R. Hall and J.J. Menn (eds.). Biopesticides: Use and Delivery. Humana Press Inc., Totowa, New Jersey.

Tillman, J.A., S.J. Seybold, R.A. Jurenka and G.J. Blomquist. 1999. Insect pheromones—an overview of biosynthesis and endocrine regulation. Insect Biochem. Mol. 29: 481–514.

Torre-Bueno, J.R. and G.S. Tulloch. 1962. A Glossary of Entomology. Brooklyn Entomological Society, Brooklyn, New York.

Turchin, P. and P. Kareiva. 1989. Aggregation in *Aphis varians*: an effective strategy for reducing predation risk. Ecology 70: 1008–1016.

van Tol, R.W.H.M., H.H.M. Helsen, F.C. Griepink and W.J. de Kogel. 2009. Female-induced increase of host-plant volatiles enhance specific attraction of aphid male *Dysaphis plantaginea* (Homoptera: Aphididae) to the sex pheromone. Bull. Entomol. Res. 99: 593–602.

Vandermoten, S., F. Francis, E. Haubruge and W.S. Leal. 2011. Conserved odorant-binding proteins from aphids and eavesdropping predators. PLoS one 6: e23608.

Vandermoten, S., M.C. Mescher, F. Francis, E. Haubruge and F.J. Verheggen. 2012. Aphid alarm pheromone: an overview of current knowledge on biosynthesis and functions. Insect Biochem. Mol. 42: 155–163.

Verheggen, F.J., Q. Fagel, S. Heuskin, G. Lognay, F. Francis and E. Haubruge. 2007. Electrophysiological and behavioral responses of the multicolored Asian lady beetle, *Harmonia axyridis* Pallas, to sesquiterpene semiochemicals. J. Chem. Ecol. 33: 2148–2155.

Verheggen, F.J., M.C. Mescher, E. Haubruge, C.M. De Moraes and E.G. Schwartzberg. 2008a. Emission of alarm pheromone in aphids: a non-contagious phenomenon. J. Chem. Ecol. 34: 1146–1148.

Verheggen, F.J., L. Arnaud, S. Bartram, M. Gohy and E. Haubruge. 2008b. Aphid and plant secondary metabolites induce oviposition in an aphidophagous hoverfly. J. Chem. Ecol. 34: 301–307.

Verheggen, F.J., Q. Capella, J.P. Wathelet and E. Haubruge. 2008c. What makes *Episyrphus balteatus* (Diptera: Syrphidae) oviposit on aphid infested tomato plants? Comm. Agr. Appl. Biol. Sci. 73: 371–381.

Verheggen, F.J., E. Haubruge, C.M. De Moraes and M.C. Mescher. 2009. Social environment influences aphid production of alarm pheromone. Behav. Ecol. 20: 283–288.

Verheggen, F.J., E. Haubruge and M.C. Mescher. 2010. Chapter 9: Alarm pheromones—Chemical signaling in response to danger. Vitamins and Hormones 83: 215–240.

Verheggen, F.J., E. Haubruge, C.M. De Moraes and M.C. Mescher. 2013. Aphid responses to volatile cues from turnip plants (*Brassica rapa*) infested with phloem-feeding and chewing herbivores. Arthropod-Plant Inte. 7: 567–577.

Vet, L.E.M. and M. Dicke. 1992. Ecology of infochemical use by natural enemies in a tritrophic context. Annu. Rev. Entomol. 37: 141–172.

Wadhams, L.J. 1990. The use of coupled GC-electrophysiological techniques in the identification of insect pheromones. pp. 289–298. *In*: A.R. McCaffery and I.D. Wilson (eds.). Chromatography and Isolation of Insect Hormones and Pheromones. Plenum Press, New York.

Way, M. 1973. Population structure in aphid colonies. pp. 76–84. *In*: A.D. Lowe (ed.). Perspectives in Aphid Biology. Entomological Society of New Zealand, Auckland.

Webster, B. 2012. The role of olfaction in aphid host location. Physiol. Entomol. 37: 10–18.

Webster, B., T.J.A. Bruce, S. Dufour, C. Birkemeyer, M. Birkett, J. Hardie and J.A. Pickett. 2008a. Identification of volatile compounds used in host location by the black bean aphid, *Aphis fabae*. J. Chem. Ecol. 34: 1153–1161.

Webster, B., T.J.A. Bruce, J.A. Pickett and J. Hardie. 2008b. Olfactory recognition of host plants in the absence of host-specific volatile compounds. Comm. Integr. Biol. 1: 167–169.

Webster, B., T.J.A. Bruce, J.A. Pickett and J. Hardie. 2010. Volatiles functioning as host cues in a blend become nonhost cues when presented alone to the black bean aphid. Anim. Behav. 79: 451–457.

Witzgall, P., P. Kirsch and A. Cork. 2010. Sex pheromones and their impact on pest management. J. Chem. Ecol. 36: 80–100.

Wohlers, P. and W.F. Tjallingii. 1983. Electroantennogram responses of aphids to the alarm pheromone (E)-β-farnesene. Entomol. Exp. Appl. 33: 79–82.

Wu, G.M., G. Boivin, J. Brodeur, L.A. Giraldeau and Y. Outreman. 2010. Altruistic defence behaviours in aphids. BMC Evol. Biol. 10: 19.

Xiangyu, J.G., F. Zhang, Y.L. Fang, W. Kan, G.X. Zhang and Z.N. Zhang. 2002. Behavioral response of aphids to the alarm pheromone component (E)-b-farnesene in the field. Physiol. Enotomol. 27: 307–311.

Yan, F.S. and J.H. Visser. 1982. Electroantennogram response of the cereal aphid *Sitobion avenae* to plant volatile components. pp. 387–388. *In*: J.H. Visser and A.K. Minks (eds.). Proceedings of the Fifth International Symposium on Insect-Plant Relationships. Pudoc, Wageningen, Netherlands.

Zhang, F. and Z.N. Zhang. 2000. Comparative study on the antennal sensilla of various forms of *Myzus persicae*. Acta Entomol. Sin. 143: 131–137.

Zhu, J., J.J. Obrycki, S.A. Ochieng, T.C. Baker, J.A. Pickett and D.W.M. Smiley. 2005. Attraction of two lacewing species to volatiles produced by host plants and aphid prey. Naturwissenschaften 92: 277–281.

Zhu, J., A. Zhang, K.C. Park, T.C. Baker, B. Lang, R. Jurenka, J.J. Obrycki, W.R. Graves, J.A. Pickett and D.W.M. Smiley. 2006. Sex pheromone of the soybean aphid, *Aphis glycines* Matsumura, and its potential use in semiochemical-based control. Environ. Entomol. 35: 249–257.

10

Aphid Honeydew: Rubbish or Signaler#

Klaus H. Hoffmann

Introduction

Parking a car under a broadleaf in the middle of summer often results in a sticky coat on the roof and hood, as though sugar water had been poured over them. The culprits can be found in the tree above the parked vehicle, aphids sucking on the phloem of plants and excreting a sugar-rich fluid that settles like dew, the honeydew. In biblical times, it was probably the sugar excretion products of a scale bug that was the manna, which saved the Israelits from starvation while crossing the Sinai desert (Hoffmann et al. 2003). In Greek mythology, drips from the Manna-ash (*Fraxinus ornus*) nursed the infant god Zeus on the island of Crete.

Besides aphids and some scale insects, honeydew is produced by some caterpillars of butterflies (Lycaenidae) and some moths. The honeydew excreted by hemipterans is the result of feeding on the phloem sap of plants, which has very high sugar content (mainly sucrose/saccharose), which produces a high osmotic pressure (Ammar et al. 2013). Although honeydew is primarily a waste product allowing phloem feeders to dispose of the excess sugars in their diet (Wilkinson et al. 1997), it serves as a food source for many animals. Honeydew is collected by some birds, wasps and bees. Honeybees process it into a dark strong honey, the honeydew honey. Honeydew of psyllids and other homopterans attracts many ant species. The ants use the sugar for brood care and many also protect the aphids from predators, such as ladybird beetles or parasitoids. Ant attendance also prevents self-contamination with sticky honeydew and the fungal growth, thereby maintaining good hygiene in the aphid colony and

Animal Ecology I and BayCEER, University of Bayreuth, 95440 Bayreuth, Germany.
E-mail: klaus.hoffmann@uni-bayreuth.de

Dedicated to my former colleague and friend Dr. Wolfgang Völkl (passed away April 2015)

clearing pathways for walking in aphid colonies (Fig. 1) (Wimp and Whitham 2001). Many ants satisfy the carbohydrate needs of their entire colony with honeydew, which they usually collect directly from the anus of the sucking aphids ("milking").

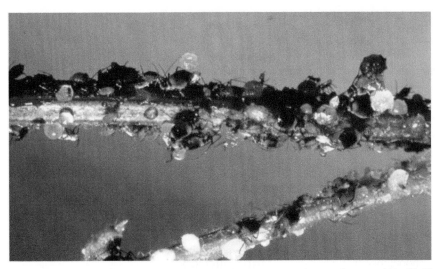

Fig. 1. Without the support of ants (*Lasius niger*), the aphid *Metopeurum fuscoviride* cannot rid itself of its excess honeydew. The large honeydew drops gather at the aphid's anus causing a life-threatening pollution. From Hoffmann et al. (2003) (Photo by W. Völkl).

Aphids frequently reach pest status in agricultural ecosystems and threaten crop production. Honeydew is often the predominant sugar source in agro ecosystems. From the perspective of the plants, ant-aphid associations can be either harmful or beneficial, depending on the ratio of the direct cost of feeding by the aphids to the indirect benefit of increased ant suppression of other herbivores (Buckley 1987). In the latter case, the aphid-ant association can be thought as part of a biological control function (Jha et al. 2012).

Here, we pursue the question whether aphid honeydews contain distinct signal compounds, which, for example, may attract consumer communities.

Ecology and Evolution of Aphid-Ant Mutualism

The trophobiotic relationship between aphids and ants is generally called mutualism. Mutualisms have been traditionally recognized as stable reciprocal interactions, in which beneficial services are exchanged between the participating species (Yao 2014). The mutualistic interaction between aphids and ants was already well documented 50 years ago (Way 1963), although it is still unclear whether some relationships are truly mutualistic or intermediate between mutualism and exploitation (Novgorodova 2004). Over the last two decades, considerable progress has been made in understanding physiological adaptations, the ecological context, and the evolutionary constrains acting on both partners when entering the association (Stadler and Dixon 2005, Yao 2014).

Given the ecological success of both ants and aphids, it is difficult to understand why only a few species of aphids have evolved a close relationship with ants. In central Europe, only about one third of aphid species are obligate myrmecophiles (Stadler and Dixon 2005). Placing aphid-ant relationships in a continuum from highly mutualistic to antagonistic, a mutualistic relationship is expected if aphids and ants achieve higher population growth rates and larger colony size. Aphids should be able to modify their honeydew to make it more attractive to ants, and ants should be able to modify their foraging behavior so that they can effectively harvest this energy-rich resource (Stadler and Dixon 2005). The chemical composition of honeydew seems to play an important role in aphid-ant mutualism (see next paragraph).

Jha et al. (2012) studied the specific mechanisms of aphid-ant mutualism and provided evidence that mutualism is driven by the direct enhancement of growth rates and fecundity of the aphids, as opposed to the frequently assumed indirect mechanism, via reduction of the parasites and predators of the aphids. Ants may also protect aphids they are tending from obligate fungal pathogens (Nielsen et al. 2010). *Formica podzolica* ants tending milkweed aphids, *Aphis asclepiadis*, can recognize and remove aphids that are covered with conidia or have been killed by an entomopathogenic fungus, *Pandora neoaphidis*. This indicates that ants may perform cleaning and quarantining services, which would enhance the aphid-ant mutualism.

The potential costs of ant attendance to aphids include the physiological costs of ant attendance through honeydew production, ant predation on aphids, competition between aphid species for mutualistic ants, the mediation of host plants on aphid-ant interactions, and the parasitism by wasps (Yao 2014). However, in the various forms of aphid-ant mutualism, the costs are influenced by spatial and temporal factors. For example, Tegelaar et al. (2012) demonstrated that the costs of ant attendance vary over 13 generations in the relationship of *Aphis fabae* and *Lasius niger*. Host alternating aphids, such as *A. fabae*, hatch on the winter host plant and produce winged offspring (alates) that may migrate and colonize summer hosts (Dixon 1998). During summer, colonies of wingless multipliers are produced (parthenogenesis), followed by a subsequent production of winged emigrants. Taylor et al. (2012) used the output of emigrating alates per colonizing alate as a useful measure of the productivity of the summer phase of the life cycle and found that the presence of ants decreased colony productivity of the aphids. This example demonstrates that aphids do not always benefit from the presence of ants, but under some conditions pay a cost in the form of reduced dispersal and colony productivity. Yao (2012) demonstrated that ant attendance directly reduces flight muscle and wing size in the aphid *Tuberculatus quercicola*. *T. quercicola* increase honeydew excretion under ant attendance at the expense of resource investment in the flight apparatus, thereby limiting their dispersal.

Moreover, aphid-ant interactions can be affected by several other factors. For example, bottom-up effects like the distribution and abundance of the host plants of aphids or the community composition of soil invertebrates (see below) may affect the growth rates of aphids. Top-down effects of natural enemies (predators and parasitoids) are also likely to shape the strength of aphid-ant relationships (Stadler and Dixon 2005). Mooney (2011) provided first evidence for genetic variation in aphids for attractiveness to ants.

Aphid tending ants not only collect honeydew and protect aphids from natural enemies, but often prey on the aphids they attend (Endo and Itino 2013). Aphids, therefore, may have evolved chemical mimicry as an anti-predation strategy. Aphids can produce ant-like cuticular hydrocarbons, which protect them from predation. Aphid cuticular hydrocarbons mainly consist of n-alkanes, with low amounts of branched alkanes and n-alkenes. Myrmecophilous and non-myrmecophilous species may differ in their cuticular hydrocarbons, especially in their n-alkane composition (Lang and Menzel 2011). This enables ants to classify aphids into potential trophobionts and potential prey.

A true mutualism between myrmecophilous aphids and ants is mainly based on the rich food supply that honeydew represents and on the protection that ants provide against natural enemies of the aphids. While predators and parasitoids actively forage by using aphid semiochemicals, scouts of aphid tending ants either use honeydew components or aphid pheromones for locating aphids. For example, *L. niger* scouts can use low levels of *A. fabae* alarm pheromone, i.e., (E)-ß-farnesene, as a cue indicating the presence of aphid colonies (Verheggen et al. 2012) (see Chapter 9 "Chemical Ecology of Aphids (Hemiptera: Aphididae)" in this volume). To understand the role of honeydew as mediator in aphid-ant interactions, the chemical composition of aphid honeydews was investigated utilizing recent progress in analytical techniques [derivatization of sugars and amino acids, gas chromatography (GC)-mass spectrometry (MS), HPLC-MS], but also aphid genomics data (see Chapter 3 "Functional and Evolutionary Genomics in Aphids" in this volume). Such progress in techniques has created new opportunities to improve our understanding of aphid-ant mutualism.

The Chemistry of Phloem Sap and Honeydew

Phloem Sap Composition

Sugar-rich diets taste good to many animals (Douglas 2006). Plants provide three types of sugar-rich food. Nectar, fleshy fruits, and phloem sap. Only one group of animals, insects of the order Hemiptera, contains species that use phloem sap as their dominant or sole food source. A few other animals (e.g., some lepidopterans, hummingbirds, and primates) consume phloem sap occasionally, but it is not a vital component of their diet. Plant phloem sap is rich in nutrients; it contains high quantities of sugars, mainly sucrose/saccharose up to a concentration of 1 M (Lohaus and Schwerdtfeger 2014), up to ten amino acids in varying amounts with the non-essential amino acids asparagine and glutamine as the most abundant ones, organic acids, vitamins, and inorganic acids, and is free of feeding deterrents and toxins (Dinant et al. 2010). The concentration of total amino acids in phloem sap is generally in the range of 60–200 nmol μL^{-1}, whereas concentration of total amino acids in the honeydew is still lower (20–100 nmol μL^{-1}) (Woodring et al. 2004). Variations in phloem sap composition occur in both diurnal and seasonal rhythms and in response to plant age (Taylor et al. 2012).

Does Aphid Honeydew Reflect the Host Plant Phloem Sap Composition?

Plants provide aphids with an unbalanced diet of high sucrose content, but low amounts of amino acids in the phloem sap. The sugar composition of honeydew generally reflects the composition of phloem sap, but various other mono-, di, and oligosaccharides are synthesized by the aphid, mainly from the surplus of sucrose. The amino acid composition of the honeydew also corresponds to the phloem sap content; asparagine and glutamine were reported as major amino acids in the honeydew of many aphid species, and also in phloem saps (Sasaki et al. 1990, Leroy et al. 2011b). Aphid infestation can modify the amino acid composition of the plant exudates; specifically enhancing the concentration of glutamine and asparagine, whereby a "specialist" for the host plant induces more important modifications in the host plant amino acid composition than a "generalist" (Leroy et al. 2011b).

Mooney and Agarwal (2008) demonstrated variations among genotypes of the common milkweed (*Asclepias syriaca*) in *per capita* ant attendance of aphids (and also in their honeydew composition) and suggested that such variations result from genetically determined differences in plant phloem composition. Plant-derived differences in the composition of aphid honeydew and their effects on colonies of aphid-tending ants were also observed by Pringle et al. (2014). They compared the chemical composition of honeydew produced by *Aphis nerii* aphid clones on two milkweed congeners, *Asclepias curassavica* and *A. incarnate*, and the response of *Linepithema humile* ant colonies to these honeydews. Despite performing better when feeding on the *A. incarnata* honeydew, ant workers marginally preferred honeydew from *A. curassavica* to honeydew from *A. incarnate* when given a choice.

Wool et al. (2006) demonstrated that seasonal variation in host-plant sap quality is reflected in changes of aphid honeydew sugar content among samples taken at different times of the year. Elevated carbon dioxide (CO_2) levels in the atmosphere lowered the amount of amino acids found in cotton phloem sap, but the content of amino acids in the honeydew of *Aphis gossypii* fed on elevated CO_2-grown cotton was not affected (Sun et al. 2009). Instead, more phloem sap was ingested by those aphids to satisfy their nutritional requirement and a larger amount of honeydew was produced.

When soybean aphids (*Aphis glycines*) were inoculated onto two related strains of soybean plants, a nodulating strain that associates with rhizobia and a non-nodulating strain that does not harbor nitrogen-fixing bacteria, aphids feeding on the nodulating strains were found to produce honeydew with significantly different sugar profile to that for aphids feeding on non-nodulating plants, but no differences in total amino acids-N content of the honeydew was observed (Whitaker et al. 2014).

Other factors that influence the chemical composition of aphid honeydew are the aphid species *per se*, the development stage and age of the aphids, the rate and duration of aphid infestation of plants, the presence of ants (mutualism), the presence of bacterial intracellular symbionts in the aphids (see chapter 5 "Bacterial symbionts of aphids (Hemiptera: Aphididae)" in this volume), the parasitized state of the aphids, and the presence of secondary plant metabolites (Leroy et al. 2011a).

Honeydew Chemical Composition and Honeydew Excretion

A common explanation for the excretion of honeydew is that the aphids must ingest large quantities of sugar-rich phloem sap to extract the necessary amino acids they need for growth and reproduction, and the surplus of sugars are excreted (Wool et al. 2006). Aphid honeydews usually contain two or three sugars in high amounts, and up to 20 in lower concentrations, the 10 essential amino acids, and up to 15 non-essential amino acids, in much lower concentrations (Dhami et al. 2011). To analyze utilization of the phloem sap and to study the chemical composition of the honeydew in more details, the honeydew has to be collected from the insects without evaporation of water. Aphids producing high amounts of honeydew, such as the ant attended *Metopeurum fuscoviride*, can be "milked" by gentle pressure on the body and the honeydew can be collected with a pipette (Woodring et al. 2007). Honeydew from less productive aphid species has to be collected with a piece of aluminum foil placed under a colony and rinsed by water. Therefore, in these species, only the percent of each compound in the honeydew can be calculated. Honeydew from galling aphids, such as in the subfamily Fordinae, has to be collected from aphids in opened galls following tactile stimulation of their tergum (Wool et al. 2006).

Dhami et al. (2011) detected over 75 distinct compounds from honeydew of scale insects by using a novel method of methylchloroformate derivatization and GC-MS analysis. They found 36 trimethylsilyl derivatives (carbohydrates) and 41 methylchloroformate derivatives (amino- and non-amino organic acids). Carbohydrates can represent up to 88% of the honeydew dry matter (Lamb 1959), usually including the photosynthate sucrose, its component monosaccharides fructose and glucose, and a variable number of oligosaccharides. Phloem feeding insects generally reduce the sucrose content of ingested phloem sap by polymerization processes that yield oligosaccharides, thus regulating gut osmotic pressure. To examine this process, the effect of the α-glucosidase inhibitor acarbose on the sugar ratios and on the osmolality of aphids was explored by Karley et al. (2005) *in vitro* and *in vivo*. Acarbose inhibited sucrase activity in gut homogenates as well as the amounts of monosaccharides and oligosaccharides in the honeydew of living aphids. Fructose as one end product from sucrase activity was absorbed and used for the energy demand of the aphids (Ashford et al. 2000), which in itself helps to reduce the osmolality in the gut. Much of the glucose cleaved from the sucrose was incorporated into oligosaccharides by transglucosidase activity (Woodring et al. 2007). Thus, the transglucosidase activity is also essential to gut osmolality (Douglas 2006). Another common and efficient method for restoring and maintaining osmoregulation and water balance is that aphids drink periodically from the xylem of the host plant (Will and Van Bel 2006).

The trisaccharide melezitose is one of the transglucosidase products found in the gut of many aphid species. The role of melezitose has received special focus, because this sugar is not present in the host plants, but synthesized by the aphids and excreted with the honeydew. Fischer and Shingleton (2001) demonstrated that honeydew sugar composition, and especially the melezitose content, are host plant- and ant-dependent. Honeydew of two *Chaitophorus* aphids, *C. populialbae* and *C. populeti* contained high proportions of melezitose and these species are typically tended by ants. In contrast, *C. tremulae* honeydew contained low proportions of melezitose and this species is typically untended. *C. populeti* and *C. populialbae* reduced their production of melezitose when reared in the absence of ants, and the reverse was true in *C. tremulae*. The results clearly show that ant tending itself may influence the honeydew sugar composition. Moreover, the honeydew melezitose content varies not only among different species on the same host plant, but also among different host plants for the same species. *C. populeti* and *C. populialbae* produced higher proportions of melezitose when reared on *Populus tremula* than on *P. alba*.

The black bean aphid, *A. fabae*, is polyphagous and regularly ant tended (Fischer et al. 2005). The honeydew production of *A. f. fabae* differs considerably between host plants. Aphids feeding on two annual plants, *Vicia faba* and *Chenopodium album* produced less than half the amount of honeydew per aphid and hour than conspecifics on the perennial host plant *Tanacetum vulgare*. *A. f. fabae* virginoparae on their winter host *Evonymus europaeus* also produced high amounts of honeydew. All honeydew samples were dominated by melezitose, except for *A. f. fabae* on *E. europaeus*, whereas the total sugar concentrations in the honeydew did not differ between the various hosts.

Woodring et al. (2007) compared the rate of oligosaccharide synthesis in the ant attended aphid *M. fuscoviride* and the unattended aphid *Macrosiphoniella tanacetaria*, both feeding on tansy host plant, *T. vulgare*. The sugar composition of the honeydew of *M. fuscoviride* feeding on *T. vulgare* indicated a rapid digestion of sucrose into glucose and fructose, and the simultaneous synthesis of considerable amounts of melezitose and some trehalose. The sugar composition of the honeydew of *M. tanacetaria* in contrast showed only traces of trehalose and melezitose, but up to 20% erlose in plant-fed aphids. These authors developed an artificial phloem sap (APS) for the two aphid species with the same sugar composition as the natural phloem sap of *T. vulgare*, containing sucrose as the only sugar. The honeydew of *M. fuscoviride* fed on APS contained high amounts of melezitose, whereas the honeydew of *M. tanacetaria* fed on APS was almost free of melezitose but contained some erlose. Generally, however, the concentrations of melezitose and trehalose (or erlose) were lower in aphids fed APS compared with those fed on the host plant. Feeding the aphids on APS through a membrane seems to be stressful and somehow inhibited the synthesis of oligosaccharides. Incubation of the isolated digestive tracts of the two aphid species in APS confirmed the results that *M. fuscoviride* shows a steady high rate of melezitose synthesis by gut enzymes, whereas in *M. tanacetaria* guts only a moderate rate of erlose synthesis and no detectable melezitose or trehalose synthesis was observed. The inclusion of tetracycline in the incubation medium eliminated any possible role of contaminating bacteria in the digestion and synthesis of sugars.

The black bean aphid *A. fabae* is a facultative ant mutualist which represents a complex of at least four subspecies, which share the same primary hosts. Vantaux et al. (2011, 2012) studied the clonal variation in the composition of honeydew of *A. fabae* and found large interclone differences in melezitose secretion. In all cases, however, the relative melezitose concentration was significantly positively correlated with the total sugar concentration and significantly negatively correlated with the relative glucose, fructose, trehalose and erlose concentrations. Moreover, their results suggest that ants were more likely to collect honeydew from the high-melezitose secreting clones.

In conclusion, the presented data provide increasing evidence that melezitose in the honeydew is a meaningful signal molecule for aphid-ant interactions (see also below).

Woodring et al. (2004) identified a total of 18 amino acids in the honeydew of eight species of aphids feeding on *T. vulgare*, with the non-essential amino acids asparagine, glutamine, glutamic acid and serine as the predominant. Honeydew of ant attended species was generally richer in amino acids compared with honeydew from unattended aphids. A somewhat unexpected result was the finding that those aphids with the highest total amino acid concentration in the honeydew always had the highest concentration of sugars. There was no evidence that any single amino acid or group of amino acids in the honeydew acts as an attractant for ant attendance in the aphid species studied. The amino acids in the honeydew of aphids are not only influenced by the diet, but also by the action of endosymbionts (*Buchnera aphidicola*), which are localized in mycetocytes or bacteriocytes (Buchner 1965, Douglas 1989, Sasaki et al. 1990, Munson et al. 1991, see Chapter 5 "Bacterial Symbionts of Aphids (Hemiptera: Aphididae)" in this volume). Endosymbionts supply aphids with vitamins, sterols, but also with up to ten essential amino acids, especially tryptophan and leucine (Sasaki et al. 1990, Bernays and Klein 2002, Leroy et al. 2011b, Shigenobu and Wilson 2011). Endosymbionts, however, do not seem to affect the sugar composition of honeydews (Katayama et al. 2013a).

Besides sugars and amino acids, aphid honeydew contains a diverse number of proteins. A proteomic investigation of honeydew from the pea aphid *Acyrthosiphon pisum*, revealed an unexpected diversity of more than 140 proteins, with a total concentration close to 5 µg/µL (Sabri et al. 2013). The protein diversity of aphid honeydew originates from several sources (i.e., the host aphid and its microbiota, including endosymbiotic bacteria and gut flora). Sixty proteins matched to insect database sequence resources, while 36 proteins were identified to be homologous to bacterial sequences. Almost half of the bacterial identified proteins were homologous to bacterial sequences associated with aphid endosymbiotic bacteria. Most of the bacterial proteins identified in honeydew (27.8%) were related to the genetic information process, while 20% of the bacterial symbiont proteins were related to the amino acid metabolism. The proteomic approach also allowed the identification of some proteins that might act as mediators in the plant-aphid interaction. The proteins flagellin and elongation factor Tu, identified from the pea aphid honeydew, are known to act as inducers of defenses in many plant species (see below).

Honeydew as a Food Source for Predators and Parasitoids

Many adult predators and parasitoids rely primarily or exclusively on carbohydrates as a source of energy required during flight. Carbohydrate feeding can increase reproductive fitness of parasitoids by increasing their longevity, fecundity and/or parasitism rate (Hogervorst et al. 2007). To meet these carbohydrate requirements, they may exploit a broad range of plant substrates, including nectar, plant sap exudates, but also the honeydew from phloem feeding arthropods (Wäckers et al. 2008). In comparison to other sugar sources, honeydew is often an inferior source. Honeydew is viscous, making it difficult to access, and its particular chemical composition limits its nutritional suitability (Wäckers 2005).

The suitability of a particular food source not only depends on food source characteristics but also on how well consumers are attuned to exploit the food. Wäckers et al. (2008), however, did not find that aphidophagous parasitoids and predators are more inclined to exploit honeydew over other carbohydrate sources like nectar in comparison to predators and parasitoids of non-honeydew producers.

The most abundant and important aphid parasitoids belong to the family Aphidiidae (Hymenoptera). Host feeding is not common among Aphidiidae and adults feed mostly on honeydew from their aphid hosts (Starý 1970). Hogervorst et al. (2007) studied the effects of honeydew sugar composition on the longevity of the aphidiid parasitoid *Aphidius ervi*. Compared to females of *A. ervi* that only had access to water, the additional access to honeydew increased the mean parasitoid lifespan by a factor 4.4 to 6.3 depending on the aphid and plant species. Differences in parasitoid longevity could to some extent be explained by carbohydrate composition of the honeydew. The plant derived sugars sucrose, glucose, and fructose generally had a higher nutritional quality, whereas the aphid synthesized oligosaccharides such as melezitose, erlose, raffinose, and trehalose lowered the nutritional value of the honeydew (Wäckers 2000).

The predatory gall midge *Aphidoletes aphidimyza* also uses aphid honeydew as an energy source. Watanabe et al. (2014) investigated the influence of sugars, including three artificial honeydews, six sugar components, and distilled water, on the longevity of unmated *A. aphidimyza*. Both females and males attained the greatest longevity on sucrose and artificial honeydew of *Aphis gossypii*, whereas longevity was shortest when they were provided only with water.

Honeydew from *A. gossypii* feeding on cotton can be used as a carbohydrate source by the aphid parasitoid *Lysiphlebus testaceites* and the whitefly parasitoid *Eretmocerus eremicus*. However, when fed on honeydew their longevity was shorter than when fed on a sucrose solution (Hagenbucher et al. 2014). Neither the genetic transformation of *Bt* cotton nor the induced resistance of *Bt* cotton against caterpillar damage affected the honeydew quality and, thereby, the fitness of the parasitoids.

Larvae of the common green lacewing *Chrysoperla carnea* are predacious and feed on a wide range of small arthropods, including aphids. In addition to their feeding on insects to cover their requirements for growth and development, they consume non-prey foods such as honeydew (Hogervorst et al. 2008). This non-prey food is not only important during a period of prey scarcity, but forms part of the diet even when suitable prey is available.

Honeydew has a second function for parasitoids and predators, as it is used as a kairomone to locate hosts and prey. Females of the aphidophagous gall midge *A. aphidimyza* showed an olfactory response to honeydew excreted by the green peach aphid, *Myzus persicae*, under laboratory conditions (Choi et al. 2004). The response was only elicited by treatments with honeydew, whereas aphids or host plant leaves were not attractive to the midges.

Honeydew Kairomone Function

Chemical cues in the honeydew that mediate aphid location by natural enemies were first evidenced by Hagen et al. (1971), who showed that deposition of aphid honeydew components attracted the ladybird beetle, *Coccinella septempunctata*. In later studies, it was claimed that the amino acid tryptophan plays an important role as an attractant (Ben Saad and Bishop 1976). Recent investigations have shown that plots treated with sucrose were significantly more attractive to ladybird beetles than untreated plots (Evans and Gunther 2005). The use of aphid honeydews as a host finding kairomone seems to be common also amongst aphid parasitoids. Several studies have proven that aphidiids use honeydew as a kairomone for host location (Rehman and Powell 2010).

Hatano et al. (2008) used Vinson's (1976) separation of the host selection process into habitat location, host location and host acceptance for both parasitoids and predators. For habitat location (i.e., detection of the host plant), volatiles emitted by plants after aphid attack may be used. These synomones indicate not only the presence of an aphid host plant, but also the presence of aphids. Host location on the food plant is guided by semiochemicals that mostly originate from the aphids, in particular aphid alarm pheromone [(E)ß-farnesene], honeydew, or the smell of the aphid itself (see Chapter 9 "Chemical Ecology of Aphids (Hemiptera: Aphididae)" in this volume). Finally, host acceptance is guided by contact chemicals. For parasitoids, this final step has been divided into host recognition and host acceptance. A host may be recognized visually or by antennal contact with chemical cues on the aphid cuticle. Host acceptance depends on an assessment of host quality made during ovipositor probing.

Although, honeydew is considered to be an important kairomone for habitat location (Pettersson et al. 2008), more recent studies with various parasitoids and predators have shown that it often acts as an arrestant, i.e., it increases the time that natural enemies search for aphids on plants (Blande et al. 2008). Often, the natural enemy needs physical contact with the honeydew. Aphid honeydew may also act as an oviposition stimulus for parasitoids. Leroy et al. (2014) demonstrated that honeydew excreted from the pea aphid *A. pisum* acts as an arrestant and a contact kairomone for young larvae and adults of a common predatory hoverfly, *Episyrphus balteatus*. First and second instar larvae increased their foraging behavior in honeydew-treated areas. When plants were sprayed with crude honeydew, the speed of movements of adult females was higher than in controls, resulting in a longer period of time spent on treated plants and laying eggs. Larvae of the ladybird beetle, *C. septempunctata*, also use honeydew as a contact kairomone when foraging for aphids, *A. pisum* and *Aphis craccivora* (Ide et al. 2007). More larvae responded by climbing the plants with aphids than to plants without aphids. When the plants were replaced with sticks, in order to exclude visual and olfactory cues from plants and aphids, more larvae climbed sticks above the area that contained honeydew than sticks above an area without honeydew. Moreover, *C. septempunctata* larvae can distinguish honeydew of the two aphid species and respond more strongly to *A. craccivora* honeydew than to *A. pisum* honeydew. Buitenhuis et al. (2004) demonstrated that even hyperparasitoids may use aphid honeydew as an infochemical to locate their hosts, a conspicuous cue from the second trophic level.

Merivee et al. (2012) investigated the stimulatory effect of various aphid honeydew sugars and sugar alcohols on chemoreceptor neurons of the antennal chaetoid gustatory sensilla in the carabid beetle *Anchonemus dorsalis*, an important aphid predator. They showed that maltose was the most stimulating sugar for *A. dorsalis*. Maltose is one of the predominant sugars in the honeydew of a number of aphids. Its content in the honeydew of *M. fuscoviride, Brachycaudus cardui* and *A. fabae* is 27.6%, 8.9% and 26.8%, respectively (Völkl et al. 1999).

Honeydew deposition on plant leaves offers an excellent substrate for fungal epiphytes that can serve as a food source for adult ladybird beetles. Thus honeydew may not only be seen as a sign of aphid presence for predators, but also as a cue for other potential food sources. Leroy et al. (2011a) isolated a bacterium from the pea aphid *A. pisum* honeydew, *Staphylococcus sciuri*, which acts as a kairomone attracting and enhancing the efficiency of aphid natural enemies. *S. sciuri* has been isolated from the gut flora of *A. pisum* and is partially excreted by the aphids. Bacterial growth on honeydew and honeydew degradation of amino acids and sugars in aphid honeydew are responsible for the emission of volatile compounds. When volatiles identified from *A. pisum* honeydew were tested individually on their impact on hoverfly *E. balteatus* foraging and oviposition behavior, three of these volatiles attracted *E. balteatus*: 3-methyl-2-butenal, 3-methylbutanoic acid and 2-methylbutanoic acid, and two of these volatiles also induced *E. balteatus* oviposition (Leroy et al. 2011a).

In areas with copious amounts of honeydew, mite-ant interactions are often observed, but rarely, if ever, do ants attack the mites. Yoder et al. (2010) studied the tritrophic interaction among red velvet mites, *Balaustium* sp. nr. *putmani*, aphids and ants (*Lasius alenius* and *Camponotus pennsylvanicus*). In this system, honeydew serves as a cue that facilitates discovery of aphid prey, but none of the honeydew sugars could be identified as an active arrestant ingredient. The honeydew also serves as an alternative and supplemental food source for the mites. The mites are not attacked by ants defending the honeydew resources because they release an alarm pheromone, neryl formate (Yoder et al. 2007), that bears a strong structural resemblance to natural terpenoid pheromones produced by ants.

Aside from its role in osmoregulation and ant attraction, melezitose in the honeydew has been suggested to be beneficial in warding off predators and parasitoids. For example, for parasitoids the secretion of oligosaccharides by aphids with their honeydew has been shown to hinder their gustatory perception of mono- and disaccharides (Wäckers 2000), so that they will less likely exploit honeydew and, therefore, less parasitize aphids (Vantaux et al. 2011).

Aphid Honeydew and the Intensity of Ant Attendance

Aphids benefit from the mutualism with ants by reduced predation and parasitism and a reduced rate of fungal infection (see Chapter 8 "The Effect of Plant Within-Species Variation on Aphid Ecology" in this volume). Aphids, therefore, compete for the mutualistic services of ants. In this competition, the ants respond more intensively to more profitable honeydew resources (Woodring et al. 2004). Such differences in the reward for the foraging ant may be either larger volumes of honeydew (Bristow 1991, Völkl et al. 1999) or the presence of preferred amino acids or sugars in the honeydew (Cherix 1987, Cushman 1991, Völkl et al. 1999, Fischer et al. 2002, 2005, Detrain et al. 2010, Vantaux et al. 2011).

Woodring et al. (2004) studied the role of honeydew amino acids and sugars in the establishment of ant attendance hierarchy in eight species of aphids feeding on tansy,

T. vulgare. The honeydew of five ant attended (*L. niger*) aphid species, *M. fuscoviride, Trama troglodytes, Aphis vandergooti, B. cardui,* and *A. fabae* was rich in total amino acids compared with that from unattended aphids *M. tanacetaria, Uroleucon tanaceti,* and *Coloradoa tanacetina.* Aphid species with the highest total amino acid concentration in the honeydew also had the highest concentration of sugars. The most intensely ant attended species in this study had the highest level of melezitose. The richness of the honeydew (rate of honeydew secretion x total concentration of sugars) along with the presence of melezitose comprised the critical factors determining the extent of ant attendance of the aphids feeding on tansy. When two potentially ant attended aphid species were present in mixed colonies on the same shoot of tansy, the preferred species (e.g., *M. fuscoviride* vs. *A. fabae*) increased in numbers, while the less preferred decreased due to predation by *L. niger* (Fischer et al. 2001).

Völkl et al. (1999) tested the sugar preferences of *L. niger* both in laboratory and field studies (Fig. 2). All colonies of *L. niger* readily collected fructose, glucose, sucrose, maltose, melezitose, and raffinose, while none of them responded to the presence of xylose, trehalose or water. In choice experiments with two sugars or sugar mixtures, *L. niger* visited food sources containing trisaccharides (melezitose, raffinose) more frequently than sources without trisaccharides. Melezitose was visited more frequently than raffinose.

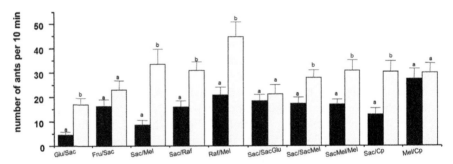

Fig. 2. Sugar preferences of the ant *Lasius niger* in paired tests (means ± SE) (n = 5 replicates per combination) in the laboratory. All tests were carried out with aquous solutions of 10% (w/v). Fru = fructose, Glu = glucose, Sac = saccharose/sucrose, Mel = melezitose, Raf = raffinose, SacGlu = saccharose: glucose 5:1, SacMel = saccharose: melezitose 5:1, Cp = crystalline honeydew from *Cinaria piceicola* (host plant: spruce). Means sharing the same letter do not differ significantly P < 0.05; Wilcoxon test. Adapted from Völkl et al. (1999).

Within the obligatory ant attended pink tansy aphid *M. fuscoviride,* honeydew production and honeydew quality changed with the age of the aphid (Fischer et al. 2002). First and second instar larvae produced only half of the amount of honeydew as older larvae and adults, but there were no differences between age classes in the total honeydew sugar concentration; melezitose was the dominant sugar in all classes. The amino acid concentration, by contrast increased with aphid age. Ant attendance correlated with the amount of honeydew produced and was much lower in colonies of first and second instar larvae than in colonies of older age classes.

In the facultative myrmecophilous aphid species *A. fabae* the honeydew sugar content is relatively low. Fischer et al. (2005) studied a potential correlation between honeydew production, honeydew sugar composition and ant attendance for *A. fabae* feeding on various host plants. *A. f. fabae* feeding on the perennial summer host, *T. vulgare*, produced twice the amount of honeydew than when feeding on the annual host plants *Vicia faba* or *Ch. album* (see above). Honeydew production of *A. f. cirsiiacanthoides* on the creeping thistle *Cirsium arvense* was the highest measured in this study. Total sugar concentration in the honeydew of *A. f. fabae* did not differ when feeding on various summer hosts, whereas the honeydew of *A. f. cirsiiacanthoides* on *C. arvense* contained a significantly higher amount of total sugars. Melezitose was the dominant sugar in all honeydew samples and the highest proportion of melezitose was found in the honeydew of *A. f. cirsiiacanthoides*. In this subspecies, the intensity of *L. niger* ant attendance was also highest. However, *L. niger* ants reduced *A. fabae* tending when they had access to a 50% cane sugar solution (Schumacher and Platner 2009). As mentioned above, *A. fabae* colonies can be made up of several clones which may display significant differences in the amount of the ant attractant sugar melezitose. These clone differences greatly impact the strength of the mutualistic interactions with *L. niger*, but also the fitness of the aphids (Vantaux et al. 2015). High-melezitose secreting clones produced fewer alates and fewer apterae in the presence of ants and hence might have a lower dispersal ability.

Detrain et al. (2010) confirmed the feeding preference shown by *L. niger* for melezitose. Regarding a recruitment trail by ant scouts, melezitose is used by the scouts as a cue indicative of a long-lasting productive carbohydrate resource that is worthy of collective exploitation.

Although, the synthesis or presence of oligosaccharides, and particularly of melezitose in aphid honeydew, has often been used to explain the preferences in ant attendance (for further references see also Vantaux et al. 2011), this cannot be the sole reason. Several species of ants have been shown not to display such a melezitose preference (Cornelius et al. 1996, Blüthgen and Fiedler 2004), especially when they do not feed solely on honeydew but on plant nectar. In addition, melezitose is produced in some insect species that are not tended by ants (Hogervorst et al. 2007).

Ant Competition for Aphid Honeydew and Plant Nectar

Nectars primarily attract pollinators. When foraging for carbohydrate source, ants may have a choice between collecting sugar provided by honeydew of aphids or by plants in nectaries and extrafloral nectaries (EFN). In both cases, there seem to be benefits for the ants' partner. Ant attendance not only reduces predation and parasitization of aphids, but decreases plant damage due to herbivory (Engel et al. 2001). In a number of cases, a positive effect of ant attendance on the plant's overall fitness has been demonstrated. A conflict situation, however, may arise when both, EFN and honeydew-producing homopterans are present on the same plant. When aphids feed on plants with EFN and aphid colony is small, ants frequently use EFN and do not tend aphids. However, as the aphid colony size increases, ants stop using EFN and strengthen their associations with aphids (Katayama et al. 2013b). Even an aphid colony containing only two individuals can supply a greater reward to ants than EFN. On the other side, small aphid colonies may also gain indirect benefits from the EFN of its host plant because ants would be attracted by the EFNs instead of the aphids (Katayama and Suzuki 2010). The paleotropic ant-plant *Humboldtia brunonis* (Fabaceae) can be occupied by up to 16 ant species, but the plant receives protection from herbivory by only one ant species and only when herbivory is very high (Chanam et al. 2015). Despite this lack of antiherbivore protection, the plant secretes EFN, which is rich in sugars and essential amino acids and potentially distracts ants from honeydew of tending Hemiptera.

Nectaries and EFN take sucrose from the phloem into the secretory tissues where it is stored as starch and/or is further processed. During active secretion, sucrose is digested by invertases, which produce hexose-rich nectars (Lohaus and Schwerdtfeger 2014). Concentrations of amino acids in nectars are generally low, but are higher in insect—than in vertebrate-pollinated flowers.

The annual broad bean, *V. faba*, bears EFN at the base of the upper leaves and is regularly infested by two aphid species, *A. fabae* and *A. pisum*. EFN and *A. fabae* are commonly attended by the ant *L. niger*, while *A. pisum* usually remains unattended (Engel et al. 2001). The sugar concentration in EFN (mainly glucose) is much higher than in the honeydew of *A. fabae* (mainly melezitose), and does not change after aphid infestation. The presence of small *A. fabae* colonies does not affect ant attendance of the EFN. The higher sugar concentration in EFN obviously outweighs the higher quality (because of the melezitose) and quantity of aphid honeydew.

Martinez et al. (2011) studied whether EFN of different plants have the same impact on the relations between an aphid species feeding on those plants and their attending ant. When examining the attraction of *Tapinoma erraticum* scout ants to honeydew from the aphid *A. gossypii* feeding on two different plants, *Prunus amygdalus* and *Mentha piperita* in the presence of artificial EFN, they found that the scouts were more attracted to artificial nectar dispensed on *P. amygdalus* leaves than on *M. piperita*, or aphids on both plants or on water. The scout ants neglected aphids in the presence of EFN on *M. piperita* but not on *P. amygdalus*. Only on *P. amygdalis* did the aphids develope fully to the winged form.

Effects of Aphids and Aphid Honeydew on Host Plant Phenotype and Plant Defense

Studies concerning direct or indirect effects of aphid colonies and other herbivores on the host plant phenology are rare. Neves et al. (2011) studied tritrophic level interactions between the aphid *Uroleucon erigeronensis*, its host plant *Baccharis dracunculifolia* (Asteraceae) and other insect herbivores. *U. erigeronensis* forms dense colonies on the apical shoots of the host plant and the honeydew produced by these aphids attracts several species of ants that might interfere with other herbivores. Their results showed that herbivore abundance was lower on shoots with aphids than on shoots without aphids, and even lower on shoots with aphids and ants. Aphids alone not only diminished the abundance of other herbivores, but also affected negatively the host plant shoot growth.

Feeding of the aphid, *Myzus persicae*, on *Arabidopsis thaliana* foliage induced the formation of a deterrent indole glucosinolate with a great defensive benefit (Kim and Jander 2007). Moreover, feeding of *M. persicae* on *A. thaliana* induced a density-dependent and systemic accumulation of trehalose in the phloem sap of the plant (Hodge et al. 2013). This trehalose was taken up by the aphids via the phloem sap, and was excreted into the aphid honeydew. It is not yet clear, however, whether this trehalose accumulation in *Arabidopsis* has a direct role or a signaling function in plant tolerance of, or resistance to, aphid feeding.

Plants respond to insect herbivory by increasing the production of defense-related jasmonic acid (JA) (Howe and Jander 2008). Schwartzberg and Tumlinson (2014) tested the ability of pea aphids, *A. pisum*, to suppress the accumulation of damage-induced JA in broad bean plants, *V. faba*. They showed that both aphid feeding and exogenous application of aphid honeydew induced an accumulation of the defense hormone salicylic acid (SA) within plant tissues and suppressed the accumulation of JA in response to leaf damage. Aphid honeydew, therefore, can suppress induced plant defense and effect herbivore recognition (Jaouannet et al. 2014). As shown above, bacterial proteins in the honeydew may contribute to prevent the activation of defense responses.

Cleland and Ajami (1974) demonstrated that the aphid *Dactynotus ambrosiae* takes up SA from the phloem when feeding on vegetative or flowering plants of the short-day plant *Xanthium strumarium*. SA is transferred into the honeydew from where it could be extracted without loss of its flower inducing activity. By analogy, other plant hormones such as cytokinins, gibberelins, indolacetic acid, and abscisic acid could be isolated from aphid honeydew without loss in biological activity (Cleland 1974). VanDoorn et al. (2015) recently detected that in whitefly honeydew, SA is converted to its glycoside (SAG). This manipulation of SA levels within whitefly secretions may influence the defense response in those plant parts that come in contact with honeydew.

Some social aphids (subfamily Hormaphidinae) form completely closed galls, wherein hundreds to thousands of insects grow and reproduce for several months in isolation. In the closed galls, the insects produce considerable amount of honeydew and the question arises, why the aphids are not drowned by the accumulated honeydew. Kutsukake et al. (2012) reported a sophisticated biological solution to the waste problem, whereby honeydew is removed via the plant vascular system. The inner surface of the gall is specialized for water and honeydew absorption. This mechanism of gall cleaning can be regarded as an "extended phenotype" and a "plant-mediated social behavior" of social aphids with an imprisoned lifestyle.

Honeydew as Carbon Source for Soil Organisms

Aphid populations on trees may lose considerable amounts of honeydew even when ant attended. The honeydew may have one of two possible fates (Dighton 1978). Firstly, honeydew may impinge on the surface of leaves of the lower canopy and become available as an energy source for the phylloplane microflora (bacteria, yeasts, filamentous fungi) (Stadler and Müller 1996, Stadler et al. 2001). Secondly, it may fall directly or as rain washings from the leaves to the soil below the tree. Dighton (1978) studied the effect of aphid honeydew on the populations of soil organisms in a woodland and a grassland soil. In the woodland soil a 30% increase in fungal population resulted from the honeydew dropping and up to a threefold increase in bacterial numbers occurred. In the grassland soil no increase in microbial population was observed, but soil respiration was 75% greater in a honeydew-treated plot than in an untreated one. In neither soil type were microarthropod populations influenced by the addition of honeydew.

Another increase in microbial biomass following honeydew input was observed by Seeger and Filser (2008) on two ruderal sites differing in soil organic matter content and vegetation cover. Honeydew treatment also raised activity densities of an epigeic Collembola taxon, the Bourletiellidae, but not of the dominant *Hemisotoma thermophila*. Ant consumption reduced the amount of honeydew reaching the soil surface by 50%. The activity density of *H. thermophila* was negatively related to ants, suggesting a top-down control.

In general, at least under moderate to low levels of aphid infestation, the effects of honeydew input seems to be obscured in the soil through buffering biological processes. Only little differences between the chemistry of soil solutions collected from the forest floor beneath infested and uninfested trees were observed (Stadler et al. 2001).

Use of Honeydew in Insect Pest Control and Human Medicine

The population dynamics of insect pests in agro ecosystems is often linked to those of other phytophagous species through the foraging activities of natural enemies (Evans and England 1996). These indirect interactions provide both opportunities and challenges for biological control. Honeydew produced by the pea aphid, *A. pisum*, in alfalfa fields can be an important source of nutrition for adults of the wasp *Bathyplectes curculionis* and may therefore enhance parasitism of the wasp host, the alfalfa weevil, *Hypera postica*. Increased parasitoid wasp density is not only observed in response to increased pea aphid density and its honeydew production, but also after field application of a sugar spray at a proper time (Jacob and Evans 1998). The presence of aphids may also promote aggregation of ladybird beetles, which consume weevil larvae in alfalfa.

The usefulness of artificial honeydew in improving biological control has been repeatedly tested, for example to manipulate the spatial distribution of ladybird beetles, parasitoids, gall midges and syrphids (Hatano et al. 2008) (see also above). The Asian citrus psyllid, *Diaphorina citri*, has recently invaded as a new citrus pest in Southern California. Tena et al. (2013a) tested whether mutualistic relationships between the Argentine ant, *Linepithema humile* and various honeydew producers (whiteflies, citrus aphids, scales and psyllids) hindered or favored the establishment of *D. citri* and its biological control with the parasitoid *Tamarixia radiata* (Hymenoptera: Eulophidae). The presence of other honeydew-producing species in the citrus trees reduced not only the intensity of the mutualistic interaction of *L. humile* with *D. citri*, but also the density of *D. citri*. Consequently, indirect competition between honeydew producers for *L. humile* may hinder the establishment of *D. citri* and facilitate its biological control. The same authors found that a parasitoid (*Aphytis melinus*), whose host does not produce honeydew, can discriminate between honeydews with high and poor nutritional quality (Tena et al. 2013b). When *A. melinus* had continuous access to the different honeydews, its fitness entirely depended on the honeydew quality. These results indicate that the presence of different honeydews in agro ecosystems should be taken into account when designing biological control programs.

Almost all honeydew producers feed on the phloem sap of the host plant and several compounds are not degraded during passage through the digestive tract (see above). This applies in particular to proteins because phloem sucking insects possess low proteolytic activity in their gut (Rahbé et al. 1995). A similar unaltered passing of insecticidal proteins expressed by genetically modified plants into honeydew would be expected (Hogervorst et al. 2009). The authors demonstrated that snowdrop lectin (*Galanthus nivalis* agglutinin) in honeydew negatively affects honeydew consuming parasitoids not only directly due to the action of the insecticidal compound, but also indirectly through an altered honeydew composition. When honeydew can be an important route of exposure to transgenic products for a wide range of nontarget organisms, this should be considered in risk assessment studies.

Commercially available "Biocontrol Honeydew™" [Ladies in Red; Organic Pest Control; USA] simulates a mixture of honeydew, nectar, and pollen and attracts many beneficial insects. The main attractants are sugars and the amino acid tryptophan. The high protein diet stimulates egg production in ladybird beetles, lace wings and other natural predators that will help to control plant pest infestations.

An antibacterial action of bee honey was reported as early as 1892 (Van Ketel 1892). Honey inhibits the growth of many microorganisms, viruses, parasites and fungi (Dinkov 2013). The main antibacterial substances in honey are sugars, which by their osmotic effect exert an antibacterial action. Honey also contains antioxidant activity factors, including glucose oxidase, catalase, ascorbic acid, flavonoids, phenolic acids, carotenoid derivatives, organic acids, amino acids and proteins (Otilia et al. 2008). Dinkov (2013) demonstrated that in honeydew honeys the antibacterial and antioxidant activities may be even higher than in blossom honeys. By using comprehensive two-dimensional gas chromatography coupled to time-of flight mass spectrometry Janoskova et al. (2014) detected up to 300 volatile organic compounds in honeydew honeys, belonging to various chemical classes such as hydrocarbons, alcohols, aldehyds, ketones, terpenes and benzene derivatives. Mayer et al. (2014) recently confirmed the potential of honeydew honey to be of medical grade for treatment of non-healing leg ulcers.

Conclusions

Recent biochemical and proteomic analyses on aphid honeydews provide new insights into a substance previously considered as a waste product. Aphids primarily excrete honeydew to get rid of the high amounts of sucrose/saccharose taken up with the phloem sap. Honeydew sugars and amino acids generally reflect the phloem sap composition, but various compounds in the honeydew are synthesized by the sap feeders or their microbial endosymbionts. Aphid honeydews contain distinctive compounds attracting consumer communities like mutualistic ants, parasitoids and predators. The trisaccharide melezitose in the honeydew seems to be an important signal molecule in many but not all aphid-ant mutualistic interactions. The carbohydrate polymerization process also helps to reduce the osmotic pressure in the gut of the aphids. Many parasitoids and predators of aphids feed on the honeydew and use honeydew compounds as kairomones to locate hosts and prey. Melezitose, however, seems to ward off parasitoids and predators. When extrafloral nectaries (EFN) and honeydew producing aphids occur simultaneously on the same plant, a conflict situation for tending ants can arise. The higher sugar concentration in EFN may then outweigh a higher quality and quantity of honeydew. Aphids feeding on their host plant, or the exogenous application of aphid honeydew, can induce salicylic acid (SA) accumulation within plant tissues and suppresses induced plant defense. Much progress has been made during the last decade in using honeydew in insect pest biocontrol and in human medicine.

The answer to the question of honeydew as rubbish or signaler is that the excrete honeydew serves many organisms as food, whereby specific compounds in the honeydew can act as attractant or arrestant in multitrophic systems (Hoffmann et al. 2006).

Keywords: Honeydew, sugars, amino acids, signal compounds, kairomone, tritrophic interactions, aphid-ant mutualism, osmoregulation, pest biocontrol

References

Ammar, E.-D., R. Alessandro, R.G. Shatters, Jr. and D.G. Hall. 2013. Behavioral, ultrastructural and chemical studies on the honeydew and waxy secretions by nymphs and adults of the Asian citrus psyllid *Diaphorina citri* (Hemiptera: Psyllidae). PlosOne 8(6): e64938.
Ashford, D.A., W.A. Smith and A.E. Douglas. 2000. Living on a high sugar diet: the fate of sucrose ingested by a phloem-feeding insect, the pea aphid *Acyrthosiphon pisum*. J. Insect Physiol. 46: 335–341.
Ben Saad, A. and G.W. Bishop. 1976. Attraction of insects to potato plants through use of artificial honeydews and aphid juice. Entomophaga 21: 49–57.
Bernays, E.A. and B.A. Klein. 2002. Quantifying the symbiont contribution to essential amino acids: the importance of tryptophan for *Uroleucon ambrosiae*. Physiol. Entomol. 27: 275–284.
Blande, J.D., J.A. Pickett and G.M. Poppy. 2008. Host foraging for differentially adapted *Brassica*-feeding aphids by the braconid parasitoid *Diaeretiella rapae*. Plant Signal. Behav. 3: 580–582.
Blüthgen, N. and K. Fiedler. 2004. Preferences for sugars and amino acids and their conditionality in a diverse nectar-feeding ant community. J. Anim. Ecol. 73: 155–166.
Bristow, C. 1991. Why are so few aphids ant-tended? pp. 104–109. *In*: C.R. Huxley and D.F. Cutler (eds.). Ant-plant Interactions. Oxford University Press, Oxford.
Buchner, P. 1965. Endosymbiosis of Animals with Plant Microorganisms. Interscience Publisher, New York.
Buckley, R.C. 1987. Interactions involving plants, Homoptera, and ants. Annu. Rev. Ecol. Syst. 18: 111–35.
Buitenhuis, R., J.N. McNeil, G. Boivin and J. Brodeur. 2004. The role of honeydew in host searching of aphid hyperparasitoids. J. Chem. Ecol. 30: 273–285.
Chanam, J., S. Kasinathan, G.K. Pramanik, A. Jagdeesh, K.A. Joshi and R.M. Borges. 2015. Foliar extrafloral nectar of *Homboldtia brunonis* (Fabaceae), a paleotropic ant-plant, is richer than phloem sap and more attractive than honeydew. Biotropica 47: 1–5.
Cherix, D. 1987. Relation between diet and polyethism in *Formica* colonies. Experientia Suppl. 54: 93–115.
Choi, M.Y., B.D. Roitberg, A. Shani, D.A. Raworth and G.H. Lee. 2004. Olfactory response by the aphidophagous gall midge, *Aphidoletes aphidimyza* to honeydew from green peach aphid, *Myzus persicae*. Entomol. Exp. Appl. 111: 37–45.
Cleland, C.F. 1974. Isolation of flower-inducing and flower inhibiting factors from aphid honeydew. Plant Physiol. 54: 899–903.
Cleland, C.F. and A. Ajami. 1974. Identification of the flower-inducing factor isolated from aphid honeydew as being salicylic acid. Plant Physiol. 54: 904–906.

Cornelius, M.L., J.K. Grace and J.R. Yates. 1996. III. Acceptability of different sugars and oils to three tropical ant species (Hymen., Formicidae). Anz. Schädlingskunde, Pflanzenschutz, Umweltschutz 69: 41–43.

Cushman, J.H. 1991. Host plant mediation of insect mutualism: variable outcomes in herbivore-ant interactions. Oikos 61: 138–144.

Detrain, C., F.J. Verheggen, L. Diez, B. Wathelet and E. Haubruge. 2010. Aphid-ant mutualism: how honeydew sugars influence the behavior of ant scouts. Physiol. Entomol. 35: 168–174.

Dhami, M.K., R. Gardner-Gee, J. Van Houtte, S.G. Villas-Bôas and J.R. Beggs. 2011. Species-specific chemical signatures in scale insect honeydew. J. Chem. Ecol. 37: 1231–1241.

Dighton, J. 1978. Effects of synthetic lime aphid honeydew on populations of soil organisms. Soil Biol. Biochem. 10: 369–376.

Dinant, S., J.L. Bonnemain, C. Girousse and J. Kehr. 2010. Phloem sap intricacy and interplay with aphid feeding. C. R. Biol. 333: 504–515.

Dinkov, D.H. 2013. Correlation between antibacterial and antioxidant activity in oak honeydew and Acacia (*Robinia pseudoacacia* I.) bee honeys. The V. International Scientific and Practical Conference on Current State and Perspectives of Food Industry and Catering Development. Chelyabinsk, Russia 1: 265–277.

Dixon, A.F.G. 1998. Aphid Ecology. Chapman & Hall, London.

Douglas, A.E. 1989. Mycetocyte symbiosis in insects. Biol. Rev. 64: 409–434.

Douglas, A.E. 2006. Phloem-sap feeding by animals: problems and solutions. J. Exp. Bot. 57: 747–754.

Endo, S. and T. Itino. 2013. Myrmecophilous aphids produce cuticular hydrocarbons that resemble those of their tending ants. Popul. Ecol. 55: 27–34.

Engel, V., M.K. Fischer, F.L. Wäckers and W. Völkl. 2001. Interactions between extrafloral nectaries, aphids and ants: are there competition effects between plant and homopteran sugar sources? Oecologia 129: 577–584.

Evans, E.W. and S. England. 1996. Indirect actions in biological control of insects: pests and natural enemies in alfalfa. Ecol. Appl. 6: 920–930.

Evans, E.W. and D.I. Gunther. 2005. The link between food and reproduction in aphidophagous predators: a case study with *Harmonia axyridis* (Coleoptera: Coccinellidae). Eur. J. Entomol. 102: 423–430.

Fischer, M.K. and A.R. Shingleton. 2001. Host plant and ants influence the honeydew sugar composition of aphids. Funct. Ecol. 15: 544–550.

Fischer, M.K., K.H. Hoffmann and W. Völkl. 2001. Competition for mutualists in ant-homopteran interaction mediated by hierarchies of ant attendance. Oikos 92: 531–541.

Fischer, M.K., W. Völkl, R. Schopf and K.H. Hoffmann. 2002. Age-specific patterns in honeydew production and honeydew composition in the aphid *Metopeurum fuscoviride*: implications for ant-attendance. J. Insect Physiol. 48: 319–326.

Fischer, M.K., W. Völkl and K.H. Hoffmann. 2005. Honeydew production and honeydew sugar composition of polyphagous black bean aphid, *Aphis fabae* (Hemiptera: Aphididae) on various host plants and implications for ant-attendance. Eur. J. Entomol. 102: 155–160.

Hagen, K.S., R.L. Tassan and E.F. Sawall, Jr. 1971. The use of food sprays to increase effectiveness of entomophagous insects. Proc. Tall Timbers Conf. Ecol. Anim. Control Habitat Manag. 3: 59–81.

Hagenbucher, S., F.L. Wäckers and J. Romeis. 2014. Aphid honeydew quality as a food source for parasitoids is maintained in Bt cotton. PlosOne 9(9): e107806.

Hatano, E., G. Kunert, J.P. Michaud and W.W. Weisser. 2008. Chemical cues mediating aphid location by natural enemies. Eur. J. Entomol. 105: 797–806.

Hodge, S., J.L. Ward, M.H. Beale, M. Bennett, J.W. Mansfield and G. Powell. 2013. Aphid-induced accumulation of trehalose in *Arabidopsis thaliana* is systemic and dependent upon aphid density. Planta 237: 1057–1064.

Hoffmann, K.H., M. Fischer, W. Völkl and M.W. Lorenz. 2003. Of aphids and ants. German Research-Magazine of the Deutsche Forschungsgemeinschaft 3/2003: 20–23.

Hoffmann, K.H., K. Dettner and K.-H. Tomaschko. 2006. Chemical signals in insects and other arthropods: from molecular structure to physiological functions. Physiol. Biochem. Zool. 79: 344–356.

Hogervorst, P.A.M., F.L. Wäckers and J. Romeis. 2007. Effects of honeydew sugar composition on the longevity of *Aphidius ervi*. Entomol. Exp. Appl. 122: 223–232.

Hogervorst, P.A.M., F.L. Wäckers, A.C. Carette and J. Romeis. 2008. The importance of honeydew as food for larvae of *Chrysoperla carnea* in the presence of aphids. J. Appl. Entomol. 132: 18–25.

Hogervorst, P.A.M., F.L. Wäckers, J. Woodring and J. Romeis. 2009. Snowdrop lectin (*Galanthus nivalis* agglutinin) in aphid honeydew negatively affects survival of a honeydew-consuming parasitoid. Agric. Forest Entomol. 11: 161–173.

Howe, G.A. and G. Jander. 2008. Plant immunity to insect herbivores. Annu. Rev. Plant Biol. 59: 41–66.

Ide, T., N. Suzuki and N. Katayama. 2007. The use of honeydew in foraging for aphids by larvae of the ladybird beetle, *Coccinella septempunctata* L. (Coleoptera: Coccinellidae). Ecol. Entomol. 32: 455–460.

Jacob, H.S. and E.W. Evans. 1998. Effects of sugar spray and aphid honeydew on field populations of the parasitoid *Bathyplectes curculionis* (Hymenoptera: Ichneumonidae). Environm. Entomol. 27: 1563–1568.

Janoskova, N., O. Vyviurska and I. Spanik. 2014. Identification of volatile organic compounds in honeydew honeys using comprehensive gas chromatography. J. Food Nutr. Res. 53: 353–362.

Jaouannet, M., P.A. Rodriguez, P. Thorpe, C.J.G. Lenoir, R. MacLeod, C. Escudero-Martinez and J.I.B. Bos. 2014. Plant immunity in plant-aphid interactions. Front. Plant Sci. 5: 663.

Jha, S., D. Allen, H. Liere, I. Perfecto and J. Vandermeer. 2012. Mutualisms and population regulation: mechanism matters. PlosOne 7(8): e43510.

Karley, A.J., D.A. Ashford, L.M. Minto, J. Pritchard and A.E. Douglas. 2005. The significance of gut sucrose activity for osmoregulation in the pea aphid, *Acyrthosiphon pisum*. J. Insect Physiol. 51: 1313–1319.

Katayama, N. and N. Suzuki. 2010. Extrafloral nectaries indirectly protect small aphid colonies via ant-mediated interactions. Appl. Entomol. Zool. 45: 505–511.

Katayama, N., T. Tsuchida, M.K. Hojo and T. Ohgushi. 2013a. Aphid genotype determines intensity of ant attendance: do endosymbionts and honeydew composition matter? Ann. Entomol. Soc. Amer. 106: 761–770.

Katayama, N., D.H. Hembry, M.K. Hojo and N. Suzuki. 2013b. Why do ants shift their foraging from extrafloral nectar to aphid honeydew? Ecol. Res. 28: 919–926.

Kutsukake, M., X.Y. Meng, N. Katayama, N. Nikoh, H. Shibao and T. Fakatsu. 2012. An insect-induced novel plant phenotype for sustaining social life in a closed system. Nature Commun. 3: 1187.

Kim, J.H. and G. Jander. 2007. *Myzus persicae* (green peach aphid) feeding on *Arabidopsis* induces the formation of a deterrent indole glucosinolate. Plant J. 49: 1008–1019.

Lamb, K.B. 1959. Composition of the honeydew of the aphid *Brevicoryne brassicae* (L.) feeding on swedes (*Brassica napobrassica* DC.). J. Insect Physiol. 3: 1–13.

Lang, C. and F. Menzel. 2011. *Lasius niger* ants discriminate aphids based on their cuticular hydrocarbons. Anim. Behav. 82: 1245–1254.

Leroy, P.D., A. Sabri, S. Heuskin, P. Thonart, G. Lognay, F.J. Verheggen, F. Francis, Y. Brostaux, G.W. Felton and E. Haubruge. 2011a. Microorganisms from aphid honeydew attract and enhance the efficacy of natural enemies. Nat. Commun. 2: 348.

Leroy, P.D., B. Wathelet, A. Sabri, F. Francis, F.J. Verheggen, Q. Capella, P. Thonart and E. Haubruge. 2011b. Aphid-host plant interactions: does aphid honeydew exactly reflect the host plant amino acid composition? Arthropod Plant Interact. 5: 193–199.

Leroy, P.D., R. Almohamad, S. Attia, Q. Capella, F.J. Verheggen, E. Haubruge and F. Francis. 2014. Aphid honeydew: an arrestant and a contact kairomone for *Episyrphus balteatus* (Diptera: Syrphidae) larvae and adults. Eur. J. Entomol. 111: 237–242.

Lohaus, G. and M. Schwerdtfeger. 2014. Comparison of sugars, iridoid glycosides and amino acids in nectar and phloem sap of *Maurandya barclayana*, *Lophospermum erubescens*, and *Brassica napus*. PlosOne 9(1): e87689.

Martinez, J.J.I., M. Cohen and N. Mgocheki. 2011. The response of an aphid tending ant to artificial extra-floral nectaries on different host plants. Arthropod Plant Interact. 5: 185–192.

Mayer, A., V. Slezak, P. Takac, J. Olejnik and J. Majtan. 2014. Treatment of non-healing leg ulcers with honeydew honey. J. Tissue Viability 23: 94–97.

Merivee, E., A. Must, E. Tooming, I. Williams and I. Sibul. 2012. Sensitivity of antennal gustatory receptor neurons to aphid honeydew sugars in the carabid *Anchomenus dorsalis*. Physiol. Entomol. 37: 369–378.

Mooney, K.A. 2011. Genetically based population variation in aphid association with ants and predators. Arthropod Plant Interact. 5: 1–7.

Mooney, K.A. and A.A. Agarwal. 2008. Plant genotype shapes ant-aphid interactions: implications for community structure and indirect plant defense. Am. Nat. 171(6): E195–E205. doi: 10.1086/587758.

Munson, M.A., P. Baumann, M.A. Clark, L. Baumann, N.A. Moran, D.J. Voegtlin and Campell B.C. 1991. Evidence for the establishment of aphid-eubacterium endosymbiosis in an ancestor of four aphid families. J. Bacteriol. 173: 6321–6324.

Neves, F.D., M. Fagundes, C.F. Sperber and G.W. Fernandes. 2011. Tri-trophic level interactions affect host plant development and abundance of insect herbivores. Arthropod Plant Interact. 5: 351–357.

Nielsen, Ch., A.A. Agrawal and A.E. Hajek. 2010. Ants defend aphids against lethal disease. Biol. Lett. 6: 205–208.

Novgorodova, T.A. 2004. The symbiotic relationships between ants and aphids. Zhurn. Obsh. Biol. 65: 153–166.

Otilia, B., L. Maregithas, I.K. Rindt, M. Niculae and D. Dezmirean. 2008. Honeydew honey: correlations between chemical composition, antioxidant capacity and antibacterial effect. Zootehnie si Biotehnologii 41: 271–277.

Pettersson, J., V. Ninkovic, R. Glinwood, S. Al Abassi, M. Birkett, J. Pickett and L. Wadhams. 2008. Chemical stimuli supporting foraging behavior of *Coccinella septempunctata* L. (Coleoptera: Coccinellidae): volatiles and allelobiosis. Appl. Entomol. Zool. 43: 315–321.

Pringle, E.G., A. Novo, I. Ableson, R.V. Barbehenn and R.L. Vannette. 2014. Plant-derived differences in the composition of aphid honeydew and their effects on colonies of aphid-tending ants. Ecol. Evol. 4: 4065–4079.

Rahbé, Y., N. Sauvion, G. Febvay, W.J. Peumans and A.M.R. Gatehouse. 1995. Toxicity of lectins and processing of ingested proteins in the pea aphid *Acyrthosiphon pisum*. Entomol. Exp. Appl. 76: 143–155.

Rehman, A. and W. Powell. 2010. Host selection behavior of aphid parasitoids (Aphidiidae: Hymenoptera). J. Plant Breed. Crop Sci. 2: 299–311.

Sabri, A., S. Vandermoten, P.D. Leroy, E. Haubruge, T. Hance, P. Thonart, E. De Pauw and F. Francis. 2013. Proteomic investigation of aphid honeydew reveals an unexpected diversity of proteins. PlosOne 8(9): e74656.

Sasaki, T., T. Aoki, H. Hayashi and H. Ishikawa. 1990. Amino acid composition of the honeydew of symbiotic and aposymbiotic pea aphids *Acyrthosiphon pisum*. J. Insect Physiol. 36: 35–40.

Schumacher, F. and C. Platner. 2009. Nutrient dynamics in a tritrophic system of ants, aphids and beans. J. Appl. Entomol. 133: 33–46.

Schwartzberg, E.G. and J.H. Tumlinson. 2014. Aphid honeydew alters plant defence responses. Funct. Ecol. 28: 386–394.

Seeger, J. and J. Filser. 2008. Bottom-up down from the top: honeydew as a carbon source for soil organisms. Eur. J. Soil Biol. 44: 483–490.

Shigenobu, S. and A.C.C. Wilson. 2011. Genomic revelations of a mutualism: the pea aphid and its obligate bacterial symbiont. Cell. Mol. Life Sci. 68: 1297–1309.

Stadler, B. and T. Müller. 1996. Aphid honeydew and its effect on the phyllosphere microflora of *Picea abies* (L.) Karst. Oecologia 108: 771–776.

Stadler, B. and A.F.G. Dixon. 2005. Ecology and evolution of aphid-ant interactions. Annu. Rev. Ecol. Evol. Syst. 36: 345–372.

Stadler, B., K. Fiedler, T.J. Kawecki and W.W. Weisser. 2001. Costs and benefits for phytophagous myrmecophiles: when ants are not always available. Oikos 92: 467–78.

Starý, P. 1970. Biology of aphid parasitoids (Hymenoptera: Aphidiidae) with respect to integrated control. Ser. Entomol. 6: 1–643.

Sun, Y.C., B.B. Jing and F. Ge. 2009. Response of amino acid changes in *Aphis gossypii* (Glover) to elevated CO_2 levels. J. Appl. Entomol. 133: 189–197.

Taylor, S.H., W.E. Parker and A.E. Douglas. 2012. Patterns on aphid honeydew production parallel diurnal shifts in phloem sap composition. Entomol. Exp. Appl. 142: 121–129.

Tegelaar, K., M. Hagman, R. Glinwood, J. Pettersson and O. Leimar. 2012. Ant-aphid mutualism. The influence of ants on the aphid summer cycle. Oikos 121: 61–66.

Tena, A., C.D. Hoddle and M.S. Hoddle. 2013a. Competition between honeydew producers in an ant-hemipteran interaction may enhance biological control of an invasive pest. Bull. Entomol. Res. 103: 714–723.

Tena, A., E. Llácer and A. Urbaneja. 2013b. Biological control of a non-honeydew producer mediated by a distinct hierarchy of honeydew quality. Biol. Contr. 67: 117–122.

VanDoorn, A., M. de Vries, M.R. Kant and R.C. Schuurink. 2015. Whitefly glycosylate salicylic acid and secrete the conjugate via their honeydew. J. Chem. Ecol. 41: 52–58.

Van Ketel, B.A. 1892. Festnummer der Berichten van den Niederlandsche Maatschappij. Bevordering der Pharmacie. pp. 67–96.

Vantaux, A., W. Van den Ende, J. Billen and T. Wenseleers. 2011. Large interclone differences in melezitose secretion in the facultatively ant-tended black bean aphid *Aphis fabae*. J. Insect Physiol. 57: 1614–1621.

Vantaux, A., T. Parmentier, J. Billen and T. Wenseleers. 2012. Do *Lasius niger* ants push low-quality black bean aphid mutualists? Anim. Behav. 83: 257–262.

Vantaux, A., S. Schillewaert, T. Parmentier, W. Van den Ende, J. Billen and T. Wenseleers. 2015. The cost of ant attendance and melezotose secretion in the black bean aphid *Aphis fabae*. Ecol. Entomol. 40: 511–517.

Verheggen, F.J., L. Diez, L. Sablon, C. Fischer, S. Bartram, E. Haubruge and C. Detrain. 2012. Aphid alarm pheromone as a cue for ants to locate aphid partners. PlosOne 7(8): e41841.

Vinson, S.B. 1976. Host selection by insect parasitoids. Annu. Rev. Entomol. 21: 109–133.

Völkl, W., J. Woodring, M. Fischer, M.W. Lorenz and K.H. Hoffmann. 1999. Ant-aphid mutualisms: the impact of honeydew production and honeydew sugar composition on ant preferences. Oecologia 118: 483–491.

Wäckers, F.L. 2000. Do oligosaccharides reduce the suitability of honeydew for predators and parasitoids? A further facet to the function of insect-synthesized honeydew sugars. Oikos 90: 197–201.

Wäckers, F.L. 2005. Suitability of (extra-) floral nectar, pollen, and honeydew as insect food sources. pp. 17–74. *In*: F.L. Wäckers, P.C.J. van Rijn and J. Bruin (eds.). Plant-Provided Food for Carnivorous Insects: A Protective Mutualism and its Applications. Cambridge University Press, Cambridge.

Wäckers, F.L., P.C.J. van Rijn and G.E. Heimpel. 2008. Honeydew as a food source for natural enemies: making the best of a bad meal? Biol. Contr. 45: 172–175.

Watanabe, H., N. Katayama, E. Yano, R. Sugiyama, S. Nishikawa, T. Endou, K. Watanabe, J. Takabayashi and R. Ozawa. 2014. Effects of aphid honeydew sugars on the longevity and fecundity of the aphidophagous gall midge *Aphidoletes aphidimyza*. Biol. Contr. 78: 55–60.

Way, M.J. 1963. Mutualism between ants and honeydew-producing Homoptera. Annu. Rev. Entomol. 8: 307–44.

Whitaker, M.R.L., N. Katayama and T. Ohgushi. 2014. Plant-rhizoba interactions alter aphid honeydew composition. Arthropod Plant Interact. 8: 213–220.

Wilkinson, T.L., D.A. Ashford, J. Pritchard and A.E. Douglas. 1997. Honeydew sugars and osmoregulation in the pea aphid *Acyrthosiphon pisum*. J. Exp. Biol. 200: 2137–2143.

Will, T. and A.J.E. Van Bel. 2006. Physical and chemical interactions between aphids and plants. J. Exp. Bot. 57: 729–737.

Wimp, G.M. and T.G. Whitham. 2001. Biodiversity consequences of predation and host plant hybridization on ant-aphid mutualism. Ecology 82: 440–452.

Woodring, J., R. Wiedemann, M.K. Fischer, K.H. Hoffmann and W. Völkl. 2004. Honeydew amino acids in relation to sugars and their role in the establishment of ant-attendance hierarchy in eight species of aphids feeding on tansy (*Tanacetum vulgare*). Physiol. Entomol. 29: 311–319.

Woodring, J., R. Wiedemann, W. Völkl and K.H. Hoffmann. 2007. Oligosaccharide synthesis regulates gut osmolality in the ant-attended aphid *Metopeurum fuscoviride* but not in the unattended aphid *Macrosiphoniella tanacetaria*. J. Appl. Entomol. 131: 1–7.

Wool, D., D.L. Hendrix and O. Shukry. 2006. Seasonal variation in honeydew sugar content of galling aphids (Aphidoidea: Pemphigidae: Fordinae) feeding on *Pistacia*: Host ecology and aphid physiology. Basic Appl. Ecol. 7: 141–151.

Yao, I. 2012. Ant attendance reduces flight muscle and wing size in the aphid *Tuberculatus quercicola*. Biol. Lett. 8: 624–627.

Yao, I. 2014. Costs and constraints in aphid-ant mutualism. Ecol. Res. 29: 383–391.

Yoder, J.A., D.D. Mowrey, E.J. Rellinger, J.L. Tank, P.E. Hanson and R.W. York. 2007. Detection of the mite alarm pheromone neryl formate in the velvet mite, *Balaustium* sp. (Parasitengona: Erythraeidae). Int. J. Acarol. 33: 73–78.

Yoder, J.A., M.R. Condon, C.E. Hart, M.H. Collier, K.R. Patrick and J.B. Benoit. 2010. Use of alarm pheromone against ants for gaining access to aphid/scale prey by the red velvet mite *Balaustium* sp. (Erythraeidae) in a honeydew-rich environment. J. Exp. Biol. 213: 386–392.

11

Function of Aphid Saliva in Aphid-Plant Interaction

Torsten Will

Introduction

Saliva is important in the interaction of many insect species and their plant hosts. The composition and function of aphid saliva was until yet the subject of more than 100 research articles. While early studies used enzymatic assays to determine enzyme activity, recently published studies use proteomic approaches, *in planta* protein expression and RNA interference to study the function of salivary proteins. The primary components of aphid saliva originate in the salivary glands, a pair of organs located in the dorsal metathorax (Ponsen 1972). Each half of the salivary gland pair consists of two components, a large principal gland that is often bilobed and a smaller accessory gland. The two subunits join in the salivary canal that leads to the stylet tips, where it unites with the nutrition channel to form the so called common duct (Uzest et al. 2010). The principal gland is innervated by nerves and includes eight secretory cells, whereas the accessory gland is not innervated and the cells show no differentiation. The contribution of the two subunits to the saliva has been suggested largely through plant virus transmission studies (Gray and Gildow 2003). Persistent and circulative viruses that infect the phloem are transferred from the aphid hemolymph into saliva through the accessory gland, indicating that this structure might be responsible for the production of some watery saliva. Hints exist that different salivary glands produce specific proteins, demonstrated in *Schizaphis graminum* by the use of an antibody against a 154 kDa protein (Cherqui and Tjallingii 2000), and in more detail that different subsets of secretory cells produce different proteins, observed in principal salivary glands of *Acyrthosiphon pisum* for the effectors C002

Institute for Insect Biotechnology, Justus-Liebig-University, Heinrich-Buff-Ring 26-32, 35394 Giessen, Germany.
E-mail: torsten.will@agrar.uni-giessen.de

(Mutti et al. 2008) and Armet (Wang et al. 2015). Wang and colleagues suggest that a preferential expression of a protein in a subset of secretory cells of the salivary gland may be a hallmark of saliva proteins. However, to date not enough data exist to determine the overall contribution of the accessory and principal glands to saliva production.

Epidermis Penetration and Host Plant Recognition

After reaching a potential host plant, aphids start penetrating the epidermis but before doing so they secrete a small droplet of gel saliva, forming a salivary flange on the epidermal surface at the linkage of two cells. This salivary flange is presumably needed to stabilize the stylet during penetration of the epidermis (Miles 1999), appearing to be reasonable because the epidermis is covered with a layer of waxes, composed of a mixture of hydrophobic compounds (Walton 1990). This cuticula covers all aerial plant surfaces and prevents wetting of the leave surface as well as uncontrolled evaporation. In addition, a negative correlation of wax layer thickness and aphid infestation was observed by Tsumuki et al. (1989). The authors concluded that on one hand the composition of the wax layer may be related with an increased resistance of barley against the aphid species *Rhopalosiphum padi* as already observed by other authors (Lowe et al. 1985). On the other hand, Tsumuki et al. (1989) suggested that at least part of the resistance to cereal aphids shown by some barley lines may be attributed to the wax quantity on the epidermal surface. One may think about problems of attaching the salivary flange on the cuticula in the presence of specific wax compounds and furthermore the penetration of a thicker wax layer might be more difficult to aphids but no studies were conducted so far that proof this hypothesis. However, beside their function in plant defence cuticular waxes appear to be involved in host-plant recognition (Powell et al. 1999).

The first few stylet penetrations initiated after plant contact are in general short events (< 1 min) that are limited to the epidermis. Alates often land on the upper leaf surface and make one or more probes before either taking flight (Nault and Styer 1972, Powell and Hardie 2000) or moving to the lower leaf surface, their preferred feeding site (Calabrese and Edwards 1976, Wensler 1962). Aphid stylet and labium apparently lack external contact chemoreceptors (Tjallingii 1978, Wensler 1974, 1977), but plant cell sap is ingested during brief cell penetrations and is analysed at the gustatory organ in the epipharyngeal area (Wensler and Filshie 1969). Stylet penetrations longer than approximately 1 min usually reach tissues beyond the epidermis (Bradley 1952, Nault and Gyrisco 1966), the mesophyll and parenchyma. These pathway activities include further ingestion of small quantities of cell sap during regular cell penetrations and the secretion of gel saliva into the apoplast.

The Stylet Route through the Plant

First studies on the stylet route inside the plant towards the sieve elements implied that the stylet track proceeds extra- or intracellular or mixed depending upon the aphid species (Pollard 1973). Spiller et al. (1985) were the first that complemented light microscopic observation by transmission electron microscopy and their observations were concordant with previous studies. Advanced sample preparation for transmission electron microscopic accompanied by subsequent reconstruction of up to 1200 serial sections revealed, by using the aphid species *Aphis fabae*, that the stylet pathway is entirely intercellular (Tjallingii and Hogen Esch 1993). After initial epidermis penetration and host plant acceptance, the predominant stylet path proceeds along the secondary cell wall. Furthermore, stylet pathway goes through the primary cell wall, i.e., the middle lamella, through intercellular air spaces or between plasmalemma and cell wall. Individual cells (mesophyll, parenchyma and companion cells) are briefly punctured along the stylet route towards the sieve elements. The stylet route can be circuitous, commonly involving dead-ends, direction reversal, and aborted sieve-element punctures (Tjallingii and Hogen Esch 1993). Overall, it can take from 30 minutes to several hours for the aphid to locate a suitable feeding site.

During stylet movement through the apoplast an aphid secretes gel saliva that hardens and envelops the stylet like a sheath. The salivary sheath was already observed in 1891 by Büsgen who showed that it is of proteinaceous origin and Arnaud (1918) compared the salivary sheath with the padding of an oil borehole and suggested that the sheath prevents sap leaking from the stylet as well as the ingress of undesirable plant material into the feeding channel. Most phytophagous hemiptera (aphids, white flies, plant hoppers) form a salivary sheath during feeding, emphasizing its biological relevance (Freeman et al. 2001, Wang et al. 2008, Morgan et al. 2013) (see Chapter 1 "Phylogeny of the Aphids" in this volume).

Formation and Function of the Salivary Sheath

The sheath is formed by successive events of gel saliva secretions and stylet movement. This can easily be derived from the sheath structure *in vitro* (artificial diet or *in aera*) where its shape is comparable to a pearl necklace (Will et al. 2012, Morgan et al. 2013). McLean and Kinsey (1965) suggested that an initial secretion of a droplet of gel saliva is followed by secretion of a small volume of watery saliva that serves to inflate the droplet to a spherical structure, creating a spherical cavity inside the droplet of gel saliva. The distinct tubular structure inside the salivary sheath observed by confocal-laser-scanning microscopy (Will et al. 2012), conflicts with this proposition and indicates that all saliva material hardens after being pierced by the stylet. Hardening of the salivary sheath was assumed to be associated with the oxidation of protein sulphydryl groups leading to the formation of disulfide bridges (Miles 1965, Tjallingii 2006). Will et al. (2012) observed in this context that salivary sheath formation is disturbed under reducing conditions in the presence of dithiothreitol. A novel protein ("sheath protein" (SHP)) with a molecular weight of approximately 154 kDa and a high content of cysteine was first identified in saliva of the pea aphid *A. pisum* (Carolan et al. 2009) and was later identified in the cereal aphid species *Sitobion avenae* and *Metopolophium dirhodum* (Rao et al. 2013). Disulphide bridge formation is potentially catalysed by the enzyme disulphide isomerases and several disulphide isomerases were identified in the pea aphid salivary gland secretome (Carolan et al. 2011). Silencing the expression of shp in the aphid species *A. pisum* by injection of double stranded RNA (Will and Vilcinskas 2015) showed that sheath formation was inhibited and thus demonstrated that SHP is a major structural protein of the salivary sheath.

It is suggested that gel saliva functions as a lubricant to facilitate stylet movement and that the sheath protects the stylet against mechanical forces and chemicals (Miles 1999). In addition, Will and van Bel (2006) postulated that the salivary sheath prevents the induction of defense responses in sieve elements by sealing the stylet piercing site in the plasmamembrane of sieve elements. To study the effect of an intact salivary sheath on aphid feeding, SHP specific double stranded RNA was applied to *A. pisum* by micro injection to knockdown SHP expression by RNA interference (Will and Vilcinskas 2015). Inhibition of sheath hardening by shp silencing allowed a very first functional analysis of the salivary sheath (Will and Vilcinskas 2015). Data that were obtained by behavior observation, using the electrical penetration graph technique (see Chapter 13 "Aphid Techniques" in this volume), indicate that an intact salivary sheath is necessary for sustained feeding (> 10 minutes) from sieve elements. These findings indicate that the salivary sheath prevents the induction of plant defense responses as previously suggested (Will and van Bel 2006). Aphids that lack sheath hardening show an increase of watery salivation events without subsequent ingestion (Will and Vilcinskas 2015). Latter aphids must somehow be confronted with plant defence responses that cannot be effectively overwhelmed by the secretion of watery saliva, which is as well associated with the suppression of plant defence (Will et al. 2013). A similar dsRNA sequence was applied to *S. avenae* by transgenic barley were a negative effect on ingestion is indicated by decreased reproduction and reduced survival in comparison to control plants (Abdellatef et al. 2015). Thus, it can be suggested that sealing of the stylet piercing site by gel saliva appears to be important for different aphid species on different host plants, dicots as well as monocots.

After stylet retraction from the plant the salivary sheath remains in the plant (Tjallingii and Hogen Esch 1993) and the stylet canal is filled with gel saliva material as shown for the aphid species *Megoura viciae* by scanning electron microscopy (Will et al. 2012). The question arises why additional protein material is invested to seal the salivary sheath and if this is done by all aphid species. Stylet canal filling could lead to a facilitated withdrawal of the stylet from the sheath by repulsion. This would eliminate possible suction effects inside the thin stylet canal of the salivary sheath. However this is only speculation.

Cell Wall Degrading Enzymes

Tjallingii and Hogen Esch (1993) conclude from their micrographs that stylet movement along the primary and secondary cell wall might be explained mechanically. However, earlier studies reveal the presence of cellulase (Adams and Drew 1965) and pectinase activity in salivary secretions of different aphid species like *Myzus persicae* and *S. graminum* (Adams and McAllan 1956, Ma et al. 1990, Cherqui and Tjallingii 2000) that could facilitate stylet movement along cell walls and the middle lamella (Adams and Drew 1963, Miles 1999). Nevertheless, Miles (1999) raises concerns that secreted cell wall degrading enzymes do have any effect on this process because movement of an aphid's stylet appears to be too fast.

In addition to its speculative function during stylet movement, Miles (1999) and later Will and van Bel (2008) suggested that cellulase and pectinase activity in salivary secretions of aphids may release so called pathogen induced molecular patterns (PIMPs) by cell wall degradation. The most characteristic component found in all plant cell walls is cellulose, build of β-1,4-linked glucan chains that interact with each other via hydrogen bonds to form a crystalline microfibril (Somerville 2006). In addition, plant cell walls contain matrix polysaccharides that are grouped into two general categories: (i) pectic polysaccharides that include homogalacturonan, and rhamnogalacturonan I and II (Harholt et al. 2010) and (ii) hemicellulosic polysaccharides that include xyloglucans, glucomannans, xylans, and mixed-linkage glucans (Scheller and Ulvskov 2010). Fragments of cellulose (cellodextrins) and pectin (oligoglalcturonides) act as PIMPs and trigger calcium-dependent and calcium-independent signaling pathways that are both involved in plant defense responses (Moscatiello et al. 2006).

Cell Probing

Along the stylet pathway, cells are regularly pierced by the aphid's stylet, indicated by a rapid and short "potential drop" of 5–15 seconds during electrical penetration graph recordings (Tjallingii 1985). Potential drops can be analysed in detail and are divided into three sub-waveforms: II-1, II-2 and II-3. With regard to secretion of saliva into penetrated cells EPG sub-waveform II-1 is of interest (Martin et al. 1997). Martin et al. (1997) correlated this waveform with the secretion of saliva by studying the transfer of nonpersistent phytopathogenic viruses, whose transmission is necessarily linked with salivation, while virus particles are acquired together with cell sap during waveform II-3 (Martin et al. 1997, Collar and Fereres 1998). In context with these short cell penetrations it can be suggested that saliva suppresses plant defence responses by effectors as observed in sieve elements (Will et al. 2013) but no data are available so far. Potential functions of salivary enzymes and effectors are discussed in the following text. It has to be mentioned at this point that saliva, which is collected by using artificial diet (see Chapter 13 "Aphid Techniques" in this volume), represents a mixture of saliva from different salivation events, occurring in planta in different environments. This includes secretion of gel saliva as well as secretion of watery saliva into the apoplast and intracellular into cortex cells and sieve elements.

Secretion of Watery Saliva into Sieve Elements

Sieve elements belong to the phloem and are highly specialized cells that possess a couple of unique features. They possess a high turgor pressure from approx. four bar (*Vicia faba*; Fisher 1978) up to 12 bar (*Hevea brasiliensis*; An et al. 2012), most of their organelles are reduced or absent (van Bel 2003, Cayla et al. 2015), they have a very low calcium resting level of approx. 50 nM in the cytosol (Furch et al. 2009) and they are equipped with specific callose and protein based defence mechanisms (e.g., Furch et al. 2010). Directly after piercing a sieve element with its stylet an aphid secretes a small volume of saliva before it takes up a small volume of cell sap, which is comparable to regular cell penetrations along the stylet track (Prado and Tjallingii 1994). Observations by electrical penetration graph technique indicate that aphid behaviour subsequently switches to secretion of watery saliva for several minutes (Prado and Tjallingii 1994). While secreting watery saliva into the sieve element lumen the precibarial valve blocks the nutrition channel and no sieve element sap is ingested by the aphid. Subsequent ingestion of sieve element sap can be interrupted by secretion of watery saliva, e.g., by triggering sieve-element occlusion (Will et al. 2007, 2009). Watery saliva is composed of a number of different proteins (Carolan et al. 2009) and it appears that saliva form different aphid species differs in its respective composition, indicated by the molecular weight of detected protein bands (Madhusudhan and Miles 1998, Cherqui and Tjallingii 2000, Will et al. 2007, 2009, Cooper et al. 2011). Differences were confirmed by proteomic approaches using 1D and 2D gel electrophoresis and mass spectrometry, available for *A. pisum* (Carolan et al. 2009), *M. persicae* (Harmel et al. 2008), some cereal aphid species (Rao et al. 2013) and *Macrosiphum euphorbiae* (Chaudhary et al. 2015). However, overlaps in protein composition were observed in these studies for the different aphid species as well.

Suppression of Sieve-Element Occlusion

Specific proteinacious mechanisms as well as callose deposition are suggested to act in occlusion of sieve tubes and seal these in response to wounding to prevent nutrition loss (e.g., Knoblauch and van Bel 1998, Furch et al. 2010, Jekat et al. 2013). Factors that can trigger protein based occlusion mechanisms include calcium influx (Knoblauch et al. 2001), turgor loss (Ehlers et al. 2000), and variations in redox potential (Leineweber et al. 2000). Calcium also triggers callose dependent sieve plate occlusion (Furch et al. 2010) by activating the phloem located callose synthase CalS7 (Kauss 1986, Xie et al. 2011). Furthermore, a calcium influx into the sieve element lumen may induce additional plant defence responses (Lecourieux et al. 2006).

Aphids are confronted with sieve-element occlusion mechanisms and it was suggested several times that these might act against sieve element sucking insects (Knoblauch et al. 2001, Tjallingii 2006, Will and van Bel 2006, Jiang and Walker 2007, Zhu et al. 2011). If this is the case one would suggest that occlusion mechanisms are triggered during stylet penetration of the sieve element plasma membrane. This would directly interfere with the ingestion of sieve element sap by reducing turgor pressure in sieve elements (Gould et al. 2004) and inhibiting mass flow (Peuke et al. 2006), leading to deprivation of nutrition supply to the aphid (Will and van Bel 2006). In this context it is suggested that aphids perceive sieve-element occlusion by a decrease of turgor pressure in the sieve-element lumen (Will et al. 2008).

A plant family with a unique occlusion mechanism that is easy to study by microscopy are Fabaceae, possessing a spindle like protein in each sieve element termed forisome (Knoblauch and van Bel 1998), which is built of different proteins that belong to the sieve-element-occlusion protein family (Müller et al. 2014). A first *in vivo* study that studied the relation of forisome based sieve-element occlusion and aphid ingestion demonstrated that *A. pisum*, which is highly adapted to its host plants of the family Fabaceae, does not trigger sieve element occlusion (Walker and Medina-Ortega 2012). This contradicted previously made hypotheses (Will and van Bel 2006, Will et al. 2007). In contrast to *A. pisum*, less-adapted generalist feeders like *M. persicae* and *M. euphorbiae* trigger sieve element occlusion in *V. faba*, impairing sieve element accession (Medina-Ortega and Walker 2015). Infiltration of the apoplast by a calcium channel blocker prevents sieve element occlusion by calcium responsive specialized occlusion proteins, so called forisomes, and facilitates accession to sieve elements for *M. persicae* (Medina-Ortega and Walker 2015). Thus, latter study indicates for the first time that forisomes, act in sieve tube located plant defence in *V. faba* and demonstrated a high grade of adaptation of specialized feeders to specific plant defence responses.

Detoxifying Enzymes

The detoxification of phenols by the secretion of polyphenoloxidase and peroxidase was reported for *Sitobion avenae* (Madhusudhan and Miles 1998, Urbanska et al. 1998, Chaudhary et al. 2015). The degradation of hydrogen peroxide could most likely interfere with plant defense signaling because hydrogen peroxide represents an activator of Ca^{2+} channels in the plasma membrane of plant cells (Lecourieux et al. 2006). Further detoxifying enzymes were identified in aphid saliva by enzymatic assays and proteomic approaches.

Proteins that may directly interfere with plant defense related signaling are glucose dehydrogenase and glucose oxidase that are present in saliva of the aphid species *M. persicae* and *A. pisum* (Harmel et al. 2008, Carolan et al. 2011). Both potentially interfere with jasmonic acid (JA)-regulated defense responses that were shown to be induced during infestation of Arabidopsis by the aphid species *Brevicoryne brassicae* (Kusnierczyk et al. 2011). Takemoto et al. (2013) noticed that endogenous JA production decreased in *A. pisum* infested *V. faba*. Furthermore, aphids appear to be able to modulate genes in the salicylic acid (SA) pathway (Zhu-Salzman et al. 2004). Cross-talk between JA and SA defense pathways (Pieterse et al. 2012) may allow aphids to suppress specific plant defense responses as has been previously described for whiteflies by Zarate et al. (2007). A detailed view on the role of SA and JA in plant-aphid interaction is given by Louis and Shah (2013).

Proteases

Whereas a first approach to identify salivary proteases by Cherqui and Tjallingii (2000) was unsuccessful, a proteomic study by Carolan et al. (2009) provided evidence for two proteases in the saliva of *A. pisum*, a M1 zinc metalloprotease and an angiotensin-converting enzyme (M2 metalloprotease). Metalloproteases have also been found in the saliva of other insects such as the phytophagous thrips *Frankliniella occidentalis* (Stafford-Banks et al. 2014). However, homologues of the detected proteases (Carolan et al. 2009) were not detected in the saliva of the two cereal aphid species *S. avenae* and *M. dirhodum* (Rao et al. 2013).

Salivary proteases are believed to counteract protein-mediated defense responses of insects host plants (Francischetti et al. 2003, Carolan et al. 2009, Stafford-Banks et al. 2014). In detail, the breakdown of sieve-tube sap proteins involved in defense, signaling, and/or occlusion (Kehr 2006) may lead to suppression of plant defense (Carolan et al. 2009). Furch et al. (2015) used a functional assay to verify the presence of protease activity in *A. pisum* and detected protease activity in saliva of *M. euphorbiae* as well. Since latter was not detected by a recent proteomic approach (Chaudhary et al. 2015) this is a good example how functional approaches may supplement modern techniques and *vice versa*. In a subsequent experiment Furch et al. (2015) mixed saliva of both species and phloem sap from *Cucurbita maxima in vitro*. A high grade of degradation of the PP1 protein, which is involved in sieve element occlusion (Gaupels et al. 2008), was observed in sieve element sap samples in the presence of saliva from *M. euphorbiae* less degradation was observed when sieve element sap was mixed with saliva from *A. pisum*. Furch et al. (2015) suggested that this can be seen as an adaptation of *M. euphorbiae* to its host plant whereat *A. pisum* does not feed on cucurbits.

Beside their function in suppression of plant defense salivary proteases are suggested to function in degradation of sieve-tube proteins, allowing aphids to access an amino acid source (Carolan et al. 2009) to supplement the limited amount of essential amino acids in sieve-element sap (Gündüz and Douglas 2009). The breakdown strategy might work best for dicotyledons, containing high protein levels in their sieve elements (e.g., *Ricinus communis* 2–5 mg/ml (Schobert et al. 1998), *Cucurbita maxima* 19–100 mg/ml (Richardson et al. 1982, Schobert et al. 1998, Zimmermann et al. 2013)), while the protein content in sieve element sap of monocotyledonous is below 1 mg/ml (Fisher et al. 1992, Schobert et al. 1998, Gaupels et al. 2008). Although this function of salivary proteases from aphids seems plausible, results from Furch et al. (2015) does not support this hypothesis because degradation was mainly observed for the PP1 protein.

Calcium-binding Proteins

A first proof that salivary proteins interact with calcium as an important second messenger and trigger compound of some sieve-element occlusion mechanisms and further plant defence responses was presented by Will et al. (2007). In their approach they used isolated forisomes from *V. faba* and concentrated saliva from the aphid species *M. viciae*. As mentioned earlier, forisomes act in sieve element occlusion by changing their shape from a low-volume to a high volume state in response to a calcium influx (Knoblauch et al. 2001). This does not only work *in vivo* but was also shown *in vitro* were forisomes were used as a gate in a microfluidic system (Knoblauch et al. 2012). Aphid saliva changes forisome shape from the high-volume state back into its low-volume state. The absence of forisome transformation in the presence of proteolytically digested salivary proteins indicates an active role for salivary proteins in sequestering calcium to prevent occlusion. By using different marker techniques on western blot (e.g., $^{45}Ca^{2+}$) seven calcium-binding proteins of different molecular weight were identified in saliva of *M. viciae* (Will et al. 2007).

Another putative candidate that acts in calcium-binding was identified in the saliva of *A. pisum* (Carolan et al. 2009). This regucalcin-like protein has a molecular mass of 43 kDa, which is comparable in size to a calcium-binding protein that was previously detected in aphid saliva (Will et al. 2007). Regucalcin is a member of the senescence marker protein-30 (SMP-30) family that functions by sequestering signaling molecules such as calcium (Fujita et al. 1992, Shimokawa and Yamaguchi 1993). Furthermore, regucalcin acts in maintaining intracellular calcium homeostasis by activating calcium-pumps in the plasma membrane, endoplasmic reticulum, and mitochondria of many animal cell types (Yamaguchi 2000) and has an inhibitory effect on calcium/calmodulin-dependent enzymes and protein kinase C

(Yamaguchi 2005). There are indications that the endoplasmatic reticulum inside the highly specialized sieve elements (Sjolund and Shih 1983, Arsanto 1986) functions as an internal calcium store (Furch et al. 2009), thus representing a potential target for regucalcin.

Wang et al. (2015) recently characterized the Armet protein, which was discovered in saliva of *A. pisum* (Carolan et al. 2011). Armet is bifunctional in mammals, where it intracellular acts as a component of the unfolded protein response in the lumen of the endoplasmic reticulum (Mizobuchi et al. 2007, Apostolou et al. 2008). Extracellular it functions as a neurotrophic factor (Petrova et al. 2003, Palgi et al. 2009). In invertebrates, published work on Armet is limited to Drosophila, documenting comparable functions to mammals in that species (Palgi et al. 2009, 2012). The Armet transcript level in the salivary gland of *A. pisum* is much higher than in the rest of the organism (Carolan et al. 2011) and Armet was shown to be secreted when expressed in *Spodoptera frugiperla* Sf9 cells, indicating that it is a component of saliva and is secreted during the feeding process (Wang et al. 2015). When Armet specific dsRNA is injected in *A. pisum*, behavior observation by EPG demonstrates that ingestion is decreased while aphids have to secrete more watery saliva into pierced sieve elements. Although specific calcium-binding motifs are absent in Armet it binds calcium. Thus, the possibility should be considered that Armet might conceivably take part in counteracting calcium-triggered occlusion mechanism in plants (Wang et al. 2015).

However, any physiologically significant effect of calcium binding by Armet and other calcium-binding salivary proteins in sieve element sap would have to be highly localized because the molar concentrations of individual aphid proteins in phloem sap are undoubtedly very low, and thus the effect of calcium binding by such a protein on global calcium concentration in phloem sap would be as well (Wang et al. 2015). Hence, calcium binding capacity of saliva might be at a maximum during the saliva secretion process at the beginning of sieve element penetration in the close vicinity to the aphid stylet. In this context it has to be taken into account that calcium diffusion is decreased due to binding to cytoplasmic sites, leading to local calcium hotspots (large local concentrations) just under the cells plasmamembrane (Stockbridge and Moore 1984), as observed, e.g., in stomatal guard cells (Gilroy and Trewavas 2001). Taken together it can be suggested that a local increase in calcium (at the stylet penetration site) can be buffered by short term aggregations of calcium binding salivary proteins.

Salivary Effectors

Effectors are, e.g., known from different bacterial pathogens and are generally defined as proteins and/or small molecules that alter host cell structure and function (Hogenhout et al. 2009). During the last years several effectors were identified in aphids as well and current techniques that are used for identification of salivary effectors are RNA interference, mediated by injection or via transgenic plants and the transgenic expression of effectors *in planta* (see Chapter 12 "Biotechnological Approaches to Aphid Management" in this volume). Both approaches are combined with bioassays, e.g., observation of survival and reproduction, and often with observation of aphid behavior by EPG technique (see Chapter 13 "Aphid Techniques" in this volume).

C002 can be seen as the first identified effector in aphid saliva (Mutti et al. 2006, 2008). C002, whose production takes place in the principal salivary glands of *A. pisum*, plays a significant role in successful aphid feeding (Mutti et al. 2008). Suppression of C002 transcripts by RNA interference in *A. pisum* led to a reduced life span and has a negative impact on the ability of an aphid to locate/reach sieve elements during probing (Mutti et al. 2006). In a few successful penetrations of sieve elements, aphids were unable to sustain ingestion of sieve-element sap for longer than 30 minutes (Mutti et al. 2008). When *M. persicae* fed on MpC002-expressing plants, an enhanced fecundity was observed (Bos et al. 2010), while the reproduction rates of *M. persicae* feeding on plants that express C002 from *A. pisum* are not influenced (Pitino and Hogenhout 2013). This observation indicates that the function of salivary effectors is specific to aphid species. Interestingly *A. pisum* and *M. persicae* C002 are divergent in amino acid sequence (Mutti et al. 2006, Bos et al. 2010), which may explain effector specificity.

Further effectors with a beneficial effect on reproduction and thus on aphid colonization are PIntO1 and PIntO2, orthologues of C002. PIntO1 and PIntO2 were detected in salivary gland transcriptome of multiple aphid species and appear to be specific for the respective aphid species (Pitino and Hogenhout 2013). A study of the effector Me23 demonstrates that effectors are specific not only to aphid species but also to host plants by demonstrating that the fecundity of *M. euphorbiae* was enhanced when aphids fed on Me23 expressing *Nicotiana benthamiana* and not on Me23 expressing tomato that both represent suitable host plants to this aphid species (Atamian et al. 2013). As described for different fungi (Stergiopoulos and de Wit 2009) above described effectors may be able to facilitate ingestion by interfering with plant signal cascades, leading to suppressing of plant defense responses and thus appear to contribute to aphid-plant compatibility (Pitino and Hogenhout 2013). Beside the effectors that promote aphid infestation, some effectors induce plant defense responses. Thus they act as pathogen associated molecular patterns (PAMPs). MP10 and MP42 were shown to reduce fecundity when expressed in plants (Bos et al. 2010). These effectors possibly interacting with plant receptors of the NBS-LRR superfamily and by doing so trigger plant defense responses (Bos and Hogenhout 2011). The identified aphid resistance genes Mi-1.2 in tomato (Martinez de Ilarduya et al. 2003) and Vat in melon (Chen et al. 1996) belong to the NBS-LRR receptor family (Smith and Clement 2012). A comparative summary of aphid effectors and their presence in the aphid species *A. pisum*, *M. persicae* and *M. euphorbiae* is given by Chaudhary et al. (2015).

Proteins from Endosymbiotic Bacteria

Aphids are closely linked with different endosymbiotic bacteria whereat the most important one is the obligate primary symbiont *Buchnera aphidicola*, which is harbored inside bacteriocytes in the body of an aphid (see Chapter 5 "Bacterial Symbionts of Aphids (Hemiptera: Aphididae)" in this volume). To identify proteins of endosymbiont origin in the saliva of *M. euphorbiae*, Chaudhary et al. (2014) searched against *Buchnera*-predicted proteins and identified 11 proteins. Four were only detected in the gel saliva of the studied aphid species. Among the *Buchnera* proteins was the chaperonin GroEL, constituting 10% of the *Buchnera* proteins (Baumann et al. 1996).

Using antibodies against *Escherichia coli* GroEL, this type of protein was detected in aphid saliva (Filichkin et al. 1997). Furthermore, a single peptide matching to both *E. coli* and *Buchnera* GroEL has also been identified in aphid saliva (Vandermoten et al. 2014). Due to cross-reactivity of the GroEL antibody and the high similarity of the GroEL peptide between *E. coli* and *Buchnera*, a conclusive determination of the GroEL origin could not be made.

The role of GroEL in aphid-plant interaction was tested by direct infiltration into *Arabidopsis thaliana*, by expression of GroEL in *Arabidopsis* or through *Pseudomonas fluorescens*, engineered to express the type III secretion system, in *Arabidopsis* and *Solanum lycopersicum* (Chaudhary et al. 2014). All treatments triggered defense responses in plants, indicated by oxidative burst and the expression of pattern triggered immunity (PTI) marker genes. Thus, GroEL is recognized both extracellularly and intracellularly and functions as a microbe associated molecular pattern (MAMP). Where tested, induced PTI led to resistance against aphids indicated by reduced fecundity. Chaudhary and colleagues suggest that induced PTI may directly target the *Buchnera* endosymbiont to control the insect pest. *Buchnera* is an obligate endosymbiont, shown to be relevant for conversion of non-essential amino acids into essential (IAGC 2010). Removal of *Buchnera* leads to delayed development and a strong decrease of reproduction (Sasaki et al. 1991). By targeting the endosymbiont the plant immune system is taking advantage of the strict mutual dependency of a host-insect with its symbiont (Chaudhary et al. 2014).

Because *Buchnera* are housed within bacteriocytes, the question arises how bacterial proteins get into the salivary glands so that they can be part of aphid saliva. Chaudhary et al. (2014) suggest that bacterial proteins get into the hemolymph during bacteriocyte turnover or degeneration in the postreproductive aphid stage (Douglas and Dixon 1987, Nishikori et al. 2009) and furthermore speculate that movement of macromolecules from the hemocoel to the salivary glands is a potential way to eliminate macromolecules in aphids (Miles 1968, 1972, Ponsen 1972).

Secretion of Watery Saliva during Ingestion

After penetration of a sieve element, the high turgor pressure inside the sieve-element lumen drives the sieve-element sap through the stylet nutrition channel into the alimentary tract of the aphid. This passive feeding process is regulated by the precibarial valve. During the ingestion process, watery saliva is secreted in regular intervals, mixes with sieve-element sap in the nutrition channel and gets with the nutrition flow into the alimentary tract (Tjallingii 2006). The function of salivary proteins inside the aphid gut may be comparable to those inside sieve elements. Aggregation of occlusion proteins, e.g., by members of the SEO family (Jekat et al. 2013), could plug the nutrition channel inside the stylet or could form massive protein aggregations in the stomach and salivary proteins could inhibit these aggregations. However, calcium binding proteins appear to be of minor importance for this purpose since special cells in the midgut appear to take up calcium and store it as calcium carbonate (Ehrhardt 1965).

Salivary proteases (Carolan et al. 2009, Furch et al. 2015) could degrade protease inhibitors (PIs) and by doing so act in inactivating PIs that harm aphids. PIs are produced by wild type plants but are used for agrobiotechnological pest control as well (Will and Vilcinskas 2013, see Chapter 12 "Biotechnological Approaches to Aphid Management" in this volume) and potentially interact with proteases located inside the aphid midgut as identified in *A. pisum* (Cristofoletti et al. 2003). Ingested PIs negatively affect aphid fitness, indicated by a reduced reproduction rate (e.g., Tran et al. 1997), which is most likely caused by a shortage in nitrogen supply to the aphid.

Phytochemicals in the ingested sieve element sap such as phenolic compounds have to be detoxified by the aphid. This may be achieved prior to ingestion of sieve element sap or inside the alimentary tract by peroxidases, phenoloxidases (Urbanska et al. 1998) and oxidoreductases (Miles and Oertli 1993) such as those identified in the saliva of *M. persicae* and *A. pisum* (Harmel et al. 2008, Carolan et al. 2009). GMC-oxidoreductases that were observed for *A. pisum* and several cereal aphid species (Carolan et al. 2009, Nicholson et al. 2012, Rao et al. 2013) could fulfill a potential function in scavenging of reactive oxygen species and detoxification of phytochemicals within the aphid (Carolan et al. 2009, Miles and Oertli 1993). However, the pH optimum of enzymes has to be considered when one is thinking about the activity of a salivary enzyme inside sieve elements and the aphid's alimentary tract. The pH inside sieve elements of *V. faba* is approximately 7.5 (Hafke et al. 2005) while the pH inside the gut of *A. pisum* ranges from 5.5 in the midgut to 8.5 in the intestine (Cristofoletti et al. 2003). Urbanska and Niraz (1990) determined the pH optimum of a peroxidase (pH 5.0–7.0) and a phenoloxidase (pH 8.2–9.4) from saliva of *S. avenae*, indicating that enzyme activity must vary for different enzymes between the different compartments.

Keywords: Aphid saliva, plant defense, salivary sheath, sieve element, watery saliva

References

Abdellatef, E., T. Will, A. Koch, J. Imani, A. Vilcinskas and K.H. Kogel. 2015. Silencing the expression of the salivary sheath protein causes transgenerational feeding suppression in the aphid *Sitobion avenae*. Plant Biotechnol. J. doi: 10.1111/pbi.12322.

Adams, J.B. and J.W. McAllan. 1956. Pectinase in the saliva of *Myzus persicae* (Sulz.) (Homoptera: Aphididae). Can. J. Zool. 34: 541–543.

Adams, J.B. and M.E. Drew. 1963. The effects of heating on the hydrolytic activity of aphid extracts on soluble cellulose substrates. Can. J. Zool. 41: 1263.

Adams, J.B. and M.E. Drew. 1965. A cellulose-hydrolyzing factor in aphid saliva. Can. J. Zool. 43: 489–496.

An, F., D. Cahill, J. Rookes, W. Lin and L. Kong. 2012. Real-time measurement of phloem turgor pressure in *Hevea brasiliensis* with a modified cell pressure probe. Botanical Studies 55: 19.

Apostolou, A., Y. Shen, Y. Liang, J. Luo and S. Fang. 2008. Armet, a UPR-upregulated protein, inhibits cell proliferation and ER stress-induced cell death. Exp. Cell Res. 314: 2454–2467.

Arnaud, G. 1918. Les Astérinées. pp. 52–65. Annales de l'Ecole d'Agriculture. Montpellier, France.

Arsanto, J.P. 1986. Ca^{2+}-binding sites and phosphatase activities in sieve element reticulum and P-protein of chick-pea phloem: a cytochemical and X-ray microanalysis survey. Protoplasma 132: 160–171.

Atamian, H.S., R. Chaudhary, V.D. Cin, E. Bao, T. Girke and I. Kaloshian. 2013. In planta expression or delivery of potato aphid *Macrosiphum euphorbiae* effectors Me10 and Me23 enhances aphid fecundity. Mol. Plant Microbe Interact. 26: 67–74.

Baumann, P., L. Baumann and M.A. Clark. 1996. Levels of *Buchnera aphidicola* chaperonin groEL during growth of the aphid *Schizaphis graminum*. Curr. Microbiol. 32: 279–285.

Bos, J.I.B. and S.A. Hogenhout. 2011. Effector proteins that modulate plant-insect interactions. Curr. Opin. Plant Biol. 14: 1–7.

Bos, J.I.B., D. Prince, M. Pitino, M.E. Maffei, J. Win and S.A. Hogenhout. 2010. A Functional genomics approach identifies candidate effectors from the aphid species *Myzus persicae* (green peach aphid). PLoS Genetics 6: e1001216. doi: 10.1371/journal.pgen.1001216.

Bradley, R.H.E. 1952. Studies on the aphid transmission of a strain of Henbane mosaic virus. Ann. Appl. Biol. 39: 78–97.

Büsgen, M.D.H. 1891. Studien an Pflanzen und Pflanzenläusen. Jenaische Zeitschrift für Naturwissenschaft 25: 340–428.

Calabrese, E.J. and L.J. Edwards. 1976. Light and gravity in leaf-side selection by the green peach aphid, *Myzus persicae*. Ann. Entomol. Soc. Am. 69: 1145–1146.

Carolan, J.C., C.I.J. Fitzroy, P.D. Ashton, A.E. Douglas and T.L. Wilkinson. 2009. The secreted salivary proteome of the pea aphid *Acyrthosiphon pisum* characterised by mass spectrometry. Proteomics 9: 2457–2467.

Carolan, J.C., D. Caragea, K.T. Reardon, N.S. Mutti, N. Dittmer, K. Pappan, F. Cui, M. Castaneto, J. Poulain, C. Dossat, D. Tagu, J.C. Reese, G.R. Reeck, T.L. Wilkinson and O.R. Edwards. 2011. Predicted effector molecules in the salivary secretome of the pea aphid (*Acyrthosiphon pisum*): a dual transcriptomic/proteomic approach. J. Proteome Res. 10: 1505–1518.

Cayla, T., B. Batailler, R. Le Hir, F. Revers, J.A. Anstaed, G.A. Thompson, O. Grandjean and S. Dinant. 2015. Live imaging of companion cells and sieve elements in Arabidopsis leaves. PLoS One 10: e0118122.

Chaudhary, R., H.S. Atamian, Z. Shen, S.P. Briggs and I. Kaloshian. 2014. GroEL from the endosymbiont *Buchnera aphidicola* betrays the aphid by triggering plant defense. Proc. Natl. Acad. Sci. USA 111: 8919–8924.

Chaudhary, R., H.S. Atamian, Z. Shen, S.P. Briggs and I. Kaloshian. 2015. Potato aphid salivary proteome: enhanced salivation using resorcinol and identification of aphid phosphoproteins. J. Proteome Res. 14: 1762–1778.

Chen, J.Q., B. Delobel, Y. Rahbé and N. Sauvion. 1996. Biological and chemical characteristics of a genetic resistance of melon to the melon aphid. Entomol. Exp. Appl. 88: 250–253.

Cherqui, A. and W.F. Tjallingii. 2000. Salivary proteins of aphids, a pilot study on identification, separation, and immunolocalisation. J. Insect Physiol. 46: 1177–1186.

Collar, J.L. and A. Fereres. 1998. Nonpersistent virus transmission efficiency determined by aphid probing behavior during intracellular punctures. Environ. Entomol. 27: 583–591.

Cooper, W.R., J.W. Dillwith and G.J. Puterka. 2011. Comparison of salivary proteins from five aphid (Hemiptera: Aphididae) species. Environ. Entomol. 40: 151–156.

Cristofoletti, P.T., A.F. Ribeiro, C. Deraison, Y. Rahbé and W.R. Terra. 2003. Midgut adaptation and digestive enzyme distribution in a phloem feeding insect, the pea aphid *Acyrthosiphon pisum*. J. Insect Physiol. 49: 11–24.

Douglas, A.E. and A.F.G. Dixon. 1987. The mycetocyte symbiosis in aphids: variation with age and morph in virginoparae of *Megoura viciae* and *Acyrthosiphon pisum*. J. Insect Physiol. 33: 109–113.

Ehlers, K., M. Knoblauch and A.J.E. van Bel. 2000. Ultrastructural features of well-preserved and injured sieve elements: minute clamps keep the phloem transport conduits free for mass flow. Protoplasma 214: 80–92.

Ehrhardt, P. 1965. Magnesium and calcium-containing granules in the midgut cells of aphids. Experientia 21: 337–338.

Filichkin, S.A., S. Brumfield, T.P. Filichkin and M.J. Young. 1997. *In vitro* interactions of the aphid endosymbiotic SymL chaperonin with barley yellow dwarf virus. J. Virol. 71: 569–577.

Fisher, D.B. 1978. An evaluation of the Munch hypothesis for phloem transport in soybean. Planta 139: 25–28.

Fisher, D.B., Y. Wu and M.S.B. Ku. 1992. Turnover of soluble proteins in the wheat sieve tube. Plant Physiol. 100: 1433–1441.

Francischetti, I.M.B., T.N. Mather and J.M.C. Ribeiro. 2003. Cloning of a salivary gland metalloprotease and characterization of gelatinase and fibrin(ogen)lytic activities in the saliva of the Lyme Disease tick vector *Ixodes scapularis*. Biochem. Biophys. Res. Commun. 305: 869–875.

Freeman, T.P., J.S. Buckner, D.R. Nelson, C.-C. Chu Chang and T.J. Henneberry. 2001. Stylet penetration by *Bemisia argentifolii* (Homoptera: Aleyrodidae) into host leaf tissue. Ann. Entomol. Soc. Am. 94: 761–768.

Fujita, T., K. Uchida and N. Maruyama. 1992. Purification of senescence marker protein-30 (SMP30) and its androgen-independent decrease with age in the rat liver. Biochim. Biophys. Acta 1116: 122–128.

Furch, A.C.U., A.J.E. van Bel, M.D. Fricker, H.H. Felle, M. Fuchs and J.B. Hafke. 2009. Sieve-element Ca²⁺ channels as relay stations between remote stimuli and sieve-tube occlusion in *Vicia faba*. Plant Cell 2: 2118–2132.

Furch, A.C.U., T. Will, M.R. Zimmermann, J.B. Hafke and A.J.E. van Bel. 2010. Remote-controlled stop of phloem mass flow by biphasic occlusion in *Cucurbita maxima*. J. Exp. Bot. 61: 3697–3708.

Furch, A.C.U., A.J.E. van Bel and T. Will. 2015. Aphid salivary proteases are capable of degrading sieve-tube proteins. J. Exp. Bot. 66: 533–539.

Gaupels, F., T. Knauer and A.J.E. van Bel. 2008. A combinatory approach for analysis of protein sets in barley sieve-tube samples using EDTA-facilitated exudation and aphid stylectomy. J. Plant Physiol. 165: 95–103.

Gilroy, S. and A. Trewavas. 2001. Signal processing and transduction in plant cells: the end of a beginning? Nature Rev. Mol. Cell Biol. 2: 307–314.

Gould, N., P.E.H. Minch and M.R. Thorpe. 2004. Direct measurements of sieve element hydrostatic pressure reveal strong regulation after pathway blockage. Funct. Plant Biol. 31: 987–993.

Gray, M.S. and F.E. Gildow. 2003. Luteovirus-aphid interactions. Annu. Rev. Phytopathol. 41: 539–566.

Gündüz, E.A. and A.E. Douglas. 2009. Symbiotic bacteria enable insects to use a nutritionally inadequate diet. Proc. R. Soc. Lond. B Biol. Sci. 276: 987–991.

Hafke, J.B., J.-K. van Amerongen, F. Kelling, A.C.U. Furch, F. Gaupels and A.J.E. van Bel. 2005. Thermodynamic battle for photosynthate acquisition between sieve tubes and adjoining parenchyma in transport phloem. Plant Physiol. 138: 1527–1537.

Harholt, J., A. Suttangkakul and H.V. Scheller. 2010. Biosynthesis of pectin. Plant Physiol. 153: 384–395.

Harmel, N., E. Létocart, A. Cherqui, P. Giordanengo, G. Mazzucchelli, E. De Pauw, E. Haubruge and F. Francis. 2008. Identification of aphid salivary proteins: a proteomic investigation of *Myzus persicae*. Insect Mol. Biol. 17: 165–174.

Hogenhout, S.A., R.A.L. van der Hoorn, R. Terauchi and S. Kamoun. 2009. Emerging concepts in effector biology of plant-associated organisms. Mol. Plant Microbe Interact. 22: 115–122.

Jekat, S.B., A.M. Ernst, A. von Bohl, S. Zielonka, R.M. Twyman, G.A. Noll and D. Prüfer. 2013. P-proteins in *Arabidopsis* are heteromeric structures involved in rapid sieve tube sealing. Front Plant Sci. 4: 225. doi: 10.3389/fpls.2013.00225.

Jiang, Y.X. and G.P. Walker. 2007. Identification of phloem sieve elements as the site of resistance to silverleaf whitefly in resistant alfalfa genotypes. Entomol. Exp. Appl. 125: 307–320.

Kauss, H. 1986. Ca²⁺ dependence of callose synthesis and the role of polyamines in the activation of 1,3-β-glucan synthase by Ca²⁺. pp. 131–137. *In*: A.J. Trewavas (ed.). Molecular and Cellular Aspects of Calcium in Plant Development. Volume 4, NATO ASI Subseries A, Plenum Press, New York.

Kehr, J. 2006. Phloem sap proteins: their identities and potential roles in the interaction between plants and phloem-feeding insects. J. Exp. Bot. 57: 767–774.

Knoblauch, M. and A.J.E. van Bel. 1998. Sieve tubes in action. The Plant Cell 10: 35–50.

Knoblauch, M., W.S. Peters, K. Ehlers and A.J.E. van Bel. 2001. Reversible calcium-regulated stopcocks in legume sieve tubes. Plant Cell 13: 1221–1230.

Knoblauch, M., M. Stubenrauch, A.J.E. van Bel and W. Peters. 2012. Forisome performance in artificial sieve tubes. Plant Cell Environ. 35: 1419–1427.

Kusnierczyk, A., D.H.T. Tran, P. Winge, T.S. Jørstad, J.C. Reese, J. Troczynska and A.M. Bones. 2011. Testing the importance of jasmonate signalling in induction of plant defences upon cabbage aphid (*Brevicoryne brassicae*) attack. BMC Genomics 12: 423.

Lecourieux, D., R. Ranjeva and A. Pugin. 2006. Calcium in plant defence-signalling pathways. New Phytol. 171: 249–269.

Leineweber, K., A. Schulz and G.A. Thompson. 2000. Dynamic transitions in the translocated phloem filament protein. Aust. J. Plant Physiol. 27: 733–741.

Louis, J. and J. Shah. 2013. *Arabidopsis thaliana–Myzus persicae* interaction: shaping the understanding of plant defense against phloem-feeding aphids. Front. Plant Sci. 4: 213. doi: 10.3389/fpls.2013.00213.

Lowe, H.J.B., G.J.P. Murphy and M.L. Parker. 1985. Non-glaucousness, a probable aphid resistance character of wheat. Ann. Appl. Biol. 106: 555–560.

Ma, R., J.C. Reese, W.C. Black IV and P. Bramel-Cox. 1990. Detection of pectinesterase and polygalacturonase from salivary secretions of living greenbugs, *Schizaphis graminum* (Homoptera: Aphididae). J. Insect Physiol. 36: 507–512.

Madhusudhan, V.V. and P.W. Miles. 1998. Mobility of salivary components as a possible reason for differences in the responses of alfalfa to the spotted alfalfa aphid and pea aphid. Entomol. Exp. Appl. 86: 25–39.

Martin, B., J.L. Collar, W.F. Tjallingii and A. Fereres. 1997. Intracellular ingestion and salivation by aphids may cause the acquisition and inoculation of non-persistently transmitted plant viruses. J. Gen. Virol. 78: 2701–2705.

Martinez de Ilarduya, O., Q. Xie and I. Kaloshian. 2003. Aphid-induced defense responses in Mi-1-mediated compatible and incompatible tomato interactions. Mol. Plant Microbe Interact. 16: 699–708.

McLean, D.L. and M.G. Kinsey. 1965. Identification of electrically recorded curve patterns associated with aphid salivation and ingestion. Nature 205: 1130–1131.

Medina-Ortega, K.J. and G.P. Walker. 2015. Faba bean forisomes can function in defense against generalist aphids. Plant Cell Environ. 38: 1167–1177.

Miles, P.W. 1965. Studies on the salivary physiology of plant-bugs: the salivary secretions of aphids. J. Insect Physiol. 11: 1261–1268.

Miles, P.W. 1968. Insect secretions in plants. Annu. Rev. Phytopathol. 6: 137–164.

Miles, P.W. 1972. The saliva of Hemiptera. Adv. Insect Physiol. 9: 183–255.

Miles, P.W. 1999. Aphid saliva. Biol. Rev. 74: 41–85.

Miles, P.W. and J.J. Oertli. 1993. The significance of antioxidants in the aphid-plant interaction: the redox hypothesis. Entomol. Exp. Appl. 67: 275–283.

Mizobuchi, N., J. Hoseki, H. Kubota, S. Toyokuni, J. Nozaki, M. Naitoh, A. Koizumi and K. Nagata. 2007. ARMET is a soluble ER protein induced by the unfolded protein response via ERSE-II element. Cell Struct. Funct. 32: 41–50.

Morgan, J.K., G.A. Luzio, E.-D. Ammar, W.B. Hunter, D.G. Hall and R.G. Shatters, Jr. 2013. Formation of stylet sheaths *in āere* (in air) from eight species of phytophagous hemipterans from six families (Suborders: Auchenorrhyncha and Sternorrhyncha). PLoS ONE 8(4): e62444.

Moscatiello, R., P. Mariani, D. Sanders and F.J. Maathuis. 2006. Transcriptional analysis of calcium-dependent and calcium-independent signalling pathways induced by oligogalacturonides. J. Exp. Bot. 57: 2847–2865.

Müller, B., S. Groscurth, M. Menzel, B.A. Rüping, R.M. Twyman, D. Prüfer and G.A. Noll. 2014. Molecular and ultrastructural analysis of forisome subunits reveals the principles of forisome assembly. Ann. Bot. 113: 1121–1137.

Mutti, N.S., Y.S. Park, J.C. Reese and G.R. Reeck. 2006. RNAi knockdown of a salivary transcript leading to lethality in the pea aphid, *Acyrthosiphon pisum*. J. Insect Sci. 6: 38.

Mutti, N.S., J. Louis, L.K. Pappan, K. Pappan, K. Begum, M.-S. Chen, Y. Park, N. Dittmer, J. Marshall, J.C. Reese and G.R. Reeck. 2008. A protein from the salivary glands of the pea aphid, *Acyrthosiphon pisum*, is essential in feeding on a host plant. Proc. Nat. Ac. Sci. USA 105: 9965–9969.

Nault, L.R. and G.G. Gyrisco. 1966. Relation of the feeding process of the pea aphid to the inoculation of pea enation mosaic virus. Ann. Entomol. Soc. Am. 59: 1185–1197.

Nault, L.R. and W.E. Styer. 1972. Effects of sinigrin on host selection by aphids. Entomol. Exp. Appl. 15: 423–437.

Nicholson, S.J., S.D. Hartson and G.J. Puterka. 2012. Proteomic analysis of secreted saliva from Russian Wheat Aphid (*Diuraphis noxia* Kurd.) biotypes that differ in virulence to wheat. J. Proteom. 75: 2252–2268.

Nishikori, K., K. Morioka, T. Kubo and M. Morioka. 2009. Age- and morph-dependent activation of the lysosomal system and Buchnera degradation in aphid endosymbiosis. J. Insect Physiol. 55: 351–357.

Palgi, M., R. Lindström, J. Peränen, T.P. Piepponen, M. Saarma and T.I. Heino. 2009. Evidence that DmMANF is an invertebrate neurotrophic factor supporting dopaminergic neurons. Proc. Natl. Acad. Sci. USA 106: 2429–2434.

Palgi, M., D. Greco, R. Lindström, P. Auvinen and T.I. Heino. 2012. Gene expression analysis of Drosophila Manf mutants reveals perturbations in membrane traffic and major metabolic changes. BMC Genomics 13: 134.

Petrova, P., A. Raibekas, J. Pevsner, N. Vigo, M. Anafi, M.K. Moore, A.E. Peaire, V. Shridhar, D.I. Smith, J. Kelly, Y. Durocher and J.W. Commissiong. 2003. MANF: a new mesencephalic, astrocyte-derived neurotrophic factor with selectivity for dopaminergic neurons. J. Mol. Neurosci. 20: 173–188.

Peuke, A.D., C. Windt and H. van As. 2006. Effects of cold-girdling on flows in the transport phloem in *Ricinus communis*: is mass flow inhibited? Plant Cell Environ. 29: 15–25.

Pieterse, C.M.J., D. van der Does, C. Zamioudis, A. Leon-Reyes and S.C.M. van Wees. 2012. Hormonal modulation of plant immunity. Annu. Rev. Cell Dev. Biol. 28: 489–521.

Pitino, M. and S.A. Hogenhout. 2013. Aphid protein effectors promote aphid colonization in a plant species-specific manner. Mol. Plant Microbe Interact. 26: 130–139.

Pollard, D.G. 1973. Plant penetration by feeding aphids (Hemiptera: Aphidoidea): a review. Bull. Entomol. Res. 62: 631–714.

Ponsen, M.B. 1972. The site of potato leaf roll virus multiplication in its vector *Myzus persicae*, an anatomical study. Meded. Landbouwhogesch Wageningen 16: 1–147.

Powell, G. and J. Hardie. 2000. Host-selection behaviour by genetically identical aphids with different plant preferences. Physiol. Entomol. 25: 54–62.

Powell, G., S.P. Maniar, J.A. Pickett and J. Hardie. 1999. Aphid response to non-host epicuticular lipids. Entomol. Exp. Appl. 91: 115–123.

Prado, E. and W.F. Tjallingii. 1994. Aphid activities during sieve element punctures. Entomol. Exp. Appl. 72: 157–165.

Rao, S.A., J.C. Carolan and T.L. Wilkinson. 2013. Proteomic profiling of cereal aphid saliva reveals both ubiquitous and adaptive secreted proteins. PLoS One 8(2): e57413.

Richardson, P.T., D.A. Baker and L.C. Ho. 1982. The chemical composition of cucurbit vascular exudates. J. Exp. Bot. 33: 1239–1247.

Sasaki, T., H. Hayashi and H. Ishikawa. 1991. Growth and reproduction of the symbiotic and aposymbiotic pea aphids, *Acyrthosiphon pisum* maintained on artificial diets. J. Insect Physiol. 37: 749–756.

Scheller, H.V. and P. Ulvskov. 2010. Hemicelluloses. Annu. Rev. Plant Biol. 61: 263–289.

Schobert, C., L. Baker, J. Szederkenyi, P. Großmann, E. Komor and H. Hayashi. 1998. Identification of immunologically related proteins in sieve-tube exudate collected from monocotyledonous and dicotyledonous plants. Planta 206: 245–252.

Shimokawa, N. and M. Yamaguchi. 1993. Expression of hepatic calcium-binding protein regucalcin mRNA is mediated through Ca2+/calmodulin in rat liver. FEBS Lett. 316: 79–84.

Sjolund, R.D. and C.Y. Shih. 1983. Freeze fracture analysis of phloem structure in plant tissue cultures. I. The sieve element reticulum. J. Ultrastruct. Res. 82: 111–121.

Smith, C.M. and S.L. Clement. 2012. Molecular bases of plant resistance to arthropods. Annu. Rev. Entomol. 57: 309–328.

Somerville, S. 2006. Cellulose synthesis in higher plants. Annu. Rev. Cell Dev. Biol. 22: 53–78.

Spiller, N.J., F.M. Kimmins and M. Llewellyn. 1985. Fine structure of aphid stylet pathways and its use in host plant resistance studies. Entomol. Exp. Appl. 38: 293–295.

Stafford-Banks, C.A., D. Rotenberg, B.R. Johnson, A.E. Whitfield and D.E. Ullman. 2014. Analysis of the salivary gland transcriptome of *Frankliniella occidentalis*. PLoS One 9: e94447. doi: 10.1371/journal.pone.0094447.

Stergiopoulos, I. and P.J. de Wit. 2009. Fungal effector proteins. Annu. Rev. Phytopathol. 47: 233–263.

Stockbridge, N. and J.W. Moore. 1984. Dynamics of intracellular calcium and its possible relationship to phasic transmitter release facilitation at the frog neuromuscular junction. J. Neurosci. 4: 803–811.

Takemoto, H., M. Uefune, R. Ozawa, G.I. Arimura and J. Takabayashi. 2013. Previous infestation of pea aphids *Acyrthosiphon pisum* on broad bean plants resulted in the increased performance of conspecific nymphs on the plants. J. Plant Interact. 4. doi: 10.1080/17429145.2013.786792.

The International Aphid Genomics Consortium. 2010. Genome sequence of the pea aphid *Acyrthosiphon pisum*. PLoS Biol. 8: e1000313. doi:10.1371/journal.pbio.1000313.

Tjallingii, W.F. 1978. Mechanoreceptors of the aphid labium. Entomol. Exp. Appl. 24: 531–537.

Tjallingii, W.F. 1985. Membrane potentials as an indication for plant cell penetrations by aphid stylets. Entomol. Exp. Appl. 38: 187–193.

Tjallingii, W.F. 2006. Salivary secretions by aphids interacting with proteins of phloem wound responses. J. Exp. Bot. 57: 739–745.

Tjallingii, W.F. and T. Hogen Esch. 1993. Fine structure of aphid stylet routes in plant tissues in correlation with EPG signals. Physiol. Entomol. 18: 317–328.

Tran, P., T.M. Cheesbrough and R.W. Keickhefer. 1997. Plant proteinase inhibitors are potential anticereal aphid compounds. J. Econ. Entomol. 90: 1672–1677.

Tsumuki, H., K. Kanehisa and K. Kawada. 1989. Leaf surface wax as a possible resistance factor of barley to cereal aphids. Appl. Entomol. Zoology 24: 295–301.

Urbanska, A. and S. Niraz. 1990. The phenol detoxifying enzymes of the grain aphid. Symposium Biologia Hungarica 39: 545–547.

Urbanska, A., W.F. Tjallingii, A.F.G. Dixon and B. Leszczynski. 1998. Phenol oxidizing enzymes in the grain aphid's saliva. Entomol. Exp. Appl. 86: 197–203.

Uzest, M., D. Gargani, A. Dombrovsky, C. Cazevieille, D. Cot and S. Blanc. 2010. The "acrostyle": a newly described anatomical structure in aphid stylets. Arthropod Struct. Dev. 39: 221–229.

van Bel, A.J.E. 2003. The phloem, a miracle of ingenuity. Plant Cell Environ. 26: 125–149.

Vandermoten, S., N. Harmel, G. Mazzucchelli, E. De Pauw, E. Haubruge and F. Francis. 2014. Comparative analysis of salivary proteins from three aphid species. Insect Mol. Biol. 23: 67–77.

Walker, G.P. and K.J. Medina-Ortega. 2012. Penetration of faba bean sieve elements by pea aphid does not trigger forisome dispersal. Entomol. Exp. Appl. 144: 326–335.

Walton, T.J. 1990. Waxes, cutin and suberin. pp. 106–158. *In*: J.L. Harwood and J. Boyer (eds.). Methods in Plant Biochemistry Vol. 4. Academic Press, London, UK.

Wang, W., H. Dai, Y. Zhang, R. Chandrasekar, L. Luo, Y. Hiromasa, C. Sheng, G. Peng, S. Chen, J.M. Tomich, J. Reese, O. Edwards, L. Kang, G. Reeck and F. Cui. 2015. Armet is an effector protein mediating aphid-plant interactions. FASEB J. 29: 2032–2045.

Wang, Y., M. Tang, P.Y. Hao, Z.F. Yang, L.L. Zhu and G.C. He. 2008. Penetration into rice tissues by brown planthopper and fine structure of the salivary sheaths. Entomol. Exp. Appl. 129: 295–307.

Wensler, R.J.D. 1962. Mode of host selection by an aphid. Nature 195: 830–831.

Wensler, R.J.D. 1974. The fine structure of distal receptors on the labium of the aphid *Brevicoryne brassicae* L. (Homoptera). Cell Tiss. Res. 181: 409–421.

Wensler, R.J.D. 1977. The fine structure of the distal receptors on the labium of the aphid *Brevicoryne brassicae* L. (Homoptera). Cell Tiss. Res. 181: 409–421.

Wensler, R.J.D. and B.K. Filshie. 1969. Gustatory sense organs in the food canal of aphids. J. Morphol. 129: 473–492.

Will, T. and A.J.E. van Bel. 2006. Physical and chemical interactions between aphids and plants. J. Exp. Bot. 57: 729–737.

Will, T. and A.J.E. van Bel. 2008. Induction as well as suppression: how aphid saliva may exert opposite effects on plant defense. Plant Signal Behav. 3: 427–430.

Will, T. and A. Vilcinskas. 2013. Aphid proofed plants: biotechnical approaches for aphid control. Adv. Biochem. Eng./Biotechnol. 136: 179–203.

Will, T. and A. Vilcinskas. 2015. The structural sheath protein of aphids is required for phloem feeding. Insect Biochem. Mol. Biol. 57: 34–40.

Will, T., W.F. Tjallingii, A. Thönnessen and A.J.E. van Bel. 2007. Molecular sabotage of plant defense by aphid saliva. Proc. Natl. Acad. Sci. USA 104: 10536–10541.

Will, T., A. Hewer and A.J.E. van Bel. 2008. A novel perfusion system shows that aphid feeding behaviour is altered by decrease of sieve-tube pressure. Entomol. Exp. Appl. 127: 237–245.

Will, T., S.R. Kornemann, A.C.U. Furch, W.F. Tjallingii and A.J.E. van Bel. 2009. Aphid watery saliva counteracts sieve-tube occlusion: a universal phenomenon? J. Exp. Biol. 212: 3305–3312.

Will, T., K. Steckbauer, M. Hardt and A.J.E. van Bel. 2012. Aphid gel saliva: sheath structure, protein composition and secretors dependence on stylet-tip milieu. PLoS ONE 7(10): e46903. doi:10.1371/journal.pone.0046903.

Will, T., A.C.U. Furch and M.R. Zimmermann. 2013. How phloem-feeding insects face the challenge of phloem-located defenses. Front. Plant Sci. 4: 336.

Xie, B., X. Wang, M. Zhu, Z. Zhang and Z. Hong. 2011. CalS7 encodes a callose synthase responsible for callose deposition in the phloem. Plant J. 65: 1–14.

Yamaguchi, M. 2000. Role of regucalcin in calcium signaling. Life Sci. 66: 1769–1780.

Yamaguchi, M. 2005. Role of regucalcin in maintaining cell homeostasis and function. Int. J. Mol. Med. 15: 371–389.

Zarate, S.I., L.A. Kempema and L.L. Walling. 2007. Silverleaf whitefly salicylic acid defense and suppresses effectual jasmonic acid defense. Plant Physiol. 143: 866–875.

Zhu, L.C., J.C. Reese, J. Louis, L. Campbell and M.S. Chen. 2011. Electrical penetration graph analysis of the feeding behavior of soybean aphids on cultivars with antibiosis. J. Econ. Entomol. 104: 2068–2072.

Zhu-Salzman, K., R.A. Salzman, J.E. Ahn and H. Koiwa. 2004. Transcript regulation of sorghum defense determinants against a phloem-feeding aphid. Plant Phsyiol. 134: 420–431.

Zimmermann, M.R., J.B. Hafke, A.J.E. van Bel and A.C.U. Furch. 2013. Interaction of xylem and phloem during exudation and wound occlusion in *Cucurbita maxima*. Plant Cell Environ. 36: 237–247.

12

Biotechnological Approaches to Aphid Management

Benjamin R. Deist and *Bryony C. Bonning**

Introduction

Aphids are among the most important pests of agriculture and horticulture and negatively impact crops and ornamentals both through feeding on phloem and through transmission of plant viruses (Emden and Harrington 2007). Aphids vector almost half of the plant viruses of significant economic importance (Miles 1989, Sylvester 1989). Aphid management is driven primarily by the application of classical chemical insecticides which can result in rapid selection for aphid resistance, and in loss of non-target organisms including honey bees.

Biotechnology has allowed for the development of alternative approaches to aphid management. As aphids are phloem feeders, delivery of bioactive molecules for suppression of aphid populations relies either on ingestion from the plant and access to target sites via the gut epithelium, or on penetration into the body cavity via the cuticle. The majority of new approaches to aphid management are focused on the production of aphid resistant transgenic plants, in line with transgenic crops being the fastest adopted technology in modern agriculture (James 2013). However, although alternative non-plant genetically modified organism (GMO)-based strategies are under investigation, these approaches have yet to be applied to target aphids. Non-plant GMO strategies include the use of (1) an engineered, debilitated plant virus to deliver a bioactive agent from the plant (Dawson and Folimonova 2013, Folimonov et al. 2007, Hajeri et al. 2014), which is under investigation for use in citrus against the Asian

Department of Entomology, Iowa State University, Ames, IA 50011, USA.
* Corresponding author: bbonning@iastate.edu

citrus psyllid (*Diaphorina citri*); (2) a transgenic entomopathogic fungus to deliver an agent through the insect cuticle, (Hokanson et al. 2014, Wang and Leger 2007); (3) recombinant insect viruses for infection via the gut and delivery of bioactives within the hemocoel, with the majority of research in this area conducted with lepidopteran pests (Kroemer et al. 2015, Federici et al. 2008).

Transgenic Approaches to Aphid Management

The focus of the current chapter is to outline biotechnological approaches that have been shown to work against aphids either in laboratory feeding- or transgenic plant-based bioassays (Table 1). The three primary gut-based approaches outlined in this chapter, namely the use of gut-active agents, hemocoel-active agents and gene silencing, are illustrated in Fig. 1.

Table 1. Representative examples of strategies used toward aphid resistant transgenic plants for suppression of aphid populations.

Strategy	Target species	Approach	Bioactive	References
Plant-derived R genes	Potato aphid, *Macrosiphum euphorbiae*	Transgenic tomato	mi-1.2	(Goggin et al. 2006)
Aphid alarm pheromone	Green peach aphid, *Myzus persicae*	Transgenic *Arabidopsis*	Aphid alarm pheromone synthase	(Beale et al. 2006) (Kunert et al. 2010)
Cysteine protease inhibitors	*M. persicae*	Transgenic oilseed rape	Oryzacystatin (OC-I)	(Rahbe et al. 2003)
	M. persicae, Potato aphid, *M. euphorbiae*	Transgenic eggplant	Oryzacystatin (OC-I)	(Ribeiro et al. 2006)
Lectins	*M. persicae*	Transgenic tobacco	*Galanthus nivalis* agglutinin (GNA)	(Hilder et al. 1995)
Modified Bt toxins	Pea aphid, *Acyrthosiphon pisum* *M. persicae*	Membrane feeding assays	Cyt2Aa modified with gut binding peptide, GBP3.1 (CGAL, CGSL constructs)	(Chougule et al. 2013)
Neurotoxins	*M. persicae*	Transgenic *Arabidopsis*	Plant virus CP-neurotoxin fusion, CP-P-Hv1a	(Bonning and Chougule 2014)
dsRNA	*M. persicae*	Transient expression in *Nicotiana benthamiana* and transgenic *Arabidopsis thaliana*	C002 and Rack-1 dsRNA	(Pitino et al. 2011)
		Transgenic *Arabidopsis*	C002, PIntO1, PIntO2 dsRNA	(Pitino and Hogenhout 2013)

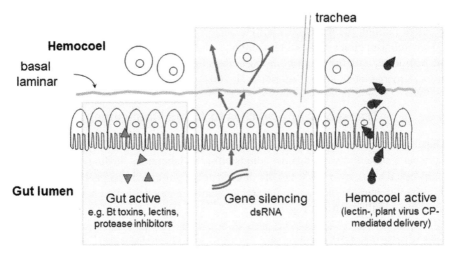

Fig. 1. Gut-mediated delivery of bioactive molecules for suppression of aphid populations. Illustrated are (1) agents that are active in the gut via disruption of gut enzymes (protease inhibitors) or the midgut epithelium (e.g., Bt toxins, lectins), (2) the use of dsRNA for silencing of gene transcripts, (3) the use of delivery proteins such as lectins and aphid transmitted plant virus-derived coat proteins for delivery of insect specific neurotoxins to the aphid hemocoel.

Plant-derived R Genes and Secondary Metabolites

Numerous R (resistance) genes that serve to protect plants against pests have been identified. These genes respond to effector proteins from plant pests and pathogens, and activate signaling cascades to activate plant defenses (Chisholm et al. 2006). The *Mi-1.2* gene, which has nucleotide binding and leucine-rich motifs typical of R genes, was identified from tomato. This gene confers resistance to aphids (Rossi et al. 1998) as well as various other plant pests including nematodes. When *Mi-1.2* was expressed in a susceptible tomato line, potato aphid populations were suppressed. However, *Mi-1.2* did not confer aphid resistance to transgenic eggplant suggesting that other factors involved in aphid resistance are not conserved between tomato and eggplant (Goggin et al. 2006). The signaling pathways involved in *Mi-1.2*-mediated aphid resistance have recently been elucidated (Wu et al. 2015). A negative aspect to the use of R genes for transgenic resistance to aphids is that these genes typically confer resistance to specific species and biotypes, with virulent biotypes likely to evolve.

Multiple plant secondary metabolites serve to protect plants against aphid infestation and there are several examples where plants have been engineered with biosynthetic enzymes for production of such metabolites. For example, chrysanthemum engineered for caffeine production was resistant to cotton aphid (Kim et al. 2011). Expression of β-glucosidase in chloroplasts in tobacco resulted in suppression of aphid populations mediated by increased density of trichomes, which physically protect the plant, and from increased sucrose esters (Jin et al. 2011). Genetic alteration of plant terpenoids can also be used for suppression of aphids (Wang et al. 2001).

Alarm Pheromone

The transgenic expression of an aphid alarm pheromone synthase in crop plants exploits aphid communication systems. Aphid alarm pheromone is primarily comprised of (E)-β-farnesene (EBF) for many aphid species (Francis et al. 2005, Vandermoten et al. 2012), and is released by aphids in imminent danger, often from natural enemies, leading to predator avoidance behavior (see Chapter 9 "Chemical Ecology of Aphids (Hemiptera: Aphididae)" in this volume). In addition, EBF has been implicated in serving as a short range attractant to aphid natural enemies (Vandermoten et al. 2012). Two early papers published on the transgenic production of EBF for aphid management showed differing results. The work of Beale et al. (2006) found that the constitutive expression of an EBF synthase gene in *Arabadopsis thaliana* resulted in the green peach aphid, *Myzus persicae* preferentially feeding on wild-type *A. thaliana* rather than the transgenic *A. thaliana*. It was also observed that the aphid parasitoid *Diaeretiella rapae* spent more time on the transgenic *A. thaliana* than on the wild-type. In contrast, Kunert et al. (2010) reported that—using the progeny of the transgenic *A. thaliana* used by Beale et al. (2006)—*M. persicae* distributed equally among transgenic *A. thaliana* and wild-type *A. thaliana* in a choice experiment.

Upon initial examination, the difference in results obtained by Beale et al. (2006) and Kunert et al. (2010) is puzzling as the wild potato (*Solanum berthaultii*) produces EBF and successfully deters aphid colonization (Gibson and Pickett 1983). However, experimental design differences between the two studies may provide an explanation for their contrasting results. Beale et al. (2006) measured *M. persicae* preference between wild-type *A. thaliana* and transgenic *A. thaliana*, at one minute and fifteen minutes while Kunert et al. (2010) measured *M. persicae* preference between wild-type and transgenic *A. thaliana* after 24 h. A longer period of time for the choice experiments likely allowed the *M. persicae* to habituate to the presence of EBF (de Vos et al. 2010). Further support for this scenario comes from Gibson and Pickett (1983) who indicated that EBF was likely released transiently due to *M. persicae* activity on *S. berthaultii*, while the transgenic *A. thaliana* used by both grops constantly released EBF (Kunert et al. 2010, Beale et al. 2006). Subsequent work showed that EBF expression in tobacco repelled aphids while attracting lacewings, natural enemies of the aphid (Yu et al. 2012).

While prolonged exposure does result in a reversible habituation of *M. persicae*— and likely other aphid species—to EBF, the transgenic expression of EBF synthases in crops could be a viable option for aphid management (de Vos et al. 2010). EBF as a short term attractant for some natural enemies of aphids provides an additional benefit (Vandermoten et al. 2012), although whether natural enemies also become habituated to EBF on continuous exposure, remains to be addressed.

Protease Inhibitors

Protease inhibitors are molecularly diverse, comprised of small molecules, peptides, and proteins united in their ability to inhibit protease activity. Of these, the protein based protease inhibitors have been grouped into 48 distinct families based on the amino acid sequence of the inhibitor region (Rawlings et al. 2004). Ingested protease inhibitors, provided they are limited to the digestive tract, have a negative impact on amino acid uptake via reduced protein digestion in aphids. Ultimately this can result in reduced size, delayed development, and death of the target aphid (Boulter 1993). For the purposes of aphid management, the cysteine protease inhibitors provide a good option as they have been shown to have negative impacts on the development and survival of various aphid species. To this end, cysteine proteases have been expressed in several plant species with deleterious effects on aphids.

The cysteine protease inhibitor, oryzacystatin I (OC-I), has been expressed in transgenic oilseed rape (*Brassica napus* L. cv. Drakkar) (Rahbe et al. 2003) and in eggplant (*Solanum melongena* L.) (Ribeiro et al. 2006). Phloem expression of OC-I resulted in depressed population growth in *M. persicae* fed on transgenic *B. napus* and *S. melongena* (Ribeiro et al. 2006, Rahbe et al. 2003). Rahbe et al. (2003) noted that *M. persicae* fed on transgenic *B. napus* had reduced fecundity and reduced biomass relative to aphids fed on wild-type *B. napus*. However, a significant difference in mortality between *M. persicae* fed on transgenic and wild-type *B. napus* was not observed (Rahbe et al. 2003). In contrast, higher mortality was observed for intermediate aged *M. persicae* fed on transgenic *S. melongena* expressing OC-I compared to those fed on wild-type *S. melongena*. In line with the work of Rahbe (Rahbe et al. 2003), transgenic *A. thaliana* expressing the cysteine protease inhibitor HvCPI-6, delayed development and reduced population growth of *M. persicae* in *in planta* bioassays, but did not result in a significant increase in aphid mortality relative to aphids fed on wild-type *A. thaliana* (Carrillo et al. 2011).

In vitro characterization of HvCPI-6 action against *M. persicae* and *A. pisum* gut enzymes showed a 70% decrease in cathepsin L-like activity and a 50% decrease in cathepsin B-like activity (Carrillo et al. 2011). Similarly, OC-I was shown to inhibit cathepsin L/H-like proteases in whole aphid extracts prepared from *M. persicae* that fed on transgenic *B. napus* (Rahbe et al. 2003). OC-I localization studies suggested that OC-I localized to the bacteriocytes and oenocytes in addition to the midgut.

While transgenic expression of cysteine protease inhibitors does not reliably induce mortality in aphids, feeding on such plants can result in a fitness cost to the aphid. Plant expression of cysteine protease inhibitors alone or in conjunction with other bioactive molecules could pose sufficient fitness cost to colonizing aphids to prevent yield loss. However, a key drawback to the use of protease inhibitors is the ability of insects to rapidly adapt by upregulating proteases insensitive to the inhibitors, or that can degrade the inhibitor (Jongsma and Beekwilder 2011).

Lectins

Lectins are carbohydrate binding proteins produced by plants and are associated with defense against pathogens and herbivores, including insects (Vandenborre et al. 2011). In aphids, lectins cause a variety of symptoms ranging from reduced feeding to death. The severity of symptoms appears to be dependent on both aphid species and lectin type, as a lectin that causes high mortality in one aphid species may depress aphid development but not cause significant mortality in a different aphid species (Vandenborre et al. 2011, Rahbe et al. 1995). Since the snowdrop lectin (GNA) was first expressed in tobacco plants (*Nicotiana tabacum*) some 20 years ago

(Hilder et al. 1995), multiple attempts have been made to transgenically express lectins for aphid suppression, using more than 13 different lectins derived from seven lectin families (Yu et al. 2014), with varying degrees of success.

Of the aphid-active lectins, GNA has been the most commonly used in transgenic crops (Table 2). Wheat (*Triticum aestivum* L. cv. Bobwhite) (Stoger et al. 1999), maize (*Zea mays* L.) (Wang et al. 2005), potato (*Solanum tuberosum*) (Down et al. 1996), and *N. tabacum* (Hilder et al. 1995) have all successfully been engineered to express GNA and all inhibited aphid colonization in laboratory experiments. It is important to note that not only has GNA been expressed in several different plant species, it has also shown activity against several aphid species including: The grain aphid (*Sitobion avenae*) (Stoger et al. 1999), corn leaf aphid (*Rhopalosiphum maidis* Fitch) (Wang et al. 2005), glasshouse potato aphid (*Aulacorthum solani*) (Down et al. 1996), and *M. persicae* (Hilder et al. 1995). Interestingly, none of the GNA-expressing plants significantly reduced aphid survival relative to that of aphids fed on control plants. However, significant fitness costs were incurred by aphid species that attempted to colonize the GNA-expressing plants, with reduced fecundity and delayed development resulting in smaller colonizing populations relative to aphids colonizing control plants (Hilder et al. 1995, Stoger et al. 1999, Down et al. 1996, Wang et al. 2005). The lack of significant mortality observed in aphids fed on the different transgenic plants was unexpected as GNA was shown to be toxic against *A. pisum* (Rahbe et al. 1995), *M. persicae* (Hilder et al. 1995), and *A. solani* (Down et al. 1996) in laboratory bioassays.

In addition to the use of GNA for development of aphid resistant transgenic plants, several other lectins have been used. *Allium sativum* leaf agglutinin (ASAL) was expressed in chickpea (*Cicer arietinum* L.) for resistance to cowpea aphid (*Aphis craccivora*), with the resulting transgenic plants inducing significant mortality and reduced fecundity in *A. craccivora* relative to control plants (Chakraborti et al. 2009). Kanrar et al. (2002) noted a slight increase in mortality and reduction

Table 2. Representative lectins expressed in plants for aphid management.

Lectin	Plant	Aphid	References
Galanthus nivalis agglutinin (GNA)	*Nicotiana tabacum* *Solanum tuberosum* *Zea mays* L. *Triticum aestivum* L. cv. Bobwhite	*Myzus persicae* *Aulacorthum solani* *Rhopalosiphum maidis* Fitch *Sitobion avenae*	(Hilder et al. 1995) (Down et al. 1996) (Wang et al. 2005) (Stoger et al. 1999)
Concanavalin A (ConA)	*Solanum tuberosum*	*Myzus persicae*	(Gatehouse et al. 1999)
Wheat germ agglutinin (WGA)	*Brassica juncea*	*Lipaphis erysimi* Kalt.	(Kanrar et al. 2002)
Allium sativum leaf agglutinin (ASAL)	*Cicer arietinum* L.	*Aphis craccivora*	(Chakraborti et al. 2009)
Allium cepa L. agglutinin (ACA)	*Brassica juncea*	*Lipaphis erysimi* Kalt.	(Hossain et al. 2006)
ACA-ASAL	*Brassica juncea*	*Lipaphis erysimi* Kalt.	(Hossain et al. 2006)

of fecundity of mustard aphid (*Lipaphis erysimi* Kalt.) fed on leaf discs cut from Indian mustard (*Brassica juncea*) expressing wheat germ agglutinin (WGA). *Allium sativum* agglutinin (ACA) and a fusion of ACA and ASAL (ACA-ASAL) were expressed in *B. juncea* and inhibited colonization by *L. erysimi*. Expression of ACA and that of ACA-ASAL in *B. juncea* resulted in significant mortality and a probable decrease in fecundity of *L. erysimi* fed on transgenic plants (Hossain et al. 2006). In contrast, concanavalin A (ConA) did not cause significant mortality when fed to *M. persicae* in artificial diet or when fed on *S. tuberosum* expressing ConA. However, similar to GNA, ConA was able to cause a significant reduction in *M. persicae* fecundity both when delivered in artificial diet, and when expressed in *S. tuberosum* (Gatehouse et al. 1999). Interestingly GNA, ASAL, ConA, WGA, and ACA display different glycan binding preferences, which may contribute to the observed differences in toxicity. GNA and ASAL only bind mannose residues, while ConA binds mannose, glucose, and N-acetyl-glucosamine. In contrast, WGA binds sialic acid and β-N-acetyl-glucosamine and ACA binds β-galactose and N-acetyl-galactosamine (Rahbe et al. 1995, Bandyopadhyay et al. 2001). While mannose is the most prevalent N-glycan in insects (Harrison and Jarvis 2006), lectin binding does not necessarily result in an insecticidal effect with other factors contributing to toxicity (Harper et al. 1995).

Lectins that reduce fecundity, delay aphid development or otherwise impede population growth can be just as valuable as lectins that induce mortality in aphids: Lectins such as ACA that induce significant mortality can be deployed as a singular strategy, while lectins that induce a fitness cost can be used in conjunction with other management strategies, transgenic or otherwise, for suppression of aphid populations.

The development of designer lectins is also possible as demonstrated by the ACA-ASAL fusion, which induced mortality in *L. erysimi* at a comparable level to ACA alone when expressed in *B. juncea* (Hossain et al. 2006). Initial studies indicate that the expression of lectins does not have a direct impact on some aphid natural predators (Couty et al. 2001, Birch et al. 1999, Down et al. 2000). The specificity of each lectin for specific sugar moieties and the extent of exposure will dictate the potential for impact on non-target organisms. Further development of aphid resistant, lectin expressing plants is warranted, but inconsistencies in performance under field conditions need to be addressed.

Modified Bt Toxins

Toxins derived from *Bacillus thuringiensis* (Bt) have been used in genetically modified crops to manage coleopteran and lepidopteran pests of agricultural significance for more than 20 years (Gatehouse et al. 2011, Sanahuja et al. 2011). Despite the extensive use of Bt toxins for pest management, the mechanisms by which Bt toxins fatally damage the gut epithelium remain unclear (Vachon et al. 2012). What is known is that upon ingestion, Bt toxins are solubilized and activated by enzymatic cleavage of an N- and/or C-terminal peptide from the protoxin. The toxin then binds to receptors on the surface of the insect gut epithelium, and a conformational change results in insertion of the toxin into the membrane. The resulting pore causes osmotic lysis and cell death (Vachon et al. 2012).

While Bt toxins display a high level of activity against Lepidoptera, Coleoptera, and Diptera, they do not display the same level of toxicity against the Hemiptera, notably aphids (Chougule and Bonning 2012, Raps et al. 2001). Some Bt toxins do display some toxicity against aphids (Porcar et al. 2009, Walters and English 1995, Loth et al. 2015), but in general, aphids are not widely susceptible. In the case of the Crystal (Cry) toxins, low susceptibility is likely associated with inefficient solubilization and activation of the Cry protoxin and low binding of the activated Cry toxin to the insect gut receptors (Li et al. 2011).

Bt toxin engineering has been used to enhance the toxicity against species with limited susceptibility, as well as to shift or broaden the specificity of toxins to other insect species. Bt toxin modification strategies center around improving toxin activation in the insect gut and either enhancing binding strength between the toxin and its receptor(s), or broadening the receptor binding profile of the toxin (Deist et al. 2014). Toxin activation can be enhanced by the replacement of native toxin enzyme cleavage sites with enzyme cleavage sites that favor the gut enzyme profile of the target insect (Walters et al. 2008). While there is one strategy for modification of Bt toxins for improved activation, there are multiple strategies for toxin modification for increased binding of Bt toxins to insect gut epithelial receptors. Bt toxin binding has been enhanced by domain swapping (Karlova et al. 2005), point modifications (Rajamohan et al. 1996), incorporation of binding motifs from unrelated toxins (Mehlo et al. 2005), the incorporation of binding motifs selected from phage display libraries (Chougule et al. 2013), and the selection of modified Cry toxins from phage display libraries (Fujii et al. 2013). Engineering Bt toxins for toxicity against hemipterans, aphids in particular, is a relatively new approach with correspondingly few publications. However, the application of methods used to engineer Bt toxins for activity against other insect species with limited susceptibility provides several strategies for enhancing aphicidal toxicity.

The earliest example of Bt toxin modification that enhanced activity of a Bt toxin against a hemipteran pest targeted the leafhopper (*Cicadulina mbila*), rather than an aphid species. Mehlo et al. (2005) fused the B-chain of the lectin ricin to Cry1Ac, broadening the activity of Cry1Ac to include *C. mbila*. It is important to note that the ricin B-chain fused to Cry1Ac is a galactose binding domain. The engineered Cry1Ac toxin, designated BtRb, reduced survival of *C. mbila* feeding on maize expressing BtRB to approximately 5% after four days (Mehlo et al. 2005). This work demonstrated that Bt toxins can be engineered for high levels of toxicity against hemipteran targets (Mehlo et al. 2005).

A cytolytic (Cyt) toxin, Cyt2Aa, has been modified for activity against the pea aphid (*Acyrthrosiphon pisum*) (Chougule et al. 2013). A short peptide, GBP3.1, selected from a phage display library for *in vivo* binding to the gut of *A. pisum* (Liu et al. 2010), was incorporated into exposed loops on the surface of Cyt2Aa, with GBP3.1 only being incorporated once for each modified toxin (Chougule et al. 2013). Several of the resulting toxins—CGAL1, CGAL3, CGAL4, CGSL1, and CGSL4—displayed a marked increase in toxicity towards *A. pisum*, with LC_{50} values decreasing from > 150 µg/ml for Cyt2Aa, to between 9.55 µg/ml and 28.74 µg/ml for the modified toxins (Chougule et al. 2013).

Three of the modified Cyt2Aa toxins—CGAL1, CGAL3, and CGAL4—also displayed increased toxicity against the green peach aphid (*Myzus persicae*), although the improvement in toxicity was not as dramatic as observed for *A. pisum* (Chougule et al. 2013). Increased toxicity against *M. persicae*, is likely due to homology between the gut protein bound by GBP3.1 (alanyl-aminopeptidase N, APN) (Linz et al. 2015) with that of *A. pisum* (Chougule et al. 2013). APN is a common gut protein in aphids (Cristofoletti et al. 2003, Liu et al. 2012) and is also bound by plant lectins (Fitches et al. 2008). While the Cyt2Aa toxin, modified with GBP3.1, impacted multiple aphid species, whether the modified toxins affect Hemiptera beyond Aphididae remains to be determined.

While Bt toxin modification is a relatively new approach for aphid management, it has proven successful for other pest insects (Deist et al. 2014). Knowledge of the physiological factors that limit the toxicity of specific toxins against aphid species is required for appropriate toxin modification to target aphids. In addition, knowledge of toxin structure to facilitate identification of sites that can be modified without destabilizing the toxin aids toxin modification to target aphids.

Neurotoxins

In addition to bioactive molecules such as lectins and protease inhibitors that target the gut of the aphid, the use of delivery systems to deliver peptide hormones and toxins that act within the hemocoel has recently been demonstrated (Bonning and Chougule 2014). In addition to direct suppression of aphid development by binding to mannose residues in the aphid gut, aphid-active lectins are also able to move from the surface of the gut epithelium into the insect hemocoel (Fitches et al. 2001). A fusion protein consisting of an insect-specific neurotoxin derived from a spider (*Segestria florentia* toxin 1, SFI1) and the garlic lectin, GNA (SFI1-GNA) was toxic to *M. persicae* in feeding assays (Down et al. 2006), with GNA delivering the toxin to its target site within the aphid hemocoel. *Arabidopsis* expression of GNA fused to a different spider venom toxin ω-ACTX-Hv1a, resulted in 40% mortality of *M. persicae* after seven days on detached leaves, but did not affect reproduction of surviving aphids. The Hv1a-GNA fusion protein was cleaved by *M. persicae* gut proteases and the linker between the two proteins appeared to be degraded in the plant (Nakasu et al. 2014). These data show promise for use of Hv1a-GNA for aphid resistant transgenic plants with an optimized linker sequence, with the added benefit that GNA can deliver toxins into the hemocoel of multiple pest insects (Nakasu et al. 2014, Bonning and Chougule 2014).

By using a completely different strategy to achieve the same end, Bonning et al. (2014), fused the coat protein of an aphid-transmitted plant virus to the insect specific neurotoxin ω-ACTX-Hv1a. Luteovirids such as Pea enation mosaic virus (PEMV) are ingested by aphids as they feed on phloem. The virions bind to receptors on the gut epithelium and are transcytosed across the epithelium (Gray and Gildow 2003). Following release into the hemocoel, virions move into the accessory salivary gland for transmission to host plants with saliva on subsequent aphid feeding. The coat protein of PEMV was shown to be sufficient to carry fused proteins into the aphid hemocoel. Delivery of the neurotoxin deleteriously impacted four aphid species in membrane feeding assays, and significantly suppressed populations of

M. persicae when the PEMV CP-P-Hv1a fusion protein was delivered from transgenic *Arabidopsis* (Bonning et al. 2014). In contrast to the GNA-based delivery system, the PEMV CP-based system is expected to confer specificity and is only expected to work with aphids; while GNA binds to mannose residues present in the guts of many arthropods, the PEMV CP interacts with a specific receptor protein, and the extent of binding in non-aphid species will be determined by the degree of similarity between the homologous receptor proteins.

RNAi

RNA interference (RNAi) is the inhibition of protein translation by RNA molecules. The three primary pathways that fall within RNAi include the cleavage of dsRNA by the enzyme Dicer into short interfering RNA (siRNA) duplexes of 21–22 nt, one strand of which is loaded onto the RNA-induced silencing complex (RISC). The RISC complex is guided by the siRNA to bind complementary sequence, which is then cleaved. This pathway represents a primary antiviral defense pathway in arthropods that is triggered by the dsRNA intermediate produced on replication of RNA viruses (Gammon and Mello 2015, Vijayendran et al. 2013). The utility of RNAi for pest management was demonstrated by plant expression of dsRNA for silencing of essential genes with complementary sequence in the western corn rootworm, *Diabrotica virgifera virgifera* LeConte (Baum et al. 2007), and by suppression of detoxification enzymes in the cotton bollworm, *Helicoverpa armigera*, such that the pest became susceptible to gossypol from the host plant (Mao et al. 2007, 2011). Since then, it has been established that the efficacy of RNAi varies greatly between insect orders, working well in Coleoptera but being highly variable in other orders, including Hemiptera (Terenius et al. 2011, Christiaens and Smagghe 2014).

Several studies have demonstrated reduction in the target transcript leading to suppression of protein translation by RNAi in aphids using feeding or injection for delivery of silencing RNA. Injection of siRNA targeting the C002 transcript in saliva of the pea aphid, *A. pisum* resulted in a two-fold reduction in the target transcript by 24 h (Mutti et al. 2006). Similarly, dsRNA injected to target calreticulin and cathepsin L resulted in 40% silencing after 5 days (Jaubert-Possamai et al. 2007, Sapountzis et al. 2014) and injection of dsRNA reduced angiotensin-converting enzyme (ACE) transcripts by 22 to 67% in pea aphid saliva (Wang, W. et al. 2015). A putative aquaporin gene (ApAQP1) (Shakesby et al. 2009), the V-ATPase gene in the pea aphid (Whyard et al. 2009) multiple target genes in the grain aphid, *Sitobion avenae* (Zhang et al. 2013, Wang, D. et al. 2015), and catalase (Deng and Zhao 2014) were suppressed by feeding of aphids on dsRNA in artificial diet. In addition, insecticide resistant adults of *A. gossypii* fed on dsRNA to target carboxylesterases implicated in insecticide resistance, showed increased sensitivity to the insecticides (Gong et al. 2014).

Gene silencing in aphids from transgenic plants has also been demonstrated, with target gene knockdown ranging from 30 to 80%: Transgenic wheat was used for knockdown of the carboxylesterase gene (CbE E4) in the grain aphid thereby increasing sensitivity to Phoxim insecticides (Xu et al. 2014), and receptor for activated kinase C (Rack-1), serine protease and hunchback were targeted in

M. persicae (Bhatia et al. 2012, Mao and Zeng 2013, Pitino et al. 2011). A side-by-side comparison of the impact of transgenic *Arabidopsis* mediated suppression of *Rack1*, *C002* and *PIntO2* in *M. persicae* showed transcript suppression of up to 70% between 4 and 8 days of aphid exposure, with C002 transcript suppression having the most dramatic effect on aphid population growth (Coleman et al. 2014).

For successful feeding, aphids produce sheath saliva (see Chapter 11 "Function of Aphid Saliva in Aphid-Plant Interaction" in this volume). The structural sheath protein (SHP) in this saliva seals around the stylet to produce a sheath that allows for sustained feeding on phloem. Transgenic barley expressing dsRNA to target SHP resulted in reduced *shp* transcript levels in aphids fed on these plants, along with reduced aphid growth, reproduction and survival rates (Abdellatef et al. 2015, Will and Vilcinskas 2015). Interestingly, transgenerational effects of feeding on SHP dsRNA were observed affecting the feeding, development and survival of progeny. Taken together, these studies support the premise that RNAi-based approaches are viable options for suppression of pestiferous aphids.

Additional Approaches

In addition to the strategies described above that have been shown to deleteriously impact aphid populations, a number of other approaches may warrant further examination for aphid management applications. Aphid myosuppressins and myosuppressin analogs are orally toxic to aphids with dose-dependent effects. The most active aphid myosuppressin, Acypi-MS resulted in 100% aphid mortality by 10 days (Down et al. 2011). The aphicidal action of these peptides may result from inhibition of visceral muscle contraction, and stimulation of enzyme secretion from the digestive tissues. The pea aphid is also susceptible to the stabilized mimics of pyrokinin/pheromone biosynthesis activating neuropeptides (PK/PBAN) and tachykinin-related peptides, which may act by interfering with gut motility (Nachman et al. 2011a, 2011b).

The primary symbiotic bacteria associated with aphids, provide an additional target for manipulation toward aphid population suppression. The antimicrobial peptide indolicidin reduced the number of bacteriocytes harboring *Buchnera* spp. in *M. persicae* resulting in impaired reproduction and survival (Le-Feuvre et al. 2007). Transgenic expression of such antimicrobial peptides to target aphid endosymbionts represents a novel approach for resistance to aphids.

Future Prospects

While significant advances have been made toward proof of concept for transgenic plant resistance to aphids (Chougule et al. 2013, Nakasu et al. 2014, Bonning et al. 2014), the ultimate adoption of any plant-based aphid resistance technology will depend on whether the industrial sector in agricultural biotechnology will invest the time and funding required to market such plants. The cost of development and regulatory requirements for transgenic plants (in the tens of millions of U.S. dollars) is prohibitive for all but the most economically important pests. It is highly likely that adoption of aphid resistant transgenic plants would reveal the pervasive nature and extent of aphid-mediated losses, which are otherwise difficult to estimate. Despite the estimated U.S. $1.6 billion lost from damage and management of the invasive soybean aphid, *Aphis glycines*, alone in the United States Midwest over a 10 year period (Kim et al. 2008), the industrial sector has shown minimal interest in developing and marketing aphid resistant transgenic crops. It seems more likely that non-transgenic plant approaches such as the use of topically applied silencing RNAs (Killiny et al. 2014) will be adopted for aphid management in the future.

Acknowledgements

This work was funded in part by the United States Department of Agriculture NIFA award number 2012-67013-19457, and by Hatch Act and State of Iowa funds.

Keywords: aphid management, transgenic approaches, R genes, secondary metabolites, alarm pheromones, protease inhibitors, lectins, Bt toxins, neurotoxins, RNAi

References

Abdellatef, E., T. Will, A. Koch, J. Imani, A. Vilcinskas and K.H. Kogel. 2015. Silencing the expression of the salivary sheath protein causes transgenerational feeding suppression in the aphid *Sitobion avenae*. Plant Biotechnol. J. doi: 10.1111/pbi.12322.

Bandyopadhyay, S., A. Roy and S. Das. 2001. Binding of garlic (*Allium sativum*) leaf lectin to the gut receptors of homopteran pests is correlated to its insecticidal activity. Plant Sci. 161(5): 1025–1033.

Baum, J.A., T. Bogaert, W. Clinton, G.R. Heck, P. Feldmann, O. Ilagan, S. Johnson, G. Plaetinck, T. Munyikwa, M. Pleau, T. Vaughn and J. Roberts. 2007. Control of coleopteran insect pests through RNA interference. Nat. Biotechnol. 25(11): 1322–1326.

Beale, M.H., M.A. Birkett, T.J.A. Bruce, K. Chamberlain, L.M. Field, A.K. Huttly, J.L. Martin, R. Parker, A.L. Phillips, J.A. Pickett, I.M. Prosser, P.R. Shewry, L.E. Smart, L.J. Wadhams, C.M. Woodcock and Y. Zhang. 2006. Aphid alarm pheromone produced by transgenic plants affects aphid and parasitoid behavior. Proc. Natl. Acad. Sci. USA 103(27): 10509–10513.

Bhatia, V., R. Bhattacharya, P.L. Uniyal, R. Singh and R.S. Niranjan. 2012. Host generated siRNAs attenuate expression of serine protease gene in *Myzus persicae*. PLoS One 7(10): e46343.

Birch, A.N.E., I.E. Geoghegan, M.E.N. Majerus, J.W. McNicol, C.A. Hackett, A.M.R. Gatehouse and J.A. Gatehouse. 1999. Tri-trophic interactions involving pest aphids, predatory 2-spot ladybirds and transgenic potatoes expressing snowdrop lectin for aphid resistance. Mol. Breed. 5(1): 75–83.

Bonning, B.C. and N.P. Chougule. 2014. Delivery of intrahemocoelic peptides for insect pest management. Trends Biotechnol. 32(2): 91–98.

Bonning, B.C., N. Pal, S. Liu, Z. Wang, S. Sivakumar, P.M. Dixon, G.F. King and W.A. Miller. 2014. Toxin delivery by the coat protein of an aphid-vectored plant virus provides plant resistance to aphids. Nat. Biotechnol. 32(1): 102–105.

Boulter, D. 1993. Insect pest control by copying nature using genetically engineered crops. Phytochem. 34(6): 1453–1466.

Carrillo, L., M. Martinez, F. Alvarez-Alfageme, P. Castañera, G. Smagghe, I. Diaz and F. Ortegoal. 2011. A barley cysteine-proteinase inhibitor reduces the performance of two aphid species in artificial diets and transgenic *Arabidopsis* plants. Transgenic Res. 20(2): 305–319.

Chakraborti, D., A. Sarkar, H.A. Mondal and S. Das. 2009. Tissue specific expression of potent insecticidal, *Allium sativum* leaf agglutinin (ASAL) in important pulse crop, chickpea (*Cicer arietinum* L.) to resist the phloem feeding *Aphis craccivora*. Transgenic Res. 18(4): 529–544.

Chisholm, S.T., G. Coaker, B. Day and B.J. Staskawicz. 2006. Host-microbe interactions: shaping the evolution of the plant immune response. Cell 124(4): 803–814.

Chougule, N.P. and B.C. Bonning. 2012. Toxins for transgenic resistance to hemipteran pests. Toxins 4(6): 405–429.

Chougule, N.P., H. Li, S. Liu, L.B. Linz, K.E. Narva, T. Meade and B.C. Bryony. 2013. Retargeting of the *Bacillus thuringiensis* toxin Cyt2Aa against hemipteran insect pests. Proc. Natl. Acad. Sci. USA 110(21): 8465–8470.

Christiaens, O. and G. Smagghe. 2014. The challenge of RNAi-mediated control of hemipterans. Curr. Opin. Insect Sci. 6: 15–21.

Coleman, A.D., M. Pitino and S.A. Hogenhout. 2014. Silencing of aphid genes by feeding on stable transgenic *Arabidopsis thaliana*. Methods Mol. Biol. 1127: 125–136.

Couty, A., G. de la Vina, S.J. Clark, L. Kaiser, M.H. Pham-Delegue and G.M. Poppy. 2001. Direct and indirect sublethal effects of *Galanthus nivalis* agglutinin (GNA) on the development of a potato-aphid parasitoid, *Aphelinus abdominalis* (Hymenoptera: Aphelinidae). J. Insect Physiol. 47(6): 553–561.

Cristofoletti, P.T., A.F. Ribeiro, C. Deraison, Y. Rahbe and W.R. Terra. 2003. Midgut adaptation and digestive enzyme distribution in a phloem feeding insect, the pea aphid *Acyrthosiphon pisum*. J. Insect Physiol. 49(1): 11–24.

Dawson, W.O. and S.Y. Folimonova. 2013. Virus-based transient expression vectors for woody crops: a new frontier for vector design and use. Annu. Rev. Phytopathol. 51: 321–337.

de Vos, M., W.Y. Cheng, H.E. Summers, R.A. Raguso and G. Jander. 2010. Alarm pheromone habituation in *Myzus persicae* has fitness consequences and causes extensive gene expression changes. Proc. Natl. Acad. Sci. USA 107(33): 14673–14678.

Deist, B.R., M.A. Rausch, M.T. Fernandez-Luna, M.J. Adang and B. Bonning. 2014. Bt toxin modification for enhanced efficacy. Toxins 6(10): 3005–3027.

Deng, F. and Z. Zhao. 2014. Influence of catalase gene silencing on the survivability of *Sitobion avenae*. Arch. Insect Biochem. Physiol. 86(1): 46–57.

Down, R.E., A.M.R. Gatehouse, W.D.O. Hamilton and J.A. Gatehouse. 1996. Snowdrop lectin inhibits development and decreases fecundity of the glasshouse potato aphid (*Aulacorthum solani*) when administered *in vitro* and via transgenic plants both in laboratory and glasshouse trials. J. Insect Physiol. 42(11-12): 1035–1045.

Down, R.E., L. Ford, S.D. Woodhouse, R.J. Raemaekers, B. Leitch, J.A. Gatehouse and A.M. Gatehouse. 2000. Snowdrop lectin (GNA) has no acute toxic effects on a beneficial insect predator, the 2-spot ladybird (*Adalia bipunctata* L.). J. Insect Physiol. 46(4): 379–391.

Down, R.E., E.C. Fitches, D.P. Wiles, P. Corti, H.A. Bell, J.A. Gatehouse and J.P. Edwards. 2006. Insecticidal spider venom toxin fused to snowdrop lectin is toxic to the peach-potato aphid, *Myzus persicae* (Hemiptera: Aphididae) and the rice brown planthopper, *Nilaparvata lugens* (Hemiptera: Delphacidae). Pest Manag. Sci. 62(1): 77–85.

Down, R.E., H.J. Matthews and N. Audsley. 2011. Oral activity of FMRFamide-related peptides on the pea aphid *Acyrthosiphon pisum* (Hemiptera: Aphididae) and degradation by enzymes from the aphid gut. Regul. Peptides 171(1-3): 11–18.

Emden, H. van and R. Harrington. 2007. Aphids as Crop Pests. CABI Publishing, London, UK.

Federici, B.A., B.C. Bonning and R.J. St. Leger. 2008. Improvement of insect pathogens as insecticides through genetic engineering. pp. 15–40. *In*: C. Hill and R. Sleator (eds.). PathoBiotechnology. Austin, TX, Landes Bioscience.

Fitches, E., S.D. Woodhouse, J.P. Edwards and J.A. Gatehouse. 2001. *In vitro* and *in vivo* binding of snowdrop (*Galanthus nivalis* agglutinin; GNA) and jackbean (*Canavalia ensiformis*; Con A) lectins within tomato moth (*Lacanobia oleracea*) larvae; mechanisms of insecticidal action. J. Insect Physiol. 47: 777–787.

Fitches, E., D. Wiles, A.E. Douglas, G. Hinchliffe, N. Audsley and J.A. Gatehouse. 2008. The insecticidal activity of recombinant garlic lectins towards aphids. Insect Biochem. Mol. Biol. 38(10): 905–915.

Folimonov, A.S., S.Y. Folimonova, M. Bar-Joseph and W.O. Dawson. 2007. A stable RNA virus-based vector for citrus trees. Virology 368(1): 205–216.

Francis, F., S. Vandermoten, F. Verheggen, G. Lognay and E. Haubruge. 2005. Is the (E)-beta-farnesene only volatile terpenoid in aphids? J. Appl. Entomol. 129(1): 6–11.

Fujii, Y., S. Tanaka, M. Otsuki, Y. Hoshino, H. Endo and R. Sato. 2013. Affinity maturation of Cry1Aa toxin to the *Bombyx mori* Cadherin-like receptor by directed evolution. Mol. Biotechnol. 54(3): 888–899.

Gammon, D.B. and C.C. Mello. 2015. RNA interference-mediated antiviral defense in insects. Curr. Opin. Insect Sci. 8: 111–120.

Gatehouse, A.M.R., G.M. Davison, J.N. Stewart, L.N. Gatehouse, A. Kumar, I.E. Geoghegan, A.N.E. Birch and J.A. Gatehouse. 1999. Concanavalin A inhibits development of tomato moth

(*Lacanobia oleracea*) and peach-potato aphid (*Myzus persicae*) when expressed in transgenic potato plants. Mo. Breeding 5(2): 153–165.

Gatehouse, A.M.R., N. Ferry, M.G. Edwards and H.A. Bell. 2011. Insect-resistant biotech crops and their impacts on beneficial arthropods. Phil. Trans. Royal Soc. B Biol. Sci. 366(1569): 1438–1452.

Gibson, R.W. and J.A. Pickett. 1983. Wild potato repels aphids by release of aphid alarm pheromone. Nature 302(5909): 608–609.

Goggin, F.L., L. Jia, G. Shah, S. Hebert, V.M. Williamson and D.E. Ullman. 2006. Heterologous expression of the Mi-1.2 gene from tomato confers resistance against nematodes but not aphids in eggplant. Mol. Plant Microbe Interact. 19(4): 383–388.

Gong, Y.H., X.R. Yu, Q.L. Shang, X.Y. Shi and X.W. Gao. 2014. Oral delivery mediated RNA interference of a carboxylesterase gene results in reduced resistance to organophosphorus insecticides in the cotton Aphid, *Aphis gossypii* Glover. PLoS One 9(8): e102823.

Gray, S. and F.E. Gildow. 2003. Luteovirus-aphid interactions. Annu. Rev. Phytopathol. 41: 539–566.

Hajeri, S., N. Killiny, C. El-Mohtar, W.O. Dawson and S. Gowda. 2014. Citrus tristeza virus-based RNAi in citrus plants induces gene silencing in *Diaphorina citri*, a phloem-sap sucking insect vector of citrus greening disease (Huanglongbing). J. Biotechnol. 176: 42–49.

Harper, S.M., R.W. Crenshaw, M.A. Mullins and L.S. Privalle. 1995. Lectin binding to insect brush border membranes. J. Econom. Ent. 88(5): 1197–1202.

Harrison, R.L. and D.L. Jarvis. 2006. Protein N-glycosylation in the baculovirus-insect cell expression system and engineering of insect cells to produce "mammalianized" recombinant glycoproteins. Adv. Virus Res. 68: 159–191.

Hilder, V.A., K.S. Powell, A.M.R. Gatehouse, L.N. Gatehouse, Y. Shi, W.D.O. Hamilton, A. Merryweather, C.A. Newell, J.C. Timans, W.J. Peumans, E. van Damme and D. Boulter. 1995. Expression of snowdrop lectin in transgenic tobacco plants results in added protection against aphids. Transgenic Res. 4(1): 18–25.

Hokanson, K.E., W.O. Dawson, A.M. Handler, M.F. Schetelig and R.J. St. Leger. 2014. Not all GMOs are crop plants: non-plant GMO applications in agriculture. Transgenic Res. 23(6): 1057–1068.

Hossain, M.A., M.K. Maiti, A. Basu, S. Sen, A.K. Ghosh and S.K. Sen. 2006. Transgenic expression of onion leaf lectin gene in Indian mustard offers protection against aphid colonization. Crop Sci. 46(5): 2022–2032.

James, C. 2013. Global status of commercialized Biotech/GM Crops: 2013. ISAAA Brief 46.

Jaubert-Possamai, S., G. Le Trionnaire, J. Bonhomme, G.K. Christophides, C. Rispe and D. Tagu. 2007. Gene knockdown by RNAi in the pea aphid *Acyrthosiphon pisum*. BMC Biotechnol. 7: 63. doi: 10.1186/1472-6750-7-63.

Jin, S.X., A. Kanagaraj, D. Verma, T. Lange and H. Daniell. 2011. Release of hormones from conjugates: chloroplast expression of β-glucosidase results in elevated phytohormone levels associated with significant increase in biomass and protection from aphids or whiteflies conferred by sucrose esters. Plant Physiol. 155: 222–235.

Jongsma, M.A. and J. Beekwilder. 2011. Co-evolution of insect proteases and plant protease inhibitors. Curr. Protein Pept. Sci. 12(5): 437–447.

Kanrar, S., J. Venkateswari, P.B. Kirti and V.L. Chopra. 2002. Transgenic Indian mustard (*Brassica juncea*) with resistance to the mustard aphid (*Lipaphis erysimi* Kalt.). Plant Cell Rep. 20(10): 976–981.

Karlova, R., M. Weemen-Hendriks, S. Naimov, J. Ceron, S. Dukiandjiev and R.A. de Maagd. 2005. *Bacillus thuringiensis* δ-endotoxin Cry1Ac domain III enhances activity against *Heliothis virescens* in some, but not all Cry1-Cry1Ac hybrids. J. Invertebr. Pathol. 88(2): 169–172.

Killiny, N., S. Hajeri, S. Tiwari, S. Gowda and L.L. Stelinski. 2014. Double-stranded RNA uptake through topical application, mediates silencing of five CYP4 genes and suppresses insecticide resistance in *Diaphorina citri*. PLoS One 9(10): e110536.

Kim, C.S., G. Schaible, L. Garrett, R. Lubowski and D. Lee. 2008. Economic impacts of the U.S. soybean aphid infestation: a multi-regional competitive dynamic analysis. Agric. Res. Econ. Rev. 37(2): 227–242.

Kim, Y.-S., S. Lim, K.-K. Kang, Y.-J. Jung, Y.-H. Lee, Y.-E. Choi and H. Sano. 2011. Resistance against beet armyworms and cotton aphids in caffeine-producing transgenic chrysanthemum. Plant Biotechnol. 28: 393–395.

Kroemer, J.A., B.C. Bonning and R.L. Harrison. 2015. Expression, delivery and function of insecticidal proteins expressed by recombinant baculoviruses. Viruses 7: 422–455.

Kunert, G., C. Reinhold and J. Gershenzon. 2010. Constitutive emission of the aphid alarm pheromone, (E)-β-farnesene, from plants does not serve as a direct defense against aphids. BMC Ecology 10: 23. doi: 10.1186/1472-6785-10-23.

Le-Feuvre, R.R., C.C. Ramirez, N. Olea and L. Meza-Basso. 2007. Effect of the antimicrobial peptide indolicidin on the green peach aphid *Myzus persicae* (Sulzer). J. Appl. Entomol. 131(2): 71–75.

Li, H., N.P. Chougule and B.C. Bonning. 2011. Interaction of the *Bacillus thuringiensis* delta endotoxins Cry1Ac and Cry3Aa with the gut of the pea aphid, *Acyrthosiphon pisum* (Harris). J. Invertebr. Pathol. 107(1): 69–78.

Linz, L.B., S. Liu, N.P. Chougule and B.C. Bonning 2015. *In vitro* evidence supports membrane alanyl aminopeptidase N as a receptor for a plant virus in the pea aphid vector. J. Virol. 89(22): 11203–11212.

Liu, S.J., S. Sivakumar, W.O. Sparks, W.A. Miller and B.C. Bonning. 2010. A peptide that binds the pea aphid gut impedes entry of Pea enation mosaic virus into the aphid hemocoel. Virology 401(1): 107–116.

Liu, S., N.P. Chougule, D. Vijayendran and B.C. Bonning. 2012. Deep sequencing of the transcriptomes of soybean aphid and associated endosymbionts. Plos One 7(9): e45161.

Loth, K., D. Costechareyre, G. Effantin, Y. Rahbé, G. Condemine, C. Landon and P. da Silva. 2015. New Cyt-like δ-endotoxins from *Dickeya dadantii*: structure and aphicidal activity. Sci. Rep. 5: 8791. doi: 10.1038/srep08791.

Mao, J. and F. Zeng. 2013. Plant-mediated RNAi of a gap gene-enhanced tobacco tolerance against the *Myzus persicae*. Transgenic Res. doi: 10.1007/s11248-013-9739-y.

Mao, Y.B., W.J. Cai, J.W. Wang, G.J. Hong, X.Y. Tao, L.J. Wang, Y.P. Huang and X.Y Chen. 2007. Silencing a cotton bollworm P450 monooxygenase gene by plant-mediated RNAi impairs larval tolerance of gossypol. Nat. Biotechnol. 25(11): 1307–1313.

Mao, Y.B., X.Y. Tao, X.Y. Xue, L.J. Wang and X.Y. Chen. 2011. Cotton plants expressing CYP6AE$_{14}$ double-stranded RNA show enhanced resistance to bollworms. Transgenic Res. 20(3): 665–673.

Mehlo, L., D. Gahakwa, P.T. Nghia, N.T. Loc, T. Capell, J.A. Gatehouse, A.M.R. Gatehouse and P. Christou. 2005. An alternative strategy for sustainable pest resistance in genetically enhanced crops. Proc. Natl. Acad. Sci. USA 102(22): 7812–7816.

Miles, P.W. 1989. Specific responses and damage caused by Aphidoidea. pp. 23–47. *In*: A.K. Minks and P. Harrewijn (eds.). Aphids. Their Biology, Natural Enemies and Control. Elsevier, Amsterdam.

Mutti, N.S., Y. Park, J.C. Reese and G.R. Reeck. 2006. RNAi knockdown of a salivary transcript leading to lethality in the pea aphid, *Acyrthosiphon pisum*. J. Insect Sci. 6: 1–7.

Nachman, R.J., M. Hamshou, K. Kaczmarek, J. Zabrocki and G. Smagghe. 2011a. Biostable and PEG polymer-conjugated insect pyrokinin analogs demonstrate antifeedant activity and induce high mortality in the pea aphid *Acyrthosiphon pisum* (Hemiptera: Aphidae). Peptides 34: 266–273.

Nachman, R.J., K. Mahdian, D.R. Nassel, R.E. Isaac, N. Pryor and G. Smagghe. 2011b. Biostable multi-Aib analogs of tachykinin-related peptides demonstrate potent oral aphicidal activity in the pea aphid *Acyrthosiphon pisum* (Hemiptera: Aphidae). Peptides 32(3): 587–594.

Nakasu, E.Y., M.G. Edwards, E. Fitches, J.A. Gatehouse and A.M. Gatehouse. 2014. Transgenic plants expressing omega-ACTX-Hv1a and snowdrop lectin (GNA) fusion protein show enhanced resistance to aphids. Front. Plant. Sci. 5: 673. doi: 10.3389/fpls.2014.00673.

Pitino, M. and S.A. Hogenhout. 2013. Aphid protein effectors promote aphid colonization in a plant species-specific manner. Mol. Plant Microbe In. 26(1): 130–139.

Pitino, M., A.D. Coleman, M.E. Maffei, C.J. Ridout and S.A. Hogenhout. 2011. Silencing of aphid genes by dsRNA feeding from plants. PLoS One 6(10): e25709.

Porcar, M., A.M. Grenier, B. Federici and Y. Rahbe. 2009. Effects of *Bacillus thuringiensis* delta-endotoxins on the pea aphid (*Acyrthosiphon pisum*). Appl. Environ. Microbiol. 75(14): 4897–4900.

Rahbe, Y., N. Sauvion, G. Febvay, W.J. Peumans and A.M.R. Gatehouse. 1995. Toxicity of lectins and processing of ingested proteins in the pea aphid *Acyrthosiphon pisum*. Entomol. Exp. Appl. 76(2): 143–155.

Rahbe, Y., C. Deraison, M. Bonade-Bottino, C. Girard, C. Nardon and L. Jouanin. 2003. Effects of the cysteine protease inhibitor oryzacystatin (OC-I) on different aphids and reduced performance of *Myzus persicae* on OC-I expressing transgenic oilseed rape. Plant Science 164(4): 441–450.

Rajamohan, F., O. Alzate, J.A. Cotrill, A. Curtiss and D.H. Dean. 1996. Protein engineering of *Bacillus thuringiensis* delta-endotoxin: mutations at domain II of CryIAb enhance receptor affinity and toxicity toward gypsy moth larvae. Proc. Natl. Acad. Sci. USA 93(25): 14338–14343.

Raps, A., J. Kehr, P. Gugerli, W.J. Moar, F. Bigler and A. Hilbeck. 2001. Immunological analysis of phloem sap of *Bacillus thuringiensis* corn and of the nontarget herbivore *Rhopalosiphum padi* (Homoptera : Aphididae) for the presence of Cry1Ab. Mol. Ecol. 10(2): 525–533.

Rawlings, N.D., D.P. Tolle and A.J. Barrett. 2004. Evolutionary families of peptidase inhibitors. Biochem. J. 378: 705–716.

Ribeiro, A.P.O., E.J.G. Pereira, T.L. Galvan, M.C. Picanço, E.A.T. Picoli, D.J.H. da Silva, M.G. Fári and W.C. Otoni. 2006. Effect of eggplant transformed with oryzacystatin gene on *Myzus persicae* and *Macrosiphum euphorbiae*. J. Appl. Entomol. 130(2): 84–90.

Rossi, M., F.L. Goggin, S.B. Milligan, I. Kaloshian, D.E. Ullman and V.M. Williamson. 1998. The nematode resistance gene Mi of tomato confers resistance against the potato aphid. Proc. Natl. Acad. Sci. USA 95(17): 9750–9754.

Sanahuja, G., R. Banakar, R.M. Twyman, T. Capell and P. Christou. 2011. *Bacillus thuringiensis*: a century of research, development and commercial applications. Plant Biotech. J. 9(3): 283–300.

Sapountzis, P., G. Duport, S. Balmand, K. Gaget, S. Jaubert-Possamai, G. Febvay, H. Charles, Y. Rahbé, S. Colella and F. Calevro. 2014. New insight into the RNA interference response against cathepsin-L gene in the pea aphid, *Acyrthosiphon pisum*: molting or gut phenotypes specifically induced by injection or feeding treatments. Insect Biochem. Mol. Biol. 51: 20–32.

Shakesby, A.J., I.S. Wallace, H.V. Isaacs, J. Pritchard, D.M. Roberts and A.E. Douglas. 2009. A water-specific aquaporin involved in aphid osmoregulation. Insect Biochem. Mol. Biol. 39(1): 1–10.

Stoger, E., S. Williams, P. Christou, R.E. Down and J.A. Gatehouse. 1999. Expression of the insecticidal lectin from snowdrop (*Galanthus nivalis* agglutinin; GNA) in transgenic wheat plants: effects on predation by the grain aphid *Sitobion avenae*. Mol. Breeding 5(1): 65–73.

Sylvester, E.S. 1989. Viruses transmitted by aphids. *In*: A.K. Minks and P. Harrewijn (eds.). Aphids; their Biology, Natural Enemies and Control. Elsevier, Amsterdam.

Terenius, O., A. Papanicolaou, J.S. Garbutt, I. Eleftherianos, H. Huvenne, S. Kanginakudru, M. Albrechtsen, C. An, J.-L. Aymeric, A. Barthel, P. Bebas, K. Bitra, A. Bravo, F. Chevalier, D.P. Collinge, C.M. Crava, R.A. de Maagd, B. Duvic, M. Erlandson, I. Faye, G. Felföldi, H. Fujiwara, R. Futahashi, A.S. Gandhe, H.S. Gatehouse, L.N. Gatehouse, J.M. Giebultowicz and I. Gómez. 2011. RNA interference in Lepidoptera: an overview of successful and unsuccessful studies and implications for experimental design. J. Insect Physiol. 57(2): 231–245.

Vachon, V., R. Laprade and J.L. Schwartz. 2012. Current models of the mode of action of *Bacillus thuringiensis* insecticidal crystal proteins: a critical review. J. Invertebr. Pathol. 111(1): 1–12.

Vandenborre, G., G. Smagghe and E.J.M. Van Damme. 2011. Plant lectins as defense proteins against phytophagous insects. Phytochemistry 72(13): 1538–1550.

Vandermoten, S., M.C. Mescher, F.F.E. Haubruge and F.J. Verheggen. 2012. Aphid alarm pheromone: an overview of current knowledge on biosynthesis and functions. Insect Biochem. Mol. Biol. 42(3): 155–163.

Vijayendran, D., P.M. Airs, K. Dolezal and B.C. Bonning. 2013. Arthropod viruses and small RNAs. J. Invertebr. Pathol. 114(2): 186–195.

Walters, F.S. and L.H. English. 1995. Toxicity of *Bacillus thuringiensis* delta-endotoxins toward the potato aphid in an artificial diet bioassay. Entomol. Exp. Appl. 77(2): 211–216.

Walters, F.S., C.M. Stacy, M.K. Lee, N. Palekar and J.S. Chen. 2008. An engineered chymotrypsin/cathepsing site in domain I renders *Bacillus thuringiensis* Cry3A active against western corn rootworm larvae. Appl. Environ. Microbiol. 74(2): 367–374.

Wang, C. and R.J. St. Leger. 2007. Expressing a scorpion neurotoxin makes a fungus hyperinfectious to insects. Nature Biotechnol. 25(12): 1455–1456.

Wang, D., Q. Liu, X. Li, Y. Sun, H. Wang and L. Xia. 2015. Double-stranded RNA in the biological control of grain aphid (*Sitobion avenae* F.). Funct. Integr. Genomics 15(2): 211–223.

Wang, E., R. Wang, J. DeParasis, J.H. Loughrin, S. Gan and G.J. Wagner. 2001. Suppression of a P450 hydrolylase gene in plant trichome glands enhances natural-product-based aphid resistance. Nature Biotechnol. 19: 371–374.

Wang, W., L. Luo, H. Lu, S. Chen, L. Kang and F. Cui. 2015. Angiotensin-converting enzymes modulate aphid-plant interactions. Sci. Rep. 5: 8885. doi:10.1038/srep08885.

Wang, Z.Y., K.W. Zhang, X.F. Sun, K.X. Tang and J.R. Zhang. 2005. Enhancement of resistance to aphids by introducing the snowdrop lectin gene gna into maize plants. J. Biosci. 30(5): 627–638.

Whyard, S., A.D. Singh and S. Wong. 2009. Ingested double-stranded RNAs can act as species-specific insecticides. Insect Biochem. Mol. Biol. 39(11): 824–832.

Will, T. and A. Vilcinskas. 2015. The structural sheath protein of aphids is required for phloem feeding. Insect Biochem. Mol. Biol. 57: 34–40.

Wu, C., C.A. Avila and F.L. Goggin. 2015. The ethylene response factor Pti5 contributes to potato aphid resistance in tomato independent of ethylene signalling. J. Exp. Bot. 66(2): 559–570.

Xu, L., X. Duan, Y. Lv, X. Zhang, Z. Nie, C. Xie, Z. Ni and R. Liang. 2014. Silencing of an aphid carboxylesterase gene by use of plant-mediated RNAi impairs *Sitobion avenae* tolerance of Phoxim insecticides. Transgenic Res. 23(2): 389–396.

Yu, X., H.D. Jones, Y. Ma, G. Wang, Z. Xu, B. Zhang, Y. Zhang, G. Ren, J.A. Pickett and L. Xia. 2012. (E)-beta-farnesene synthase genes affect aphid (*Myzus persicae*) infestation in tobacco (*Nicotiana tabacum*). Funct. Integr. Genomics 12(1): 207–213.

Yu, X., G. Wang, S. Huang, Y. Ma and L. Xia. 2014. Engineering plants for aphid resistance: current status and future perspectives. Theor. Appl. Genet. 127(10): 2065–2083.

Zhang, M., Y. Zhou, H. Wang, H.D. Jones, Q. Gao, D. Wang, Y. Ma and L. Xia. 2013. Identifying potential RNAi targets in grain aphid (*Sitobion avenae* F.) based on transcriptome profiling of its alimentary canal after feeding on wheat plants. BMC Genomics 14(1): 560. doi:10.1186/1471-2164-14-560.

13

Aphid Techniques

Torsten Will

Introduction

Aphids feed from sieve elements located deep inside plants which makes it challenging to observe details of the feeding process. The electrical penetration graph technique solves this problem and today this method is widely used to study phenomena such as plant resistance, the effects of insecticides on aphid feeding, and the relevance of individual salivary proteins on aphid–plant interactions. Further techniques have been developed to study the uptake and transmission of phytopathogenic viruses. Artificial feeding was developed in 1930 and was improved in the 1960s so that the nutritional needs of aphids could be determined. Aphids can also be used as tools to access the sieve elements of plants, e.g., to study the composition of sieve element sap and to measure electrical signals in the plant which can be triggered, e.g., by caterpillar feeding.

The Electrical Penetration Graph Technique

The electrical penetration graph (EPG) technique was introduced by McLean and Kinsey in 1964, to study the impact of saliva that aphids secrete into their host plants. Two different EPG monitors are currently in use, one based on alternating current (AC) and the other on direct current (DC). The signals recoded by the two systems were initially distinct, i.e., the AC system did not detect plant membrane potentials (Tjallingii 2000), but both systems are now comparable with regard to the recorded signals (Backus and Bennet 2009). We therefore focus on studies using the DC EPG monitor and use the term EPG monitor to refer to this particular setup, which comprises a voltage source and a fixed input resistor. An aphid (connected by a thin gold wire with a length of 1–2 cm and a diameter of ~ 20 μm via a pre-amplifier to the EPG

Institute for Insect Biotechnology, Justus-Liebig-University, Heinrich-Buff-Ring 26-32, 35394 Giessen, Germany.

E-mail: torsten.will@agrar.uni-giessen.de

monitor) and its host plant (connected by a copper wire in the soil) are made part of an electrical circuit where they together act as a variable resistor. The indirect integration of the plant using a "substrate electrode" in the soil makes it unnecessary to wound the plant, and the thin wire which is attached by electrically conductive adhesives to the aphid's dorsum allows the aphid a certain degree of mobility. A vacuum device to immobilize aphids during wiring is described by van Helden and Tjallingii (2000). The aphid stylet acts as a switch that closes the electrical circuit when penetrating the plant tissue.

Fluctuations in the variable resistor can arise from several sources (Walker 2000, Tjallingii 2000). Fluids in the apoplast and inside plant cells contain different amounts of inorganic and organic ions, meaning that the environmental surrounding of the stylet tip influences electrical conductivity (reciprocal resistance) in the system. The condition of the gel saliva, secreted during stylet movement through the apoplast, also affects electrical resistance. Although the saliva is a fluid directly after secretion, specific gel saliva proteins are subsequently oxidized, leading to formation of a solid but flexible sheath cap around the stylet tip (McLean and Kinsey 1965, Will et al. 2012). The stylet then advances through the hardening gel saliva material and the process is repeated. Whereas liquid saliva is presumed to have a low electrical resistance, hardened salivary material may have the opposite property. Opening and closing of the preciberial valve, which is located in the aphid's head, controls the flow of sap through the esophagus and also influences resistance, because chitin has a low conductivity.

The last component of the electrical circuit is an additional voltage source (electron motive force) comprising different types of potentials. First, streaming potentials occur when fluids that contain inorganic and organic ions are moved through a capillary. Adjacent to the inner surface of the nutrition channel and the salivary channel in the stylet, the charge-neutrality of the liquid is violated due to the presence of the electrical double layer: a thin layer of counter ions attracted by the charged surface. Because the flow rate of fluids is highest at the center of a capillary, cations and anions are separated and create a potential difference. Fluids with differing ionic strengths may generate different streaming potential amplitudes. As soon as the flow stops, the streaming potentials disappear. Second, electrode potential occurs when two electrodes (the substrate and insect electrodes in this case) are connected by an electrolyte, when the aphid stylet penetrates the plant. Ions are transferred across the interface and desorbed by one electrode and absorbed by the other. Because the electrode potential is stable and is not influenced by aphid activity, it has no impact on the recorded EPG signal. Third, the most significant additional voltage source is the membrane potential of the plant cells, which for sieve elements is approximately -170 mV in curcubits and -117 mV in legumes (Hafke et al. 2005). During EPG recordings, this is observed as a voltage drop during the penetration of the epidermis, mesophyll and parenchyma cells, as well as the sieve elements, by an aphid stylet. There is no drop in voltage when the xylem is penetrated because the xylem is not composed of living cells and meaning that there is no membrane potential. The electrical activity of the ciberial pump muscles does not contribute to the EPG electrical signal but the activity of the salivary pump muscles is recorded (Tjallingii 1978, 1988). A detailed physical background of the EPG technique is provided by Tjallingii (2000).

All these changes in conductivity generate waveforms that can be identified in EPG recordings. Nine different waveforms have been identified for aphids based on the patterns of voltage fluctuation and the repetition rate (Hz) of the electrical signals (Tjallingii 1995). A zero line in the EPG is detected when there are no penetration activities and this is termed waveform np (non-penetration). The initiation of stylet penetration induces a signal with high amplitude and frequency (5–10 Hz) but with a duration of only a few seconds (waveform A). Subsequent stylet movement forwards and backwards through the apoplast generates regular waves with a lower signal amplitude (waveform C) accompanied by the secretion of gel saliva (waveform B). In most cases, the detailed analysis of those initial waveforms is not needed, so waveforms A–C are therefore grouped together as waveform C (the pathway phase). The repetition rate (Hz) of this waveform is mixed. Waveform C is regularly interrupted by so-called potential drops (pd), corresponding to the penetration of epidermis, cortex, mesophyll and parenchyma cells (Tjallingii 1985). Potential drops commence with a sudden decrease in voltage lasting 5–15 s and end with a rapid increase in voltage to the original level (Tjallingii 1985). Where necessary, potential drops can be analyzed in detail and are divided into three sub-waveforms: II-1, II-2 and II-3. Waveform II-1 represents the secretion of saliva and II-3 is associated with the ingestion of sap from penetrated cells (Martin et al. 1997), but II-2 does not appear to match a specific behavior so far. Another waveform that is observed during stylet movement in the apoplast is waveform F, which represents penetration problems. It has a saw tooth like appearance and can easily be identified by its high repetition rate of 11–19 Hz (Tjallingii 1987). Penetration problems normally occur during waveform C and last from a few seconds up to several minutes. Sieve element penetration starts with a potential drop but after approximately 5 s the original voltage level does not return and the observed waveform switches to E1 (2–4 Hz), indicating the secretion of watery saliva into a sieve element, and subsequently to E2 (4–7 Hz), correlating with the ingestion of sieve element sap mixed with the secretion of saliva (Prado and Tjallingii 1994). Although the repetition rates of these waveforms overlap, they can be distinguished by characteristic peaks (Prado and Tjallingii 1994). Waveform E1 is often but not necessarily followed by E2 whereas E2 is always preceded by E1. Waveform E2 can be interrupted by E1 on resistant plants (van Helden and Tjallingii 1993). The interruption of E2 by E1 has also been observed on susceptible plants after leaf tip burning (Will et al. 2009) or caterpillar feeding (Salvador-Rectalà et al. 2014). Whereas intracellular salivation during potential drops (waveform II-1) last only a few seconds (Martin et al. 1997), E1 events may last for more than one minute. Sieve element penetration usually ends with an increase in voltage and a waveform change from E2 to C. In contrast to the ingestion of sieve element sap, the uptake of xylem components (waveform G, 4–6 Hz) is an active process, driven by the ciberial sucking pump.

Computerized recording of EPG signals was achieved more than a decade ago and software packages such as Stylet+ (EPG Systems, Wageningen, NL) now provide standard tools for signal recording and analysis. Manual EPG recordings are time consuming because 15–20 replicates are required per treatment and the recommended recording time is 8 h per aphid to detect all potential variations in behavior. However, shorter recording times are suitable for some experiments, e.g., the response of aphids to physical stimuli (Will et al. 2008). Subsequent parameter calculation can be achieved by processing the recorded data using tools such as EPG-Calc (Giordanengo

2014) and Excel workbook for automatic parameter calculation (Sarria et al. 2009). The latter allows the analysis of up to 119 different parameters representing events such as plant acceptability, virus transmission and phloem accession. Data from the EPG workbook can be processed using statistical analysis software. There have been several attempts to automate the time-consuming analysis of EPG waveforms using Fourier transform analysis, but these were inappropriate particularly due to signal variability according to the species of aphid and the data quality (Tjallingii 1988, Rahbe et al. 2000). Recent attempts at automation have been more successful, e.g., the software packages APHID-AUTOEPG (Prüfer et al. 2014) and A2EPG (Adasme-Carreno et al. 2015) address the issue of signal variability and apply promising tools for the analysis of electrically monitored aphid feeding behavior. However, the user must check the analysis, edit the results and interpret the waveforms if the signal quality is poor. Nevertheless, the analysis of a complete 4-h run does not take more than 15 min including corrections if the user is familiar with the waveforms and has adequate experience with the technique. Unfortunately, both software tools struggle to identify waveforms F (stylet penetration problems), E1 (secretion of saliva into sieve elements) and E2 (ingestion from sieve elements). Whereas E1 and E2 can be difficult to distinguish even by visual inspection, mathematical models have particular trouble with the identification of waveform F. Current software packages are only suitable for aphid EPGs but adaptations for use with whiteflies and planthoppers are under development.

Since the introduction of the EPG technique (McLean and Kinsey 1964) it has been widely used to study the interaction between aphids and their host plants (e.g., Prado and Tjallingii 1994, Martin et al. 1997, Will et al. 2007, 2009, Will and Vilcinskas 2015). The first study that focused on the identification of plant resistance to aphids (Mentink et al. 1984) and the later studies that followed (e.g., Caillaud et al. 1995, Chen et al. 1997) showed that the EPG technique has the potential to reveal the localization of plant resistance to specific tissues (e.g., van Helden and Tjallingii 1993, Alvarez et al. 2006). Plant resistance towards other sap-sucking insect pests has also been investigated by using the EPG technique, including psyllids (e.g., Bonani et al. 2010), white flies (e.g., Jiang et al. 2001) and planthoppers (e.g., Kimmins 1989). In addition to the identification of natural plant resistance (Caillaud et al. 1995), the EPG technique has also been used to study the impact of insecticides (e.g., Nauen 1995, Harrewijn and Kayser 1997, Daniels et al. 2009) and transgenic plants expressing snowdrop lectin (Rao et al. 1998) and *Bacillus thuringiensis* toxins (Bernal et al. 2002) respectively on the behavior of such insects. Furthermore, combined with RNAi-based gene silencing, the EPG technique has been used to examine the function of effectors and salivary proteins during the feeding process (e.g., Mutti et al. 2006, 2008, Wang et al. 2015, Will and Vilcinskas 2015).

Beside the great potential of the EPG technique including its broad application spectrum, the experimental setup brings with it the risk to directly affect aphid behaviour. The physical wiring of the aphid limits its mobility and thus restricts the selection of feeding sites, which may lead to enhanced probing (stylet movement through the apoplast) as observed for the aphid species *Aphis fabae* over a period of 8 h (Prado and Tjallingii 1999). Furthermore, wired aphids cannot escape from unsuitable feeding sites or from non-host plants (Tjallingii 1986, Caillaud et al. 1995). These experimental influences affecting aphid behavior must be taken into consideration when transferring experimental data to natural systems.

Aphids as Bio-Electrodes

Sieve tubes are composed of longitudinally arranged sieve element modules (Esau 1969) and serve as transport conduits for photoassimilates, amino acids, proteins, mRNAs and phytohormones. In addition, they appear to act as a kind of nerve system that extends through the plant and conducts long-distance electrical signals (Lough and Lucas 2006). These so-called electrical potential waves are composed of variation potentials (VP) and action potentials (AP), which facilitate rapid communication between distant plant organs and are triggered by events such as wounding, heat, cold and electrical shocks (Fromm and Spanswick 1993, Rhodes et al. 1996, Mancuso 1999, Furch et al. 2007).

Different technical approaches are available to record long-distance electrical signals in plants. The measurement of surface potentials by attaching electrodes to the leaf surface prevents additional wounding of the plant (Mousavi et al. 2013). Although, this makes the method gentle and protective, the recorded signals are complex because many different electrical signals from several sieve tubes and other cell types are registered in parallel. Other techniques such as "blind piercing" use electrodes that are inserted into the plant tissue without knowing the cell type they penetrate. This technique is mainly used with trees, where long electrodes pierce the stem to a selected depth such as 5 mm (Oyarce and Gurovich 2010). The precise measurement of electrical signals in sieve tubes can be achieved by substomatal measurements, in which electrodes are inserted through open stomata (e.g., Felle et al. 2000, Felle and Zimmermann 2007, Zimmermann et al. 2009), and intracellular measurements, in which electrodes penetrate individual sieve elements (Hafke et al. 2005). Both approaches are technically challenging and the latter requires the removal of several cell layers to reach the sieve tubes, thus resulting in significant tissue damage and the induction of wounding-related chemical, physical and electrical signaling.

The use of aphids as bio-electrodes has several advantages compared to the methods described above. Aphids penetrate the cuticle and maneuver their stylet through the apoplast without significant damage to adjacent cells (Tjallingii and Hogen Esch 1993). The stylet then penetrates the sieve elements (Prado and Tjallingii 1994) to establish long-term feeding sites. The first studies using aphid bio-electrodes for the detection and measurement of electrical signals in sieve elements were based on the use of dissected stylets (stylectomy) in combination with microelectrodes, whereat latter were placed into the exuding sieve element sap at the stylet end (Wright and Fisher 1981, Fromm and Fei 1998). Intact aphids were first used as living bio-electrodes by Furch et al. (2010) in combination with the DC-EPG technique described above, using the aphid species *Macrosiphum euphorbiae* feeding on *Cucurbita maxima*. Furch et al. (2009) described the recorded electrical signals as "electrical potential waves", a merging of a rapid action potential and a slower variation potential (Stankovic et al. 1998, Hafke et al. 2009). The profile of the electrical signal, induced by leaf tip burning, showed a steep transient depolarization and a long lasting repolarization when observed by apoplastic and intracellular electrical recordings (Furch et al. 2009). Electrical signals recorded by DC-EPG had a similar shape but with a smaller amplitude and a shorter length. The arrangement of up to four aphids in a row along the midrib revealed that the electrical signal declines in intensity and velocity along the pathway (Furch et al. 2009). Such intracellular multi-point electrical recording in plants are much more difficult with regular glass electrodes. The decline in the electrical potential wave between the points of recording seems to contradict the definition of an action potential, following the all-or-none principle (Zawadzki et al. 1991, Dziubinska 2003). However, Furch et al. (2009) disagree with this definition and argue that action potentials can fade along the sieve tubes, particularly due to the high symplasmic continuity with surrounding phloem parenchyma cells in cucurbits (Kempers et al. 1998). The caterpillar *Pieris brassicae* induces similar DC-EPG signals, observed with the aphid species *Brevicoryne brassicae* (Salvador-Rectalà et al. 2014) as previously observed for leaf-tip burning (Furch et al. 2009).

One potential drawback of this approach is that aphid salivary components secreted into penetrated sieve elements (see Chapter 11 "Function of aphid saliva in aphid-plant interaction" in this volume) may influence sieve element physiology, e.g., by affecting the activity of ion channels on the plasma membrane. The calcium-binding capacity of salivary proteins from the aphid species *Megoura viciae* and *Acyrtosiphon pisum* (Will et al. 2007, Wang et al. 2015) could also influence the propagation and properties of sieve element action potentials because calcium influx into the sieve element lumen is a key feature of signal propagation (Demidchik and Maathuis 2007, White 2009).

Stylectomy

The aphid stylet has two outer mandibular and two inner maxillary subunits, the latter forming a salivary channel and a nutrition channel. The part of the stylet that is pushed into the plant tissue by the aphid is enveloped by a salivary sheath, while the external part is enclosed in the aphid labium. Aphids feed passively because the high turgor inside the sieve tubes pushes the cell sap into the nutrition channel, e.g., *Vicia faba* = 3.9–4.4 bar (Fisher 1978), *Salix babylonica* = 5.1–9.3 bar (Wright and Fisher 1980) and *Hevea brasiliensis* = 8–12 bar (An et al. 2014). The high turgor inside sieve tubes is explained by the Münch hypothesis, which assumes that a gradient of photoassimilates is build up between source (mature leafs) and sink (e.g., roots), leading water influx (sink)/eflux (source) that results in mass flow between both (Münch 1930). Kennedy and Mittler (1953) introduced the stylectomy technique based on the Münch (1930) pressure flow hypothesis and previous observations by Yust and Fulton (1943) that sieve tubes sap exudes from broken coccid stylets. Kennedy and Mittler (1953) used the giant willow aphid *Tuberolachnus salignus* in this initial study because the large stylet (5.0–5.8 mm) could be cut off with a razor blade, allowing them to collect sieve tube sap from *Salix* spp. for a period of up to four days. von Soest and de Meester-Manager Cats (1956) introduced custom-made scissors to cut aphid stylets whereas Fisher and Frame (1984) used surgical micro-scissors that were modified by reducing the thickness of the blades. However, scissors cannot be used to cut the stylets of smaller aphids and anaesthetized aphids are often required. Furthermore, the large scissor blades and the robust manual movements can alarm the aphids leading them to abort ingestion and attempt to escape (Fisher and Frame 1984).

To address these issues, a radio frequency (RF) microcautery system has been developed to dissect the mouth-parts of small aphids, in which a fine tungsten needle connected to the output of the device is maneuvered with a micromanipulator directly beside the labium of an aphid sitting on the leaf surface (Downing and Unwin 1977). The application of high frequency oscillation to biological tissues causes heating, and the temperature (ΔQ) produced per tissue volume (ΔV) is proportional to the specific resistor (ρ) and the square of the current density (j) as shown in Equation 1.

Equation 1: $\Delta Q = \rho * j^2 * \Delta V * \Delta t$

The first microcautery units were low-frequency devices with power levels of 3–5 W and these often caused lesions that could block the stylet nutrition channel. Unwin (1978) suggested that such issues could be overcome by using devices with a higher frequency (27 MHz) and power up to 15 W. Fisher and Frame (1984) developed an inexpensive custom-made version of a RF microcautery instrument operating at 27 Hz and 5 W based on a citizens' band radio transmitter and other readily available components, allowing variable output power with pulse lengths of up to 0.4 s. Circuit diagrams for such instruments have been published (Downing and Unwin 1977, Unwin 1978, Fisher and Frame 1984). The devices are used with custom-made fine tungsten needles with adjustable tip thickness as explained by Brady (1965), who used such needles for insect dissection.

Another method for cutting aphid stylets using a solid-state ruby laser (Maiman 1960) was developed by Barlow and McCully (1972). The principal emission wavelength of this laser (694.3 nm) is strongly absorbed by melanin, so melanin-rich tissues heat up rapidly and vaporize. Ruby lasers have been largely superseded by more efficient systems such as the yttrium-aluminum garnet (YAG) laser which has been used for stylectomy procedures with aphids (Fisher and Frame 1984), leafhoppers and planthoppers (Kawabe et al. 1980). However, Fisher and Frame (1984) argue that laser-based stylectomy suffers from disadvantages such as the fixed light path, which makes it difficult to maneuver the aphid into the light path, the need to anaesthetize aphids with CO_2, the slow cutting process that often yields to only a few exuding stylets, and the tendency to burn surrounding plant tissue. The last point is a major concern because wounding triggers plant defense responses (Schilmiller and Howe 2005) that can affect the composition of sieve tube sap (Kehr 2006).

Whichever stylectomy method is used, it allows the collection of so-called "pure sieve-element sap", whereas facilitated exudation using detached leaves is more prone to contamination. The latter involves placing leaves in a collection buffer containing the chelator EDTA to prevent calcium-dependent sieve tube occlusion by proteins and callose (King and Zeevaart 1974). This is a simple method, but the removal of calcium ions by EDTA increases the permeability of the cell membranes and allows the sieve element sap to be compromised by the content of surrounding cells and even the apoplast (Girousse et al. 1991, van Bel and Gaupels 2004). Although stylectomy can be used to collect uncontaminated sap the collection process must be carefully monitored because the duration and rate of exudation differ significantly among plant species. For example, long periods of exudation can be achieved with the aphid species

T. salignus feeding on willow (Kennedy and Mittler 1953) whereas the duration of exudation on *V. faba* was rather brief (Fisher and Frame 1984). Exudation may be prevented by a blockage in the nutrition channel inside the aphid stylet due to the coagulation of sieve element occlusion proteins in the absence of regular secretions of watery saliva (Tjallingii and Cherqui 1999). Furthermore, exuding stylets tend to produce rather small volumes of sap for analysis, generally less than 100 nl, and less sap is exuded by monocotyledonous plants such as barley compared to dicotyledonous plants (Fisher and Frame 1984). Gaupels et al. (2008) reported exudation volumes of approximately 220 nl per stylet in aphid species *Rhopalosiphum padi*. When planthoppers were used instead of aphids, Kawabe et al. (1980) collected up to 600 nl from each exuding stylet.

To improve aphid stylectomy, Gaupels et al. (2008) arranged up to 30 barley plants in parallel with a maximum of 600 aphids (5–10 per cage) and achieved a final number of 46 exuding stylets which allowed them to obtain approximately 10 µl of sieve element sap within six h. To prevent contamination and evaporation of the small exudation volume, which would otherwise make the accurate measurement of solute concentrations impossible, they used small plastic rings that were glued to the leaf surface and filled with water-saturated silicon oil after stylectomy. The rings also doubled as aphid cages. Exuding sieve element sap can easily be collected with glass capillaries or with a pipette in air or under oil.

The sieve element sap collected by stylectomy has been analyzed using a diverse range of techniques, beginning with the use of potassium-selective microelectrodes to measure potassium concentrations (Thomas 1978) and freezing-point depression with nanoliter-sized droplets to determine the osmotic potential of the exudate (Wright and Fisher 1983, Fisher 1983). Fisher and Frame (1984) used the double-isotope dansylation procedure (Macnichol 1983) to analyze amino acids, whereas micellar electrokinetic chromatography with laser-induced fluorescence was used for this purpose by Zhu et al. (2005). Hafke et al. (2005) reported the analysis of sieve element sap using pH-sensitive microelectrodes. Proteins in the sap have been analyzed by one and two dimensional polyacrylamide gel electrophoresis as well as mass spectrometry (Aki et al. 2008, Gaupels et al. 2008). The presence of transporter mRNAs in exuded sap has been confirmed by PCR (Sasaki et al. 1998, Doering-Saad et al. 2002, Gaupels et al. 2008).

In combination with stylectomy, microelectrodes can also be used to measure the physical aspects of phloem physiology such as sieve element turgor pressure (Wright and Fisher 1980) and sieve element membrane potentials (Wright and Fisher 1981, Fromm and Fei 1998). The use of aphid stylets to access sieve elements prevents the further wounding of plants, which is inevitable when using microelectrodes that penetrate the sieve element lumen (Hafke et al. 2005). Even so, it cannot be ruled out that aphid feeding also directly influences sieve element physiology.

Artificial Feeding

Aphids passively ingest nutrients from sieve elements but also actively draw water from xylem vessels (Spiller et al. 1990). Active sucking is required (Malone et al. 1999) because water flow is driven by evaporation through leaves, so the pressure within the xylem vessels is negative and thus much lower than in the phloem (Pockman et al. 1995). This active sucking behavior is the basis for aphid feeding on artificial diets, which was introduced by Hamilton (1930) and later improved (Hamilton 1935) leading to many studies in which aphids were fed on gels and (sometimes pressurized) liquids (e.g., Day and Irzykiewicz 1953, Lindemann 1948, Maltais 1952, Pletsch 1937). Hamilton (1930) and others (e.g., Schmidt 1959, Rochow 1960) focused on the uptake and transmission of phytopathogenic viruses. Evidence for the ingestion of artificial diets was obtained from short increases in survival (Maltais 1952), the uptake of radioactive compounds (Day and Irzykiewicz 1953), or dyes (Schmidt 1959), and the transmission of a persistent virus (Rochow 1960). The first study that focussed on the nutritional needs of aphids was conducted by Mittler and Dadd (1962) who presented aphids a complex artificial diet containing neutral red or ^{32}P as markers to confirm ingestion. Diets containing 10–20% sucrose led to a higher rate of feeding in the aphid species *Myzus persicae* compared with previous studies. Mittler and Dadd (1962, 1963b) were also the first to rear aphid offspring until maturity, which was achieved by presenting a liquid diet (18% sucrose with a mixture of 10 vitamins, cholesterol, mineral salts, and 20 amino acids and amides) retained under a membrane of stretched Parafilm M, 10–20 μm thick. A similar sucrose concentration is found in the phloem sap of many plants (e.g., Fukumorita and Chino 1982). The vitamin mixture was devised for the nutrition of locusts (Dadd 1961) and the amino acid mixture was based on the analysis of pea juice (Auclair et al. 1957). With this diet, the 50 per cent survival time of apterous adults increased to 13 days, compared to 6.5 days when the diet was solely 18% sucrose and 3.5 days when the aphids were provided with water. Furthermore, more than 50 per cent of newborn nymphs survived two weeks, molting at least twice, and most nymphs surviving for 16–17 days developed into dwarfed fourth-instar nymphs or adults.

The availability of a diet that allowed nymphs to grow into diminutive adults made it possible to investigate the specific nutritional requirements of *M. persicae* by removing single components (Mittler and Dadd 1962, 1963a,b, Dadd and Mittler 1965). Trace metals (iron, copper, manganese, and zinc) were later added to the artificial diets (Dadd and Mittler 1966, Ehrhardt 1968, Akey and Beck 1971, 1972). By removing the chelator EDTA, it became possible to rear several generations of *A. pisum* on an artificial diet developed for this species (Akey and Beck 1972). However, only Dadd and Mittler (1966) reported the rearing of a permanent culture of *M. persicae*.

Artificial feeding has also been used to study the impact of nitrogen starvation on free amino acids in *M. persicae* (Strong 1964), for the collection of aphid saliva from different aphid species (e.g., Cherqui and Tjallingii 2000, Will et al. 2007, Carolan et al. 2009), for choice chamber experiments (Hewer et al. 2010) and to simulate changes in the physical properties of the sieve elements, e.g., pressure, and investigate their impact on aphid behavior (Will et al. 2008). Hewer et al. (2010, 2011) were the first to test the effects of different sugars (fructose, glucose, sucrose, raffinose, sorbitol, galactose and mannose) and pH values (pH 5–8) and thus introduced new artificial diets to study aphid orientation inside the plant. For the collection of aphid saliva, a simplified diet was developed by Cherqui and Tjallingii (2000) that only contained 15% sucrose and three amino acids (serine, methionine and aspartic acid) with a pH of 6.8. Using this diet, the authors were able to collect salivary proteins and separate them by electrophoresis. Will et al. (2007) increased the pH of this diet to 7.2 to more closely resemble the pH of sieve element sap from *V. faba* and other plant species (Hafke et al. 2005). For the collection of gel saliva, Will et al. (2012) created a diet that mimics the apoplast and thus prevents the secretion of watery saliva.

As well as artificial diets, feeding chambers have been designed for different experimental needs. Mittler and Dadd (1963a) developed a two-chamber device with an upper chamber containing 2–5 ml of artificial diet sealed on both sides with Parafilm, allowing aphids in the lower chamber to feed while preventing evaporation. Groups of 20 aphids are briefly starved and then placed in the lower chamber, with the base covered with gauze. This setup was simplified by replacing the upper diet chamber with a Parafilm sachet containing 0.1–0.2 ml artificial diet, making it more suitable for routine experiments on a larger scale (Mittler and Dadd 1964). Similar setups are still used today for the collection of aphid saliva (Carolan et al. 2009) while other researchers use Perspex blocks milled with a shallow bath (9 cm radius, 1 mm depth) containing 3 ml artificial diet covered with Parafilm (Will et al. 2007). Aphids are placed in a lower chamber formed by a metal ring and covered with Parafilm. In contrast to small chambers with Parafilm sachets that harbor approximately 100 aphids, the Perspex feeding chamber can house up to 1000 of those insects to facilitate the large-scale collection of aphid saliva. Choice chambers (Hewer et al. 2010) comprise eight single baths arranged around a central field and individual artificial diets (covered with Parafilm) can be presented in opposite baths. To prevent escape, aphids are trapped inside a tightly-fitting plastic frame covered with transparent plastic film. Due to the size of the choice chamber, aphids are located on the upper side. For testing the impact of chemical and physical changes on aphid behavior, a flow-through chamber was developed connected to an EPG device (Will et al. 2008). This is a Perspex block with a milled conduit (diameter 0.15 cm), a slot in the middle of the conduit representing an artificial sieve tube, and stainless steel tube connectors. Instead of Parafilm, the milled slot is covered with transparent plastic film that better resists the system pressure of 0.2–0.25 MPa.

Keywords: aphid behaviour, aphid-plant interaction, artificial diet, electrical penetration graph technique, nutrition, plant physiology, sieve element

References

Adasme-Carreno, F., C. Munoz-Gutierrez, J. Salinas-Cornejo and C.C. Ramirez. 2015. A2EPG: a new software for the analysis of electrical penetration graphs to study plant probing behaviour of hemipteran insects. Comput. Electron. Agric. 113: 128–135.

Akey, D.H. and S.D. Beck. 1971. Continuous rearing of the pea aphid, *Acyrthosiphon pisum*, on a holidic diet. Ann. Ent. Sot. Am. 64: 353–356.

Akey, D.H. and S.D. Beck. 1972. Nutrition of the pea aphid, *Acyrthosiphon pisum*: requirements for trace metals, sulphur, and cholesterol. J. Insect Physiol. 18: 1901–1914.

Aki, T., M. Shigyo, R. Nakano, T. Yoneyama and S. Yanagisawa. 2008. Nano scale proteomics revealed the presence of regulatory proteins including three FT-like proteins in phloem and xylem saps from rice. Plant Cell Physiol. 49: 767–790.

Alvarez, A.E., W.F. Tjallingii, E. Garzo, V. Vleeshouwers, M. Dicke and B. Vosman. 2006. Location of resistance factors in the leaves of potato and wild tuber-bearing *Solanum* species to the aphid *Myzus persicae*. Entomol. Exp. Appl. 121: 145–157.

An, F., W. Lin, D. Cahill, J. Rookes and L. Kong. 2014. Variation of phloem turgor pressure in *Hevea brasieliensis*: An implication for latex yield and tapping system optimization. Ind. Crop. Prod. 58: 182–187.

Auclair, J.L., J.B. Maltais and J.J. Cartier. 1957. Factors in resistance of peas to the pea aphid, *Acyrthosiphon pisum* (Harr.) II. Amino acids. Can. Entomol. 89: 457–464.

Backus, E.A. and W.H. Bennet. 2009. The AC–DC correlation monitor: new EPG design with flexible input resistors to detect both R and emf components for any piercing–sucking hemipteran. J. Insect Physiol. 55: 869–884.

Barlow, C.A. and M. McCully. 1972. The ruby laser as an instrument for cutting the stylets of feeding aphids. Can. J. Zool. 50: 1497–1498.

Bernal, C.C., R.M. Aguda and M.B. Cohen. 2002. Effect of rice lines transformed with *Bacillus thuringiensis* toxin genes on the brown planthopper and its predator *Cyrtorhinus lividipennis*. Entomol. Exp. Appl. 102: 21–28.

Bonani, J.P., A. Fereres, E. Garzo, M.P. Miranda, B. Appezzato-Da-Gloria and J.R.S. Lopes. 2010. Characterization of electrical penetration graphs of the Asian citrus psyllid, *Diaphorina citri*, in sweet orange seedlings. Entomol. Exp. Appl. 134: 35–49.

Brady, J. 1965. A simple technique for making very fine, durable dissecting needles by sharpening tungsten wire electrolytically. Bull World Health Organ. 32: 143–144.

Caillaud, C.M., J.S. Pierre, B. Chaubet and J.P. Di Pietro. 1995. Analysis of wheat resistance to the cereal aphid *Sitobion avenae* using electrical penetration graphs and flow chart combined with correspondence analysis. Entomol. Exp. Appl. 75: 9–18.

Carolan, J.C., C.I.J. Fitzroy, P.D. Ashton, A.E. Douglas and T.L. Wilkinson. 2009. The secreted salivary proteome of the pea aphid *Acyrthosiphon pisum* characterised by mass spectrometry. Proteomics 9: 2457–2467.

Chen, J.Q., Y. Rahbe, B. Delobel, N. Sauvion, J. Guillaud and G. Febvay. 1997. Melon resistance to the aphid *Aphis gossypii*: behavioural analysis and chemical correlations with nitrogenous compounds. Entomol. Exp. Appl. 85: 33–44.

Cherqui, A. and W.F. Tjallingii. 2000. Salivary proteins of aphids, a pilot study on identification, separation, and immunolocalization. J. Insect Physiol. 46: 1177–1186.

Dadd, R.H. 1961. The nutritional requirements of locusts. IV. Requirements for vitamins of the B-complex. J. Insect Physiol. 6: 1–12.

Dadd, R.H. and T.E. Mittler. 1965. Studies on the artificial feeding of the aphid *Myzus persicae* (Sulzer)—III. Some major nutritional requirements. J. Insect Physiol. 11: 717–743.

Dadd, R.H. and T.E. Mittler. 1966. Permanent culture of an aphid on a totally synthetic diet. Experientia 22: 832–833.

Daniels, M., J.S. Bale, H.J. Newbury, R.J. Lind and J. Pritchard. 2009. A sublethal dose of thiamethoxam causes a reduction in xylem feeding by the bird cherry-oat aphid (*Rhopalosiphum padi*), which is associated with dehydration and reduced performance. J. Insect Sci. 55: 758–765.

Day, M.F. and H. Irzykiewicz. 1953. Feeding behaviour of the aphids *Myzus persicae* and *Brevicoryne brassicae*, studied with radiophosphorus. Australian J. Biol. Sci. 6: 98–108.

Demidchik, V. and F.J.M. Maathuis. 2007. Physiological roles of nonselective cation channels in plants: from salt stress to signaling and development. New Phytol. 175: 387–404.

Doering-Saad, C., H.J. Newbury, H.J. Bale and J. Pritchard. 2002. Use of aphid stylectomy and RT-PCR for the detection of transporter mRNAs in sieve elements. J. Exp. Bot. 53: 631–637.

Downing, N. and D.M. Unwin. 1977. A new method for cutting the mouthparts of feeding aphids. Physiol. Entomol. 2: 275–277.

Dziubinska, H. 2003. Ways of signal transmission and physiological role of electrical potentials in plants. Acta Soc. Bot. Pol. 72: 309–318.

Ehrhardt, P. 1968. Die Wirkung verschiedener Spurenelemente auf Wachstum, Reproduktion und Symbionten von *Neomyzus circumjlexus* Buckt. (Aphididae, Homoptera, Insecta) bei künstlicher Ernährung. Z. Vergl. Physiol. 58: 47–75.

Esau, K. 1969. The phloem. *In*: W. Zimmermann, P. Ozenda and H.D. Wulff (eds.). Encyclopedia of Plant Anatomy, Vol. 5. Bornträger, Berlin, Germany.

Felle, H.H. and M.R. Zimmermann. 2007. Systemic signaling in barley through action potentials. Planta 226: 203–214.

Felle, H.H., S. Hanstein, R. Steinmeyer and R. Hedrich. 2000. Dynamics of ionic activities in the apoplast of the sub-stomatal cavity of intact *Vicia faba* leaves during stomatal closure evoked by ABA and darkness. Plant J. 24: 297–304.

Fisher, D.B. 1978. An evaluation of the Munch hypothesis for phloem transport in soybean. Planta 139: 25–28.

Fisher, D.B. 1983. Year-round collection of willow sieve tube exudate. Planta 159: 529–533.

Fisher, D.B. and J.M. Frame. 1984. A guide to the use of the exuding-stylet technique in phloem physiology. Planta 161: 385–393.

Fromm, J. and R. Spanswick. 1993. Characteristics of action potential in willow (*Salix viminalis* L.). J. Exp. Bot. 44: 1119–1125.

Fromm, J. and H. Fei. 1998. Electrical signaling and gas exchange in maize plants of draying soil. Plant Sci. 132: 203–213.

Fukumorita, T. and M. Chino. 1982. Sugar, amino acid and inorganic contents in rice phloem sap. Plant Cell Physiol. 23: 273–283.

Furch, A.C.U., J.B. Hafke, A. Schulz and A.J.E. van Bel. 2007. Ca2+-mediated remote control of reversible sieve tube occlusion in *Vicia faba*. J. Exp. Bot. 58: 2827–2838.

Furch, A.C.U., A.J.E. van Bel, M.D. Fricker, H.H. Felle, M. Fuchs and J.B. Hafke. 2009. Sieve-element Ca2+ channels as relay stations between remote stimuli and sieve-tube occlusion in *Vicia faba*. Plant Cell 2: 2118–2132.

Furch, A.C.U., T. Will, M.R. Zimmermann, J.B. Hafke and A.J.E. van Bel. 2010. Remote-controlled stop of phloem mass flow by biphasic occlusion in *Cucurbita maxima*. J. Exp. Bot. 61: 3697–3708.

Gaupels, F., T. Knauer and A.J.E. van Bel. 2008. A combinatory approach for analysis of protein sets in barley sieve-tube samples using EDTA-facilitated exudation and aphid stylectomy. J. Plant Physiol. 165: 95–103.

Giordanengo, P. 2014. EPG-Calc: a PHP-based script to calculate electrical penetration graph (EPG) parameters. Arthropod Plant Int. 8: 163–169.

Girousse, C., J.-L. Bonnemain, S. Delrot and R. Bournoville. 1991. Sugar and amino acid composition of phloem sap of *Medicago sativa*: a comparative study of two collecting methods. Plant Physiol. Biochem. 29: 41–48.

Hafke, J.B., J.-K. van Amerongen, F. Kelling, A.C.U. Furch, F. Gaupels and A.J.E. van Bel. 2005. Thermodynamic battle for photosynthate acquisition between sieve tubes and adjoining parenchyma in transport phloem. Plant Physiol. 138: 1527–1537.

Hafke, J.B., A.C.U. Furch, M.D. Fricker and A.J.E. van Bel. 2009. Forisome dispersion in *Vicia faba* is triggered by Ca²⁺ hotspots created by concerted action of diverse Ca²⁺ channels in sieve elements. Plant Signal. Behav. 4: 968–972.

Hamilton, M.A. 1930. Notes on the culturing of insects for virus work. Ann. Appl. Biol. 17: 487–492.

Hamilton, M.A. 1935. Further experiments on the artificial feeding of *Myzus persicae* (Sulz.). Ann. Appl. Biol. 22: 243–258.

Harrewijn, P. and H. Kayser. 1997. Pymetrozine, a fast-acting and selective inhibitor of aphid feeding. *In-situ* studies with electronic monitoring of feeding behaviour. Pest Manag. Sci. 49: 130–140.

Hewer, A., T. Will and A.J.E. van Bel. 2010. Plant cues for aphid navigation in vascular tissues. J. Exp. Biol. 213: 4030–4042.

Hewer, A., A. Becker and A.J.E. van Bel. 2011. An aphid's odyssey—the cortical quest for the vascular bundle. J. Exp. Biol. 214: 3868–3879.

Jiang, Y.X., G. Nombela and M. Muniz. 2001. Analysis by DC–EPG of the resistance to *Bemisia tabaci* on a Mi-tomato line. Entomol. Exp. Appl. 99: 295–302.

Kawabe, S., T. Fukumorita and M. Chino. 1980. Collection of rice phloem sap from stylets of homopterous insects severed by YAG laser. Plant Cell Physiol. 21: 1319–1327.

Kehr, J. 2006. Phloem sap proteins: their identities and potential roles in the interaction between plants and phloem-feeding insects. J. Exp. Bot. 57: 767–774.

Kempers, R., A. Ammerlaan and A.J.E. van Bel. 1998. Symplasmic constriction and ultrastructure features of the sieve element/companion cell complex in the transport phloem of apoplasmically and symplasmically phloem-loading species. Plant Physiol. 116: 271–278.

Kennedy, J.S. and T.E. Mittler. 1953. A method of obtaining phloem sap via the mouthparts of aphids. Nature 171: 528.

King, R.W. and J.A.D. Zeevaart. 1974. Enhancement of phloem exudation from cut petioles by chelating agents. Plant Physiol. 53: 96–103.

Kimmins, F.M. 1989. Electrical penetration graphs from *Nilaparvata lugens* on resistant and susceptible rice varieties. Entomol. Exp. Appl. 50: 69–79.

Lindemann, C. 1948. Beitrag zur Ernährungsphysiologie der Blattläuse. Z. Vgl. Physiol. 31: 112–133.

Lough, T.J. and W.J. Lucas. 2006. Integrative plant biology: role of phloem long-distance macromolecular trafficking. Ann. Rev. Plant Biol. 57: 203–232.

Macnichol, P.K. 1983. Developmental changes in the free amino acid pool and total protein amino acids of pea cotyledons (*Pisum sativum* L.) Plant Physiol. 72: 492–497.

Maiman, T.H. 1960. Stimulated optical radiation in ruby. Nature 187: 493–494.

Malone, M., R. Watson and J. Pritchard. 1999. The spittlebug *Philaenus spumarius* feeds from mature xylem at the full hydraulic tension of the transpiration stream. New Phytol. 143: 261–271.

Maltais, J.B. 1952. A simple apparatus for feeding aphids aseptically on chemically defined diets. Can. Entomol. 84: 291–294.

Mancuso, S. 1999. Hydraulic and electrical transmission of wound-induced signals in *Vitis vinifera*. Aust. J. Plant Physiol. 26: 55–61.

Martin, B., J.L. Collar, W.F. Tjallingii and A. Fereres. 1997. Intracellular ingestion and salivation by aphids may cause the acquisition and inoculation of non-persistently transmitted plant viruses. J. Gen. Virol. 78: 2701–2705.

McLean, D.L. and M.G. Kinsey. 1964. A technique for electronically recording aphid feeding and salivation. Nature 202: 1358–1359.

McLean, D.L. and M.G. Kinsey. 1965. Identification of electrically recorded curve patterns associated with aphid salivation and ingestion. Nature 205: 1130–1131.

Mentink, P.J.M., F.M. Kimmins, P. Harrewijn, F.L. Dieleman, W.F. Tjallingii, B. van Rheenen and A.H. Eenink. 1984. Electrical penetration graphs combined with stylet cutting in the study of host plant resistance to aphids. Entomol. Exp. Appl. 35: 210–213.

Mittler, T.E. and R.H. Dadd. 1962. Artificial feeding and rearing of the aphid, *Myzus persicae* (Sulzer), on a completely defined synthetic diet. Nature 195: 404.

Mittler, T.E. and R.H. Dadd. 1963a. Studies on the artificial feeding of the aphid *Myzus persicae* (Sulzer)—I. Relative uptake of water and sucrose solutions. J. Insect Physiol. 9: 623–645.

Mittler, T.E. and R.H. Dadd. 1963b. Studies on the artificial feeding of the aphid *Myzus persicae* (Sulzer)—II. Relative survival, development, and larviposition on different diets. J. Insect Physiol. 9: 741–757.

Mittler, T.E. and R.H. Dadd. 1964. An improved method for feeding aphids on artificial diets. Ann. Ent. Soc. Am. 57: 139–140.

Mousavi, S.A.R., A. Chauvin, F. Pascaud, S. Kellenberger and E.F. Farmer. 2013. GLUTAMATE RECEPTOR-LIKE genes mediate leaf-to-leaf wound signaling. Nature 500: 422–426.

Münch, E. 1930. Die Stoffbewegung in der Pflanze. Fischer, Jena, Germany.

Mutti, N.S., Y.S. Park, J.C. Reese and G.R. Reeck. 2006. RNAi knockdown of a salivary transcript leading to lethality in the pea aphid, *Acyrthosiphon pisum*. J. Insect Sci. 6: 38.

Mutti, N.S., J. Louis, L.K. Pappan, K. Pappan, K. Begum, M.-S. Chen, Y. Park, N. Dittmer, J. Marshall, J.C. Reese and G.R. Reeck. 2008. A protein from the salivary glands of the pea aphid, *Acyrthosiphon pisum*, is essential in feeding on a host plant. Proc. Nat. Ac. Sci. USA 105: 9965–9969.

Nauen, R. 1995. Behaviour modifying effects of low systemic concentrations of imidacloprid on *Myzus persicae* with special reference to an antifeeding response. Pest Manag. Sci. 44: 145–153.

Oyarce, P. and L. Gurovich. 2010. Electrical signals in avocado trees. Plant Signal. Behav. 5: 34–41.

Pletsch, D.J. 1937. Improved device for artificial feeding of aphids. J. Econ. Entomol. 30: 211–212.

Pockman, W.T., J.S. Sperry and J.W. O'Leary. 1995. Sustained and significant negative water pressure in xylem. Nature 378: 715–716.

Prado, E. and W.F. Tjallingii. 1994. Aphid activities during sieve element punctures. Entomol. Exp. Appl. 72: 157–165.

Prado, E. and W.F. Tjallingii. 1999. Effects of experimental stress factors on probing behaviour by aphids. Entomol. Exp. Appl. 90: 289–300.

Prüfer, T., T. Thieme and W.F. Tjallingii. 2014. APHID-AUTOEPG software for analyzing electrically monitored feeding behaviour of aphids. Eur. J. Env. Sci. 4: 53–59.

Rahbe, Y., G. Febvay, B. Delobel and G. Bonnot. 2000. Amino acids and proteins as cues in interactions of aphids (Homoptera: Aphididae) and plant. pp. 212–236. *In*: G.P. Walker and E.A. Backus (eds.). Principles and Applications of Electronic Monitoring and other Techniques in the Study of Homopteran Feeding Behavior. Thomas Say Publications in Entomology, Entomological Society of America, Lanham, USA.

Rao, K.V., K.S. Rathore, T.K. Hodges, X. Fu, E. Stoger, D. Sudhakar, S. Williams, P. Christou, M. Bharathi, D.P. Bown, K.S. Powell, J. Spence, A.M.R. Gatehouse and J.A. Gatehouse. 1998. Expression of snowdrop lectin (GNA) in transgenic rice plants confers resistance to rice brown planthopper. Plant J. 15: 469–477.

Rhodes, J.D., J.F. Thain and D.C. Wilson. 1996. The pathway for systemic electrical signal conduction in the wounded tomato plant. Planta 200: 50–57.

Rochow, W.F. 1960. Transmission of barley yellow dwarf virus acquired from liquid extracts by aphids feeding through membranes. Virology 12: 223–232.

Salvador-Rectalà, V., W.F. Tjallingii and E.E. Farmer. 2014. Real-time, *in vivo* intracellular recordings of caterpillar-induced depolarization waves in sieve elements using aphid electrodes. New Phytol. 203: 674–684.

Sarria, E., M. Cid, E. Garzo and A. Fereres. 2009. Excel workbook for automatic parameter calculation of EPG data. Comput. Electron. Agric. 67: 35–42.

Sasaki, T., M. Chino, H. Hayashi and T. Fujiwara. 1998. Detection of several mRNA species in rice phloem sap. Plant Cell Physiol. 39: 895–897.

Schilmiller, A.L. and G.A. Howe. 2005. Systemic signaling in the wound response. Curr. Op. Plant Biol. 8: 369–377.

Schmidt, H.B. 1959. Beiträge zur Kenntnis der Übertragung pflanzlicher Viren durch Aphiden. Biol. Zentr. 78: 889–936.

Spiller, N.J., L. Koenders and W.F. Tjallingii. 1990. Xylem ingestion by aphids—a strategy for maintaining water balance. Entomol. Exp. Appl. 55: 101–104.

Stankovic, B., D.L. Witters, T. Zawadzki and E. Davies. 1998. Action potentials and variation potentials in sunflower: an analysis of their relationships and distinguishing characteristics. Physiol. Plantarum 103: 51–58.

Strong, F.E. 1964. The effects of nitrogen starvation on the concentration of free amino acids in *Myzus persicae* (Sulzer) (Homoptera, Aphidae). J. Insect Physiol. 10: 519–522.

Thomas, R.C. 1978. Ion-sensitive Intracellular Microelectrodes. How to Make and Use Them. Academic Press, London, New York, San Francisco.

Tjallingii, W.F. 1978. Electronic recording of penetration behaviour by aphids. Entomol. Exp. Appl. 24: 721–730.

Tjallingii, W.F. 1985. Membrane potentials as an indication for plant cell penetrations by aphid stylets. Entomol. Exp. Appl. 38: 187–936.

Tjallingii, W.F. 1986. Wire effects on aphids during electrical recording of stylet penetration. Entomol. Exp. Appl. 40: 89–98.

Tjallingii, W.F. 1987. Stylet penetration activities by aphids: new correlations with electrical penetration graphs. pp. 301–306. *In*: V. Labeyrie, G. Fabres and D. Lachaise (eds.). Insect-Plants. Junk, Dordrecht, NL.

Tjallingii, W.F. 1988. Electrical recording of stylet penetration activities. pp. 95–108. *In*: A.K. Minks and P. Harrewijn (eds.). Aphids, their Biology, Natural Enemies and Control, Vol. 2B. Elsevier, Amsterdam, The Netherlands.

Tjallingii, W.F. 1995. Electrical signals from the depth of the plant tissues: the electrical penetration graph (EPG). pp. 49–58. *In*: H. Niemeyer (ed.). Techniques in Plant-Insect Interactions and Biopesticides. Proceedings IFS Workshop, Santiago, Chile.

Tjallingii, W.F. 2000. Comparison of AC and DC systems for electronic monitoring of stylet penetration activities by homopterans. pp. 41–69. *In*: G.P. Walker and E.A. Backus (eds.). Principles and Applications of Electronic Monitoring and other Techniques in the Study of Homopteran Feeding Behavior. Thomas Say Publications in Entomology, Entomological Society of America, Lanham, USA.

Tjallingii, W.F. and T. Hogen Esch. 1993. Fine structure of aphid stylet routes in plant tissues in correlation with EPG signals. Physiol. Entomol. 18: 317–328.

Tjallingii, W.F. and A. Cherqui. 1999. Aphid saliva and aphid plant interactions. Proc. Sect. Exp. Appl. Entomol. 10: 169–174.

Unwin, D.M. 1978. A versatile high frequency radio microcautery. Physiol. Entomol. 3: 71–73.

van Bel, A.J.E. and F. Gaupels. 2004. Pathogen-induced resistance and alarm signals in the phloem. Mol. Plant Pathol. 5: 495–504.

van Helden, M. and W.F. Tjallingii. 1993. Tissue localization of lettuce resistance to the aphid *N. ribisnigri* using electrical penetration graphs. Entomol. Exp. Appl. 68: 269–278.

van Helden, M. and W.F. Tjallingii. 2000. Experimental design and analysis in EPG experiments with emphasis on plant resistance research. pp. 144–171. *In*: G.P. Walker and E.A. Backus (eds.). Principles and Applications of Electronic Monitoring and other Techniques in the Study of Homopteran Feeding Behavior. Thomas Say Publications in Entomology, Entomological Society of America, Lanham, USA.

von Soest, W. and V. de Meester-Manager Cats. 1956. Does the aphid *Myzus persicae* (Sulz.) imbibe tobacco mosaic virus? Virology 2: 411–414.

Walker, G.P. 2000. Beginner's guide to electronic monitoring. pp. 14–40. *In*: G.P. Walker and E.A. Backus (eds.). Principles and Applications of Electronic Monitoring and other Techniques in the Study of Homopteran Feeding Behavior. Thomas Say Publications in Entomology, Entomological Society of America, Lanham, USA.

Wang, D., Q. Liu, X. Li, Y. Sun, H. Wang and L. Xia. 2015. Double-stranded RNA in the biological control of grain aphid (*Sitobion avenae* F.). Funct. Integr. Genomics 15: 211–223.

White, P.J. 2009. Depolarization-activated calcium channels shape the calcium signatures induced by low-temperature stress. New Phytol. 183: 6–8.

Will, T. and A. Vilcinskas. 2013. Aphid proofed plants: biotechnical approaches for aphid control. Adv. Biochem. Eng. Biotechnol. 136: 179–203.

Will, T., W.F. Tjallingii, A. Thönnessen and A.J.E. van Bel. 2007. Molecular sabotage of plant defense by aphid saliva. Proc. Natl. Acad. Sci. USA 104: 10536–10541.

Will, T., A. Hewer and A.J.E. van Bel. 2008. A novel perfusion system shows that aphid feeding behaviour is altered by decrease of sieve-tube pressure. Entomol. Exp. Appl. 127: 237–245.

Will, T., S.R. Kornemann, A.C.U. Furch, W.F. Tjallingii and A.J.E. van Bel. 2009. Aphid watery saliva counteracts sieve-tube occlusion: a universal phenomenon? J. Exp. Biol. 212: 3305–3312.

Will, T., K. Steckbauer, M. Hardt and A.J.E. van Bel. 2012. Aphid gel saliva: sheath structure, protein composition and secretors dependence on stylet-tip milieu. PLoS ONE 7(10): e46903. doi:10.1371/journal.pone.0046903.

Will, T. and A. Vilcinskas. 2015. The structural sheath protein of aphids is required for phloem feeding. Insect Biochem. Mol. Biol. 57: 34–40.

Wright, J.P. and D.B. Fisher. 1980. Direct measurement of sieve tube turgor pressure, using severed aphid stylets. Plant Physiol. 65: 1133–1135.

Wright, J.P. and D.B. Fisher. 1981. Measurement of the sieve tube membrane potential. Plant Physiol. 67: 845–848.

Wright, J.P. and D.B. Fisher. 1983. Estimation of the volumetric elastic nodulus and membrane hydraulic conductivity of willow sieve tubes. Plant Physiol. 73: 1042–1047.

Yust, H.R. and R.A. Fulton. 1943. An exudation associated with the feeding location of the California red scale. J. Econ. Entomol. 36: 346–347.

Zawadzki, T., E. Davies, H. Dziubinska and K. Trebacz. 1991. Characteristics of action potentials in *Helianthus annuus* L. Physiol. Plantarum 83: 601–604.

Zhu, X., P.N. Shaw, J. Pritchard, J. Newbury, E.J. Hunt and D.A. Barret. 2005. Amino acid analysis by micellar electrokinetic chromatography with laser-induced fluorescence detection: application to nanolitre-volume biological samples from *Arabidopsis thaliana* and *Myzus persicae*. Electrophoresis 26: 911–919.

Zimmermann, M.R., H. Maischak, A. Mithofer, W. Boland and H.H. Felle. 2009. System potentials, a novel electrical long-distance apoplastic signal in plants, induced by wounding. Plant Physiol. 149: 1593–1600.

Index

Printed and bound by CPI Group (UK) Ltd, Croydon, CR0 4YY

01/11/2024

01782623-0004